Landscapes of Power and Identity

Landscapes of Power and Identity

Comparative Histories in the Sonoran Desert and the Forests of Amazonia from Colony to Republic

CYNTHIA RADDING

Duke University Press

DURHAM & LONDON

2005

Printed in the United States of America on acid-free paper ∞
Designed by Erin Kirk New
Typeset in Minion by Keystone Typesetting, Inc.
Library of Congress Cataloging-in-Publication Data appear on the last printed page of this book.

The author and Duke University Press gratefully acknowledge the publication subvention provided by the University of New Mexico, Albuquerque, New Mexico, through the Office of the Provost and the College of Arts and Sciences.

This book is dedicated to four extraordinary women whose loving presence has enriched my life and inspired my work.

Ettie Mae Gingrich Lowman

Vera Revson Radding

Eulalia Cabrera de Saldívar

Marcelina Saldívar de Murrieta

Contents

List of Illustrations ix

Abbreviations xiii

Preface xv

Acknowledgments xxi

Introduction Savannas and Deserts: Two Histories of Cultural Landscapes 1

Chapter 1 Ecological and Cultural Frontiers in Sonora and Chiquitos 19

Chapter 2 Political Economy: Communities, Missions, and Colonial Markets 55

Chapter 3 Territory: Community and Conflicting Claims to Property 89

Chapter 4 Ethnic Mosaics and Gendered Identities 117

Chapter 5 Power Negotiated, Power Defied: Political Culture, Governance, and Mobilization 162

Chapter 6 Priests and Shamans: Spiritual Power, Ritual, and Knowledge 196

Chapter 7 Postcolonial Landscapes: Transitions from Colony to Republic 240

Chapter 8 Contested Landscapes in Continental Borderlands 295

Notes 327

Glossary 375

Bibliography 385

Index 423

Illustrations

Figures

1 Charte von Nordamerica 2
2 Ancient fence rows along the San Miguel River, Sonora 6
3 Chiquitos landscape with Piraí River 16
4 Irrigation *acequia* in the Moctezuma River valley 28
5 Newly planted fence rows at Mazocahui, Sonora River Valley 28
6 Pre-Hispanic *trinchera* 30
7 Roof repair in San Xavier, Ñuflo de Chávez, Bolivia 42
8 Plano corográfico (1777) 60
9 Tabula America specialis geographica regni Peru, Braziliae, Terra Firme, Amazonum (1737) 94
10 Mapa de parte de los virreynatos de Buenos Aires, Lima, Santa Fé, y Capitanía General de Caracas (1796) 95
11 Carte du Mexique (1771) 122
12 Map of Mexico or New Spain, in which the motions of Cortés may be traced (1795) 123
13 Tabula geographica Peruae, Braziliae, et Amazonum regionis (1730) 128
14 Insert of Chiquitos, tabula geographica Peruae, Braziliae, et Amazonum regionis (1730) 129
15 Facade of the mission church at Opodepe, Sonora 220
16 Bell tower of mission church at Onavas, Sonora 221
17 Facade of mission church at Onavas, Sonora 221
18 Mission San Miguel, in the province of Velasco, Chiquitos 224
19 Mission San Miguel, detail from the sacristy 224
20 Baptistry, mission San Javier, in the province of Ñuflo de Chávez, Chiquitos 225
21 Detail of baptistry, mission San Javier, Chiquitos 225
22 Map of Peru and Bolivia (1834) 246
23 Upper Paraguay River, 1777–1903 303

24 Plan del Tremedal y sus cercanías 305
25 Map of Mexico and Guatemala (1834) 310
26 Insert of Sonora, map of Mexico and Guatemala (1834) 311
27 Estados Unidos de Méjico (1847) 313

Maps

1 Sonora 24
2 Bolivia 36

Tables

1 Comparative Chronology 12
2 Chiquitano Soil Classification Related to Crop Selection 39
3 Male Indian Tributaries in Santa Cruz 131
4 Chiquitos Population Data, 1767–68 139
5 Selected Chiquitano Ethnic Names and Meanings 140
6 *Parcialidades* Related to the Chiquitos Languages 141
7 *Parcialidades* Related to Non-Chiquitos Languages 142
8 Household Census for Santa Ana de Tepache, 1803 149
9 Demographic Profiles of San Miguel de Oposura, 1803 151
10 Parish Census of the Pueblo of Ónavas, 1801 157
11 Tax Revenues for the Prefecture of Santa Cruz, 1830 249
12 Censuses and Inventories for San Javier and Santa Ana 250
13 Census for the Province of Chiquitos, 1842 253
14 Spiritual and Temporal State of the Missions of Pimería Alta, 1818 257
15 Pimería Alta Mission Crops and Livestock, 1818 258
16 Local Governance in the Cantons of Chiquitos, 1835 and 1839 285
17 Population of the Department of Santa Cruz, 1835–1839 290
18 Limestone Quarry Workers Paid in Santa Cruz, 1840 291
19 Significant Turning Points in Nineteenth-Century Sonora and Mexico, 1824–1883 315

Boxes

1 Tohono O'odham Seasonal Calendar 34
2 O'odham Cosmology 213
3 Rarámuri Cosmology 214
4 Chiquitano Cosmology 218

Abbreviations

AASC Archivo Arquidiócesis de Santa Cruz (Bolivia)
ABNB Archivo y Biblioteca Nacionales de Bolivia
  EC Escribanía de Cámara
  GRM Gabriel René Moreno
  MyCH Ramo Mojos y Chiquitos
ACQ Archivo del Colegio de la Santa Cruz de Querétaro (Mexico)
ACSC Archivo de Catedral, Santa Cruz (Bolivia)
AGI Archivo General de Indias (Spain)
AGN Archivo General de la Nación (Mexico City, Mexico)
  AHH Archivo Histórico de Hacienda
  PI Ramo de Provincias Internas
AGNA Archivo General de la Nación (Buenos Aires, Argentina)
AHES Archivo Histórico del Estado de Sonora (Hermosillo, Mexico)
AHGES Archivo Histórico del Gobierno del Estado de Sonora (Hermosillo, Mexico)
  TP Títulos Primordiales
AIPJ Archivo de Instrumentos Públicos de Jalisco (Guadalajara, Mexico)
  LGA Libros de Gobierno de la Audiencia
  RTA Ramo de Tierras y Aguas
AMH Archivo de la Mitra (Hermosillo, Mexico)
  AD Archivo Diocesano
  AS Archivo de la Parroquia del Sagrario
ANO Archivo General de Notarías (Hermosillo, Mexico)
BL Bancroft Library, University of California-Berkeley
  HHB Hubert Howe Bancroft Collection
BN Biblioteca Nacional (Buenos Aires, Argentina)
BNFF Biblioteca Nacional Fondo Franciscano
BPEJ Biblioteca Pública del Estado de Jalisco (Guadalajara, Mexico)

ARAG	Archivo de la Real Audiencia de Guadalajara
RC	Ramo Civil
DRSW	Documentary Relations of the Southwest, University of Arizona (Tucson, renamed Office of Ethnohistorical Research in 2002)
UASC	University of Arizona Special Collections (Tucson)
UGRM	Universidad Gabriel René Moreno (Santa Cruz de la Sierra, Bolivia)
MH	Museo de Historia
UIUC	University of Illinois at Urbana-Champaign

Preface

This book is part of a personal journey that has taken me to new lands and shown me new ways of writing history. After living and working in Sonora, Mexico, for eighteen years—a long and fruitful commitment that enriched my life and informed my research on the boundaries of anthropology and history—I decided that I wanted to continue studying frontier societies and the landscapes they created on the borders of the Iberian colonies in the Americas.[1] I was eager to follow new paths, to explore the geography and history of South America. Following the advice of several friends and colleagues, I turned to the eastern lowlands of Bolivia—specifically to Chiquitos, which, like Sonora, developed historically as a Jesuit mission province on the periphery of the principal mining centers of Spanish America. My first acquaintance with the Andean cordillera and the savannas and forests of Chiquitos proved as daunting as it did exhilarating, but I soon learned that the contrasting natural environments and divergent historical processes of Sonora and Chiquitos provided a unique comparative setting for bringing into sharper focus the mutually formative meanings of landscapes and texts.

I had spent my childhood in the temperate climates and suburban surroundings of New England and the midwestern United States, where the physical environment informed my sense of beauty and provided a cyclical assurance of permanence through change in the seasonal contrasts of heat and cold, blossoming and decay. I experienced nature as a necessary component of time through these recurring rhythms of sequential seasons that flowed one on the other from spring through winter, but I did not consider it an integral part of history. My family background in agriculture inspired in me an attachment to rural landscapes that was both sentimental and intellectual, a sense of returning to visit "God's country," as my maternal great-grandmother referred to the farmland of southern Pennsylvania. It was in the aridity and intense sunlight of the Sonoran Desert, where "God's country" assumed sharply contrasting hues and proportions to the landscapes of my youth, that I learned to consider geography as a central and powerful force in shaping human culture and society.

My years in Sonora gave me a new formative experience working directly in local and state archives, conducting oral history interviews, and traveling in the region's deserts and mountainous terrain. As I learned to recognize unfamiliar species of shrubs and trees, I understood that *forest* could refer to stands of saguaro cacti, as well as to groves of oak and pine. While tracing the course of dry riverbeds by following the cottonwood trees and cultivated floodplains that spread thin green ribbons across a brown expanse of low hills and valleys, I came to comprehend that underground water tables could be more important for agriculture than the visible flow of water in surface streams. In time, as I gained experience "reading" these landscapes, I returned to the archives with a heightened sensitivity to place. My sympathy for the proud regionalism of contemporary *Sonorenses* was tempered, in part, by my own quest to understand regionality in terms of the geographical content of different spaces in the Sonoran desert and piedmont.

When I first traveled to eastern Bolivia nearly a decade ago, I confronted the need to adjust yet again to new definitions of space and to visually distinct landscapes. I had moved from the relatively open spaces of desert scrub forest, where the ground is exposed and the sky always in view, to the towering canopy and dense undergrowth of the Chiquitano semideciduous rain forest. As I struggled anew to learn the names of unfamiliar plants and animals, the strangeness of these vocabularies was matched by the novelty of the landscapes they represented. It seemed that I had traveled even farther from the temperate climate of my youth: south of the equator, the seasons were reversed and the temporal sequence of four seasons repeated in annual cycles had become collapsed into two—wet and dry. Water, although no less important than in Sonora, took on new meanings in the tropical environment of Chiquitos. Here, surface water appeared abundant, seeping through natural springs and flowing into lagoons, while there water had been precious for its scarcity, often not visible even when present. Moreover, water management produced different landscapes generated by human invention in each of these regions. Sonoran stone terraces, earthen dams, and brush weirs that modified the course of streambeds, as well as garden wells that tapped into underground aquifers, bore testimony to the constant search for water. Chiquitano landscapes, by way of contrast, were punctuated by shallowly dug ponds (*atajadas*) that provided people and livestock with water for drinking, bathing, and washing clothes. Fire demarcated space in Chiquitos more directly and pervasively than in Sonora to create grazing pastures and agricultural plots. Swidden cultivation, commonly termed "slash-and-burn," opened irregular swaths of blackened earth and tree stumps in the forest, where mixed plantings of manioc, rice, corn, and papaya grew without conforming to the fixed dimensions of plowed fields.

As this comparative project took shape in my mind, it brought together visual, sensual, and intellectual impressions of the contrasting geographies and histories of these two regions. I did not have a period of long-term residence in Chiquitos, as I did in Sonora, but my annual visits there allowed me to move between landscapes and texts, to build stories in both space and time, that helped me to reflect in new ways on my experience in Sonora. This dialogic approach to nature and history has given me the freedom to confront the often fragmentary testimonies found in archival documents with ethnographies, written historical narratives, oral interviews, and geographic descriptions and to read these verbal texts in the light of the physical environment. Equally important, my lessons in history-as-dialogue have brought me valuable acquaintances with scholars, friends, and sojourners who have helped me forge the questions that give structure and meaning to this book. As we have shared with one another our findings in the archive, our notes of past interviews, and our visits to the colonial missions and present-day communities of Sonora and Chiquitos, we bring the presence of the past into our combined research projects in ethnobotany, architecture, musicology, anthropology, and history. Our journeys are enlightened and enlivened by the ongoing conversations that return again and again to the land and the people.

Landscapes of Power and Identity undertakes two challenges that speak to the heart of the historical discipline: the comparative study of two separate colonies within the Spanish American imperium and the conjoining of environmental and cultural history. Comparative studies of imperial systems that grew out of early modern European global expansion—Portuguese, Spanish, French, British, and Dutch—have provided valuable material for collective anthologies and syntheses of world history that follow the commercial webs between metropoles and colonies. These works—some of them monumental tomes, others admirably concise syntheses—contribute large-scale imperial frameworks focused on colonial policies, economic world systems, and geopolitical spheres of influence.[2] Immanuel Wallerstein's schema of world systems, and the debates it generated, reinforced a historical view of the mutually sustaining bonds between transoceanic mercantile empires and the rise of industrial capitalism. The approaches to economic history exemplified by Eric Hobsbawm and the founding editors of *Past and Present* focused on unraveling the complexities of the transition to capitalism and modernity, extended their inquiry to territorial colonization both within and beyond the conventional boundaries of Europe.[3] Eric Wolf built his seminal work *Europe and the People without History* on the imperial overview of world-systems theorists, but challenged them by shifting his gaze from the structures of empires to the peoples who inhabit them.[4] Turning to culture and the production of knowledge, recent historical interpretations have underscored the intellectual crosscurrents between Europe and the transatlantic colonial world.[5]

This project seeks to open a middle ground between the broadly comprehensive histories of imperialism and the historical ethnographies of the colonized. It weaves together comparative stories of colonialism *within* the Ibero-American world, focused on the Spanish colonial provinces of Sonora in northwestern Mexico and Chiquitos in eastern lowland Bolivia. Its purpose is not merely to juxtapose these two case studies but rather to build specific comparisons and contrasts into the human landscapes, economic rationales, and cultures of two distinct Hispanic American frontiers. The comparison first suggests itself in the structural relationship of each of these regions to the empire, through the institutional history of the missions, the social and cultural reproduction of the indigenous communities settled within them, and their relative distance from the viceregal poles of the colonial state. Yet a deepening acquaintance with Sonora and Chiquitos brings their historical and cultural contrasts into sharper focus: in reference to the managed economy of both mission districts and their internal governance; shifting ethnic identities, gender roles, and cultural hybridity; and the quality and duration of violent conflicts.

Why compare these two regions that, at first glance, appear so dramatically distant from each other? As I pondered this question I knew that it would prove imperative for a project of this nature to choose carefully those historical problems and geographic areas that lent themselves to comparison and, in turn, became enriched through the comparative exercise. Through the research process I became convinced that Sonora and Chiquitos are not merely case studies of colonial formations in the Americas; rather, their histories together illustrate the cultural and environmental complexities of frontier societies better than each could do alone.[6] Placing their stories between the covers of the same book builds a stronger narrative of borderlands, underscoring their porous quality as zones of contact among different biomes and cultures. In order to account for both the similarities and the differences of these two regions, the first obvious explanation pointed to the environmental contrasts between the Sonoran Desert and its mountainous borderlands and the Chiquitano forests and savannas. Geographical settings do not prescribe historical transcripts, however, and the search for explanations requires multiple lenses since the environments of these two colonies were created over millennia of human interaction with the physical surroundings. Their individual stories respond to distinct climates and terrains, to be sure; yet at the same time, they comprise different constellations of power, which emerge historically in different temporal rhythms.

The second challenge this project addresses is found at the crossroads of environmental and cultural history. Its purpose is to interpret the environment as part of the social and cultural domains of historical inquiry while avoiding both geographical and cultural determinisms. It is based on the conviction that the environment does not merely provide the scenic backdrop to the human drama but that it is, rather, integral to the historical narrative. Nature comprehends the physical environment—with its constraints of topography, climate, hydrography, and biota, as well as humanly crafted landscapes and the cultural meanings they inscribe.

The methodology employed in this work builds a historical dialogue between natural and cultural processes that is attentive to changes in the environment over time. It offers a fresh interpretive approach to what are considered conventional historical sources produced during the colonial and national eras of political administration of peoples and territories in both of these colonies. Critical reading of archival documents, published contemporary reports by missionaries and travelers, and previous histories, enriched further by cultural artifacts, ecological studies, and ethnographies, yields a rich contextual framework for constructing a history that is both social and environmental. Cultural artifacts appear in archaeological remains of pre-Hispanic and postconquest origins, common dwellings and monumental architecture, swidden plots and irrigated fields, liturgical music and religious dances, and processional and roadside crosses that inscribe layers of meaning for historical landscapes. Their content is tangible and symbolic, objective and performative, leaving its imprint on the histories of both regions that unfold through a mutual reading of landscapes and texts. The narrative is organized chronologically and thematically through seven chapters that compare and contrast the regional configurations of Sonora and Chiquitos around key issues of cultural ecology, political economy, territory, ethnicity, gender, political culture, and religion. The introduction develops the ways in which this project intersects with the history and anthropology of Latin America and with broadly comparative theoretical frameworks for the Americas and other major postcolonial world regions.[7]

Acknowledgments

In the course of researching and writing a project of this scope over ten years, I have received help, encouragement, and wisdom from many friends, colleagues, and institutions. It is my pleasure to thank them at this time and to recognize the invaluable assistance they have given me. Institutional and financial support from the University of Missouri, St. Louis, and the University of Illinois, Urbana-Champaign, in addition to short-term travel grants from the Hewlett Foundation, the National Endowment for the Humanities, and the Asociación Pro-Arte y Cultura (Santa Cruz de la Sierra, Bolivia), made it possible to carry out the archival and fieldwork and to write the manuscript. I am equally pleased to express my gratitude to the Bancroft Library at the University of California, Berkeley, and to the staff at the Newberry Library in Chicago, for their help in locating primary sources. Research assistants María Tapias, Ariel Yablón, Nicole Mottier, Mark Kummerer, Nathan Clarke, Frank Alvarez, Francisco Salazar, and Margarita Ochoa helped with important tasks related to data analysis, bibliography, index, and illustrations. University of Illinois librarians Nelly González, Alven Bregman, and Jenny Johnson and University of New Mexico Centennial Science and Engineering Library map librarian Mary Wyant graciously helped me to locate and copy the maps that illustrate the book, and Nuala Bennett Koetter helped to scan and copy them. Jane Domier drew the basic placement maps of northwestern Mexico and eastern Bolivia; Robert H. Jackson generously sent me two maps that he had located in the Library of Congress.

In Bolivia, I am indebted to the professional staff of the Archivo y Biblioteca Nacionales de Bolivia under the direction of Gunnar Mendoza, Josep M. Barnadas, René Arze Aguirre, Hugo Poppe Entrambasaguas, and Marcela Inch, whose expert assistance opened to me the rich holdings of this national treasure. My visits to Sucre have been enriched by the friendship of Deysi Pacheco de Rojas and her family, for which I am very grateful. In Santa Cruz de la Sierra, Paula Peña Hasbrún and her efficient and enthusiastic staff made it possible for me to work in the Fondo Prefectural of the archive and museum of the Universidad Gabriel René-Moreno. Víctor Hugo Ramalla opened to me the archives of the Archdiocese and Cathedral of Santa Cruz and shared with me his considerable knowledge of the regional history of the Bolivian lowlands. Anthropologists and agricultural extensionists who advance the work of support and advocacy for the indigenous communities of the lowlands through Apoyo para los Campesinos del Oriente de Bolivia (APCOB), under the direction of Jürgen Riester, first made it possible for me to visit the rural communities of Lomerío and Concepción and extended the necessary contacts for oral history interviews. Father Jesús Galeote Tormo welcomed me to the parish of San Javier and shared with me his knowledge and deeply rooted commitment to the Chiquitanos' culture and contemporary communities. The friendship and generous hospitality offered to me by Paula Peña, her parents Edgar Peña and Carmen Hasbrún, Bernd Fischermann and Rosa María Quiroga, and Ana María Lema Garrett have made my visits to Santa Cruz memorable, filled with conversation, music, and good food. Together with them and Mario Arrien, Sieglinde Falkinger, Eckart Kühne, Cecila Kenning, and Alcides Parejas, we have exchanged bibliographies, reminiscences, and built networks of collaboration. They have generously introduced me to their colleagues, educated me, and welcomed me into the environment and cultures that they have dedicated their lives to studying and celebrating. I was privileged to have known Hans Roth, who initiated the restoration of the Chiquitos mission churches and inspired the renaissance of their artwork and music, and to have benefited from his monumental knowledge of the region. In San Ignacio de Velasco I found important collaborators among the teachers, students, and social workers at the school and religious community of La Granja Hogar; I thank especially Justa Morón Macoñó and her parents for their friendship and the memories they have shared with me. In La Paz I am grateful to Clara López Beltrán, Laura Escobari de Querejazu, Pedro Querejazu, and Rossana Barragán, whose hospitality, encouragement, and scholarship have informed and enriched my reading of the history of Chiquitos. Karen Vieira Powers, David Block, Thomas Abercrombie, Elizabeth Penry, and Erick Langer generously shared with me their insights into Andean culture and history.

In Mexico, I am equally indebted to a host of colleagues, friends, and institutions who have contributed in important ways to this project and to my professional formation. Directors and staff at the Archivo General de la Nación in Mexico City, the Archivo Histórico del Estado de Sonora, and the Archivo de la Mitra, both in Hermosillo, made my research pleasant and productive.

Scholars and librarians at El Colegio de Jalisco in Guadalajara, under the direction of José María Muriá, helped me to advance my work. The Instituto Nacional de Antropología e Historia, on opening a branch in Hermosillo, gave me my first professional home in its Centro Regional de Sonora, and my return visits there are always a source of pleasure and learning. I thank Julio César Montané Martí, Juan José Gracida Romo, Elisa Villalpando Canchola, and José Luis Moctezuma for sharing their research with me and for their help and encouragement. Sergio Ortega Noriega, Ignacio del Río, José Luis Mirafuentes, and Clara Bargellini are generous colleagues and esteemed friends who have nurtured my understanding and passion for *el norte mexicano*. Similarly, long-term friendships with Cecilia Rabell Romero, Carmen Ramos Escandón, and Eugenia Meyer have helped me maintain my roots in Mexico, and Antonio Escobar Ohmstede has generously included me in his innovative projects for symposia and publications.

Many colleagues have given their time and shared with me their ideas that have improved and enriched the book. David J. Weber and Susan M. Deeds, as well as the anonymous reviewers for Duke University Press, read the entire manuscript and wrote detailed critiques that proved invaluable for its completion. Jean Allman, Antoinette Burton, Clare Crowston, Donald Crummey, Harry Liebersohn, John Randolph, Frederick Hoxie, Mark Leff, James Barrett, Nils Jacobsen, Joseph Love, Kristin Hoganson, Mark Steinberg, David Prochaska, Blair Kling, Kai-wing Chow, Vernon Burton, and Peter Fritzsche, among my colleagues in the University of Illinois history department and cultural studies reading groups, challenged me to demonstrate the purpose of a comparative book and to think carefully about the multiple intersections between history, culture, and the environment. Diane Harris, D. Fairchild Ruggles, and Helaine Silverman in the Cultural Heritage Landscapes working group opened new perspectives for me on space and the cultural meaning of landscapes. Andrew Orta, Dara Goldman, Pradeep Dhillon, Peter Garrett, and Alejandro Lugo in the University of Illinois Unit for Criticism and Interpretive Theory responded to an early version of chapter 4 that stimulated me to think critically about the concept of ethnicity. William Sullivan, Frances Kuo, William Stewart, Thomas Bassett, and the affiliated faculty and students in the Human Dimensions of Environmental Systems graduate program of the University of Illinois Environmental Council provided stimulating conversations and insightful critiques of early drafts of this work. University of Illinois graduate students, as well as Christopher Boyer, Dain Borges, and other faculty and student participants in the University of Chicago Colonial Latin American Workshop, gave me valuable criticism for blending history and ethnography and for discussing political culture in chapters 5 and 6. Thomas Sheridan, William Beezley, Alice Schlegel, and other colleagues in the University of Arizona history department and the Office of Ethnohistorical Research have offered me good advice and unfailing encouragement as the project has taken shape. James Scott and the members of the Yale Agrarian Studies Seminar enthusiastically queried me on the project in its early stages and offered important theoretical and methodological suggestions. Elizabeth Kuznesof, Eric Van Young, Stuart Schwartz, and Mary Kay Vaughan have commented insightfully on portions of the manuscript and encouraged my professional growth in many important ways. Duke University Press editors Valerie Millholland and Miriam Angress have seen the book to fruition with the professional and caring guidance that have become the hallmark of the press.

To my family I owe the greatest debts of all. My parents Benjamin I. and Dorothy L. Radding patiently encouraged the writing process through repeated home visits, read several chapters, and gave me prompt and helpful advice. My son David Murrieta gave me important computer assistance, Daniel Murrieta shared my enthusiasm and concerns about the project, and my husband Xicoténcatl Murrieta provided constant support and helped me work out many of the ideas that became foundational for the book. For this and for their love, I am eternally grateful.

A todos, gracias.

INTRODUCTION

Savannas and Deserts: Two Histories of Cultural Landscapes

> The climate of this Pimería is temperate, not inclined to extremes of either heat or cold. . . . The terrain is flat, but interwoven with hills and mountain ranges that make it beautiful, even if the hills interrupt the roadways, which extend flat and straight through scrub woodlands of mesquite and other trees and desert *matorral* [brush]. Along the riverbeds grow cottonwoods, willows, walnuts, and *taray* [of European origin, *Tamarix gallica*]. . . . In the mountains there are good pine timbers for building churches in the missions that are already established. . . . The land is fertile, in some areas very abundant, while in others somewhat sterile, more because of the lack of cultivation than for the quality of the land. Those who live there, called *papabotas* [pápagos; *tohono o'odham*], "bean eating Pimas," are content with very little to get by. . . . The other products of this Pimería are maize, the *tepari* bean, and other seeds that the Pimas collect as they ripen and save for their sustenance.[1] And since they have dealt with the Spaniards, and the missionaries have begun to work with them, they have good wheat harvests, especially in the west, beans of all kinds, lentils, squashes, and melons. . . . From this we may infer that the fertility of this land is not at all inferior, and in some ways superior, to parts of New Spain.—LUIS VELARDE, *Primera relación de la Pimería Alta*, 1716

> The province we commonly call Chiquitos is . . . for the most part wooded, with thick forests that are abundant with honey and wax due to the many bees of different species [found there]. . . . The terrain, itself, is dry, but in the rainy season, which lasts from December to May, the countryside is flooded so disproportionately that trade is curtailed, forming rivers and lagoons that fill with many different kinds of fish. . . . Once winter has passed [May to August] the fields dry out, and the heavy work of cutting down tree trunks and clearing away under brush is necessary to plant the hillsides and raised ground with crops that yield good harvests of maize, which is the wheat of the Indies, rice, cotton, sugar, tobacco, and other native fruits of the land such as bananas, pineapples, peanuts, and a variety of squash that is better and more delicious than in Europe.—JUAN PATRICIO FERNÁNDEZ, *Relación historial de las misiones de indios chiquitos*, 1726

Two Jesuit missionaries, writing within a decade of each other, but thousands of miles apart, described the provinces of their calling in terms of the land and the peoples who inhabited them.[2] Their geographies portrayed sculpted landscapes, blessed with wild fruits but made to yield bountiful crops through

Charte von Nordamerica. Augsburg, 1807. Courtesy of Map and Geography Library, University of Illinois at Urbana-Champaign. The area north of Sonora and New Mexico is labeled "unknown Quivira tribes."

human labor and the techniques of cultivation. In both the Pimería of northern Sonora and the Chiquitanía of eastern Bolivia the mission enterprise was young, yet the European presence already formed an integral part of the landscape. Juan Patricio Fernández's erroneous statement that sugar, rice, and bananas were "native" to Chiquitos implies that by the time he came to live there, these crops had been known so long that they seemed natural, a timeless part of the scenery. For his part, Luis Velarde showed that wheat had come to Sonora through commerce with the Spaniards, extolling the fertility of the land according to the usefulness of its products. Both writers conveyed their own impressions of the exotic and unfamiliar qualities of the places they endeavored to describe, yet both took possession of these frontier provinces, comparing them favorably to New Spain, the heart of the viceroyalty in Spanish North America, and to Europe.

The texts from which these passages were taken illustrate the tension between human and natural agency that provides the foundation for this book. It is based on two simple premises: first, that people create the landscapes in which they live even as their cultures are shaped by their physical surroundings; and, second, that geographical differences matter in the course of human events because history occurs and is recorded in both space and time.[3] *Landscapes of Power and Identity* builds two comparative histories of colonial frontiers in northwestern Mexico and eastern Bolivia, where indigenous communities encountered European governors, missionaries, slave hunters, merchants, miners, and ranchers in two distinct provinces: Sonora (Mexico) and Chiquitos (Bolivia). The intersection of the human dramas of Sonora and Chiquitos on the edges of the Spanish American empire illustrates different kinds of humanly crafted landscapes and relates the organization of space to changing ethnic identities among the Sonoran and Chiquitano peoples. The book weaves a complex tapestry of nature, culture, and society through the historical narratives written by diverse and opposing sets of human actors which place the environment into narratives of conquest, confrontation, and colonial readjustments.

Environmental histories have produced broad syntheses for organizing global history, employing long-term periodizations and cross-continental comparisons. These frameworks have brought historians and the reading public into an awareness of the importance of environmental issues for understanding the human drama, and they have created a context for colonialism that brings together natural and cultural history. Writ large, environmental history centers on themes of technology including societal achievements of horticulture, irrigation, and pastoralism; broadly differentiated settlement patterns; demography and epidemiology; and continental "exchanges" of flora, fauna, and pathogens in the wake of imperial conquests.[4] Environmental historians address the consequences of imperial conquests for colonized peoples and landscapes, as well as the genesis of environmental policies put into practice in European overseas colonies.[5]

Ecological approaches to narratives of colonial encounters anchor these histories in the web of multiple relations among diverse human societies and the terrestrial environments they fashion into landscapes.[6] Early theorists of ecology thought of ecosystems as networks of natural forces held in equilibrium through the economy of nature, in accord with classic authors like Linnaeus, Ernst Haeckel, and Alexander von Humboldt.[7] Following Charles Darwin's revolutionary theory of evolution through natural selection, modern ecologists have favored the concepts of disturbance and process to emphasize changes in nature and human impacts on the environment.[8] Ecologists, who approach their research from both the biological and the social sciences, echo some of the principles of environmental management and interdependency in reference to the use of resources, to the sustainability or reproduction of ecosystems—such as tropical forests, savannas, or alluvial floodplains—and to human demography and settlement patterns.[9]

Culture, defined broadly, refers to values and systems of thought, as well as to the material conditions of living that are integral to the historical construction of social and political relations. The present study relates outward expressions of culture documented in texts, imagery, pageantry, and language to the material cultures exhibited in the economy and ecology of two specific colonial societies. As used here, culture-in-practice becomes both the object of study and the means of inquiry, referring to cultural formations viewed in historical processes of change and development.[10] The cultural, in this usage, like the social and the political, is meaningful according to the particular qualities of human experience in time and space.[11] Current anthropological theories tell us that cultural systems are not closed structures that obey an internal logic. Carefully crafted studies of particular communities and societies recognize local knowledges concerning the use and conservation of natural resources, and their stories tend to dignify the means of cultural reproduction. Yet classic ethnographies of this genre have been called to task for emphasizing stability in small-scale societies more than historical change and focusing more on cultural responses to nature than on the ways in which people alter their environments.[12]

The term *cultural ecology*, as used here, explores human ecological relationships over time in two specific colonial contexts.[13] These dual histories of imperial and ecological frontiers develop the concept of cultural ecology through the associations between human societies and the material worlds they create, which are perceived in the layered moral and historical meanings of particular geographies and landscapes.[14] Culture and nature, then, are necessarily conjoined in historical processes of reciprocal adaptations. Human creativity and its consequences set in motion social and economic forces of production and reproduction, destruction and renewal, with multiple repercussions for both nature and society over time.[15] The term *social ecology*, closely linked to the broad definition of culture employed here, conveys the linkages between human and material resources *and* the political implications for their possession and use by competing social groups.[16] It is foundational for the thesis of this book, which brings together nature, culture, and political economy to argue the centrality of the environment to stories of power and colonial confrontation. These multistranded approaches to territorial conflicts and to the ethnic and gendered meanings of community comprehend both the material claims to resources and the cultural identities linked to specific localities.[17]

Landscapes and Lived Environments

Landscapes provide the unifying framework for this comparative history. Understood as lived spaces created by human labor, landscapes emerge from ecological and cultural processes that have the power to transform deserts, savannas, forests, and streams through both human and natural agency.[18] In this sense, the changing quality of cultural landscapes merges with the working concept of social ecology to provide the story line for each of the chapters along the dual axes of space and time. The built environments created in floodplain *milpas* (family plots), cleared *chacos* (swidden plots), cattle *estancias* (ranches), mission pueblos, mines, and shifting foraging grounds in deserts and forests gave rise to distinctive material cultures and contested values that cast their descriptive imprint on the regions of Sonora and Chiquitos.[19]

Western notions of landscape, highlighting its visual and pictorial qualities

Ancient fence rows along the San Miguel River, Sonora, one kilometer south of the village of Cucurpe. Photograph by the author, 1979.

and based on Renaissance and Enlightenment traditions of artistic representation, have structured the maps and reports filed by administrators and missionaries during both the colonial and republican periods, as well as personal accounts of national officials and European travelers.[20] The descriptive emphasis on landscape-as-view that emerges from these documents, characterized by some scholars as the perspective of outsiders looking in on a particular terrestrial or fluvial environment, was often linked to entrepreneurial and imperial projects in search of human and natural resources to exploit for profit. Such projects' scientific and commercial interests were exemplified in the nineteenth century by the published works of Alcides d'Orbigny, Moritz Bach, and Francis de Castelnau for Chiquitos, and by those of Ignacio de Zúñiga, Robert Hardy, and John Bartlett for Sonora.[21] Contrasting meanings of constructed landscapes from an insiders' perspective have arisen from the territorial claims advanced by socially and ethnically differentiated communities to particular spaces within these regions and from the material artifacts of their occupation. Building terraces, burning swiddens, blocking streams with dams and weirs, cutting forests, and turning savannas into grazing pastures all comprised modes of labor that changed the vegetation and even the morphology of landforms and rivers.[22]

The material and symbolic significance of landscapes develops with the productive labor that has modified the physical environment and created historical memories. For many indigenous communities of past and present times, the locations of gardens and villages evoke legends of their ancestors and stories of lived experiences in reference to migrations and to the fusion and separation of families. The places in which *chacos* were cleared and planted, and then left to fallow, in the forests and savannas of Chiquitos; in which the fruit of the saguaro cactus was gathered in the Sonoran Desert just before the onset of summer rains; or in which floodplain *milpas* were irrigated through living fence rows express both the productive and reproductive lifeways of human labor in particular communities that, in turn, constitute the social boundaries of community and the central markers of ethnicity. History, especially among unlettered peoples, comprehends human networks maintained through biological and social bonds of sexuality, nourishment, and survival—in the face of illness, violence, and death—that link together cohorts and generations of kin groups to one another and to specific places.[23] The distribution of *chacos* in different stages of primary and secondary growth, the revisiting of streams and wetland lakes and lagoons for fishing, and excursions into the forest for hunting game and gathering honey establish the spatial points of reference that substantiate the use-rights of particular families to subsistence resources in the tropical environment of Chiquitos. Similarly, in the desert scrub forests and alluvial valleys of the Sonoran highlands, the successive planting of trees to contain floodwaters during the summer rains and to create arable terraces, the location of stone *trincheras* (terraces) and corrals, and the shifting location of *rancherías* (small hamlets)—for example, the seasonal fields and wells of the Tohono O'odham ("those who emerged from the earth"[24])—create visible signs of land use and of concerted human action over time.

Landscapes produced in specific localities merge with imperial borderlands in this study of multiple frontiers at the edges of the Spanish American colonial dominions. The physical environments of Sonora and Chiquitos evoke striking contrasts between the desert and the tropics, yet they both constituted "internal provinces" distant from the centers of viceregal authority.[25] Shifting cultivation, rainfall patterns, and the seasonality of horticulture and foraging are relevant to both of these regions, but their particular ecological patterns produce distinct climatic, vegetational, and topographic frontiers. They constituted sites of multiple conquests before and after the European invasions of the Americas, forming corridors of exchange, migration, and warfare on the shifting peripheries of the Mesoamerican and Andean cultural complexes that culminated in the fifteenth-century tributary empires of Anáhuac and Tawantinsuyo.[26] Following Spanish conquest, our narrative takes up the question of what happens when similar institutions of imperial governance, commerce, and religion are planted in two very disparate physical and cultural environments.

The ecological, ethnic, and geopolitical boundaries that intersected both of these provinces belie a fixed notion of regionality and lead us to examine the historical construction of space comparatively over time.[27] Peripheral regions developed webs of exchange and dependency that tied them to colonial centers as well as to the European metropolis. Colonialism gave rise to multilayered political hierarchies because, in practice, the sites of imperial power were decentralized across distinct nodes of military, administrative, and ecclesiastical authority in the colonies. Imperial ambitions to maintain centralized control over the colonies became diffused through these multiple webs and hierarchies, contested and shaped by colonial elites and indigenous communities.[28]

Each of these provinces formed part of multiple borderlands, shaped by trade routes, territorial disputes, and their natural environments. Sonora constituted an inland frontier in relation to Guadalajara, the seat of the *audiencia* court (leading colonial governing institution), to the viceregal capital of Mexico, and to the military commandancy established in the late eighteenth century for all the northern provinces of New Spain. Chiquitos was located east of the Andean foothills and south of the Amazonian river basin, at the intersection of opposing and unstable boundaries between the Portuguese and Spanish dominions. Indigenous patterns of trade and warfare prior to colonial ambitions to open overland trade routes established foot trails and canoe portage along the rivers and through forests and swamps that formed a corridor between Chiquitos and northern Paraguay. The Portuguese province of Mato Grosso was the origin of slaving expeditions into Chiquitos and, over time, provided a ready source of contraband trade and a refuge for Chiquitanos who wearied of mission life, even as Chiquitos received runaway slaves from the Portuguese colony. Within the Spanish sphere of influence Chiquitos was linked financially and politically to the *audiencia* of Charcas, the great mining center of Potosí, and the provincial cities of Cochabamba and Tarija. These regional webs complicate the notion of an imperial center with dependent colonies. They show multiple regional centers that gave rise to different directional flows of wealth and power, operating at times against the centralizing objectives of Iberian metropoles and the mercantilist project of accumulating wealth for the benefit of the crown and European institutions and commercial interests.

Landscapes of Power and Identity opens new approaches to comparative colonialism within the Ibero-American world by highlighting the thematic connections between historical landscapes and narratives. It addresses the following questions in eight chapters that build the environmental and cultural histories of Sonora and Chiquitos. In chapter 1, it asks how the colonial encounters were conditioned by the natural settings, technologies, and cultural expectations of the Sonoran and Chiquitano peoples. Our guide to both regions is Alvar Núñez Cabeza de Vaca, whose misadventures in the borderlands of New Spain and Paraguay provide two of the earliest texts for these Spanish American colonies. Chapter 2 asks how markets operated in each of these regions and how they intersected with wider commercial networks. What were their repercussions for control of labor, technology, and distribution of wealth? In light of the greater longevity of the corporately managed economy of the Chiquitos pueblos, in comparison with those of Sonora, how did the village polities of both regions lose control of vital material resources during the nineteenth century? Through what strategies did they attempt to rebuild their communities? Chapter 3 explores the different meanings that competing social actors among colonizers and colonized ascribed to the notions of territory and property as spatial "goods" with economic and jurisdictional values. How did they create and defend cultural landscapes? Chapter 4 takes up the different social divisions in Sonora and Chiquitos, conventionally indicated by ethnicity, class, and gender. In what contrasting ways are *mestizaje* (racial mixture) and transculturation meaningful for both regions? How are social and environmental differences mutually implicated in the changing quality of ethnic identities? Chapter 5 turns to the exercise of power in the histories of local governance and during episodes of open conflict and violence in both these frontier provinces. Military narratives of contested boundaries speak to both the environmental contours and the spatial dimensions of defending an empire against elusive "enemies," while native rebels challenge the locus of authority in mission communities. Chapter 6 delves into the different meanings of religiosity in the cultural practices and historical memories of Sonoran and Chiquitano peoples. What can we learn from history and ethnography about how their spiritual landscapes relate to their social and political worlds? Chapters 7 and 8 carry these stories into the nineteenth-century formation of the Mexican and Bolivian nation-states and focus on the tension between indigenous communities and the Creole societies that assumed control over the institutions of government and on the external relations that challenged the colonial boundaries of Sonora and Chiquitos and redefined their spatial dimensions as regions, respectively.

Gender is woven into the stories presented in each of the chapters. Colonial narratives of Indian enslavement and frontier warfare distinguish between soldiers, associating military service with masculinity, and captives, consistently described as women and children. Spaces for work and worship in the missions were gendered—in the disposition of workshops for carpentry, metallurgy, spinning, and weaving, as well as in the areas of the churches assigned to men and women. Ethnic categories carried both racial and gendered connotations in the hierarchies created by the colonial regimes and in the institutions of surveillance and representation forged in these frontier provinces. The republican constitutions of nineteenth-century Mexico and Bolivia established gendered distinctions for citizenship and political participation. In the spiritual realm, shamanism and the power of native religious leaders to transcend borders between nature and culture implied as well the crossing of conventional boundaries defining gendered identities and sexuality.

The secular span of these parallel histories, approximately from 1750 to 1850, extends from the Bourbon colonial regime of Spanish America through the first half century of national formation in Mexico and Bolivia. The book aims to contribute new historical analyses for the nineteenth century, a period generally less well studied than the colonial era, in terms of political changes occurring in the interior of indigenous communities, shifting ethnic identities, and new modes of articulation with national polities and external spheres of imperial influence. The rising importance of Great Britain as the paramount global maritime power, concomitant with the dismantling of Spain's mercantilist empire, set in motion new webs of unequal transnational exchanges in which the United States played an increasingly visible role for northern Mexico, as did Brazil for eastern Bolivia.

The long, conflictive period of postcolonial realignments that characterized most of the nineteenth-century Latin American republics raised basic issues of citizenship, of inclusion and exclusion from the body politic, and of economic power that, in turn, posed new challenges for the adaptive strategies employed by indigenous communities in their negotiations and confrontations with the colonial state and regional elites.[29] In both northwestern Mexico and eastern Bolivia new definitions of property altered the social conventions of space, the distribution of material resources, and the territorial significance of frontiers as defensive borders and zones of cultural transition. Even as the newly declared nation-states of Mexico and Bolivia, arising after two decades of warfare and civil strife, undertook to sculpt "imagined communities" through national constitutions, flags, and other symbolic emblems, their boundaries were contested by the flow of contraband trade, by colonization schemes—sponsored and clandestine—and by the reassertion of ethnic polities within and across national borders.[30]

Indigenous communities of Sonora and Chiquitos challenged the tenets of modernity that underwrote the political charters of the Mexican and Bolivian republics through cultural constructs that were both conservative and innovative. They proved conservative in that they defended tenaciously the communal spaces and prerogatives whose institutional foundations lay in the colonial missions; they seemed innovative in that they sought to implement the political notions of nation and citizen in ways that pointed to alternative discourses and oppositional courses of action. Native communities' struggles led as well to internal divisions and partial alignments with regional and national centers of power. Their political and economic subjugation to the landed elites during the closing decades of the nineteenth century through military intervention by the state, the private use of force, and the circumscription of productive resources was not necessarily foreseen, nor does it mark the end of their histories. The cultural endurance of indigenous peoples and their formulation of successive political alliances punctuate later chapters in the regional economies and ethnic identities of both Sonora and Chiquitos in ways that knit together past and present.

The chronological limits established for this comparative study do not define identical periodizations for each of the regions. Rather, within a common temporal framework, Sonora and Chiquitos present different processes of historical continuity and, conversely, moments of rupture in their institutional, spatial, and cultural networks. Placing their histories in comparative perspective within the covers of the same book helps to highlight the significance of episodical changes as these occur, and to distinguish points they have in common and where they diverge (see table 1). While the dates for initial Spanish invasions of both territories correspond to the middle third of the sixteenth century, and while each of these frontiers was marked by the Jesuit

TABLE 1 Comparative Chronology

Historical Events and Processes	*Sonora*	*Chiquitos*
Spanish explorations and initial conquests	1536–64	1543–47
Founding of Santa Cruz de la Sierra		1563
Encomienda regime of servile labor		1560–1750
Repartimiento labor drafts to silver mines	1650–1740	
Jesuit mission enterprise	1591–1767	1691–1767
Bourbon administration following Jesuit expulsion		
Franciscans administer Sonoran missions	1768–1842	
Secular clergy and lay administrators in Chiquitos		1768–1860
Mexican legislation suppresses communal lands and native village autonomy in Sonora	1828–56	
Bolivian national state auctions off mission herds and grants concessions to commercial societies to open roads and promote trade between Santa Cruz and Paraguay		1840–70
Porfirian Mexico: large private land concessions; industrial mining; hydraulic technology—Sonoran Indians become small peasant farmers, dependent agricultural laborers, or a mining proletariat largely indistinguishable from the mestizo population	1880–1910	
Yaquis, Mayos, and rural mestizos participate in the northern armies of the Mexican Revolution	1910–20	
Chiquitanos subjugated to a system of forced labor (*patronato*) on low-productive agricultural estates, raising sugar cane and cattle, and drafted to gather rubber (1890s–1912; 1941–54)		1870–1954
Mexican land reforms during and after the presidency of Lázaro Cárdenas establish *ejidos* (peasant landholdings) in the fertile Yaqui and Mayo Valleys and in the piedmont of Sonora	1934–54	
Chaco War between Bolivia and Paraguay brings Andean and Chiquitano soldiers into direct contact		1935–37
Concurrent with the Bolivian Revolution of 1952 and the breakdown of the *patronato* regime, Chiquitanos migrated out of mission towns and *estancias* (ranches) to reestablish small hamlets, now the nuclei of their cultural identity		1954–70

mission enterprise, there was nearly a century's delay between the establishment of mission towns in Sonora and in Chiquitos. Furthermore, the importance of different regimes of forced labor in each province and, in Sonora, the proximity of mining *reales* (camps and centers for assaying and processing the ore) to Indian missions, established contrasting patterns of frontier society and commerce. Postindependence developments point to the prolongation of forced labor and peonage in Chiquitos, punctuated by the Amazonian quinine and rubber "booms" of the nineteenth and twentieth centuries, and to the integration of Sonoran indigenous and mestizo peasants into the regional labor force. Distinct cultural identities endured in both regions through systems of cultivation adapted to the environment and the religious expressions of folk Catholicism, traditions that are themselves historically changing and reinvented over time, serving to demarcate sacred landscapes and ethnic territories.

Theories of History and Culture

This comparative history speaks to current debates that reach across disciplines in history, geography, and anthropology concerning different methods and conceptual frameworks for writing about imperial systems and frontier societies. The shift from world-systems models to postcolonial studies as organizing principles of research has contributed to creative cross-regional discussions of imperialism, human agency, and the lived experiences of hybrid colonial societies.[31] The field of postcolonial studies developed by literary critics, anthropologists, and historians has coalesced into a common set of issues and questions particularly applicable both conceptually and historically to the British, Dutch, and French imperial spheres of South Asia, Australasia, and Africa. I refer here to the Orientalist debates, the denunciation of racializing distinctions between Europeans and colonized peoples, and to the counter-discourses questioning the intellectual legacy of the northern European Enlightenment.[32] These cultural and political challenges have led to the critique of modernity and the revisionist decentering of Europe in global and imperial contexts.[33]

Latin American(ist) scholars have broadened the discussion, employing postcolonial theories to challenge long-standing assumptions of periodization, national identity, and colonial oppositions.[34] Together with students of other imperial systems, they address problems of racializing categories and cultural misreadings across colonial frontiers. Nevertheless, the relatively greater longevity of the Ibero-American empires, spanning nearly half a millennium and grounded in settler colonies, opens alternative perspectives on the Renaissance and the Counterreformation, the imperial state, and the complexity of colonial societies.[35] During the last quarter century, punctuated by the quincentenaries of Iberian encounters with the peoples and continents of the Americas (1992–2000), the theoretical challenges of postcolonial studies have converged with the richly varied ethnohistorical literature of Latin America, inspired by the notion of "writing history from below."[36] These intellectual crosscurrents, strengthened and challenged by the political movements of contemporary indigenous peoples, highlight the need for comparative research on the Spanish and Portuguese colonial borderlands.[37]

The *post-* in *postcolonial* implies not so much a chronological break as an oppositional stance within colonial regimes in different regions and time periods. The content of postcolonial protests has germinated in the historical experiences of colonized peoples, articulating alternative political cultures forged from the vocabularies of imperial institutions and the universal values they presumed to represent.[38] Monarchy, Catholicism, and Iberian legal traditions became important symbols that buttressed the customary rights and privileges (*usos y costumbres*) tenaciously defended by both indigenous communities and colonial elites in different historical moments. Colonial subjects infused these religious and political icons with new meanings to support their claims to autonomy, material survival, and ancestral lineage. Nineteenth-century oppositional movements, arising during the wars for independence and resurfacing throughout the formative period of Latin American republics, contradicted the tenets of liberal nationalism and demanded a hearing for alternative visions of the nation-state. Their resurgence at the turn of the twenty-first century has forced a hearing of dissenting political voices and riveted national attention on distinct ethnic and cultural identities.[39] These alternative political programs bring the past into the present, integrating political economy and culture. For the historian, these two modes of inquiry are not at odds or mutually exclusive; rather, each informs the other.[40]

The historical narratives that comprise the heart of this book are constructed from multiple sources of information generated by the colonial regimes, Creole societies, and indigenous peoples of Sonora and Chiquitos. Archives record the dominant texts of authority figures—missionaries, governors, military officers—to be sure, but they also yield the actions and voices of Indians, peasants, and slaves whose labor supported the imperial enterprise. Their testimonies reveal, if only partially, the contours of their cultural identities as they reconstructed their communities and placed limits on the material and spiritual exactions of the colonial order. The territorial defense of irrigated cropland in Sonora, or of rain forest and cattle reserves in Chiquitos, raises new questions about the environment as a resource for survival and about landscape as constructed space.[41] These texts are examined through internal comparisons, noting points of agreement and contradiction, in combination with performative and artistic cultural artifacts, observed landscapes, and contemporary ethnographies.

Landscapes of Power and Identity takes up the challenge to build an integrated narrative that is attentive to long-term processes of change, while avoiding triumphalist epics or foregone conclusions. This book weaves together multiple story lines of diverse (and warring) Indian societies and of their rival colonial overlords in order to show the specific conditions of cultural and material production that work changes in human and natural environments. The political economy of mining that demanded a labor force and agricultural surpluses; the conversion of territorial spaces into private or communal tenure of specifically measured pieces of land; the institutionalization of indigenous authorities within the mold of Spanish councils and tribunals; the redefinition of gender roles in relation to work, technology, and the biological and cultural reproduction of domestic communities; new forms of military confrontation and warfare; and conflicting standard-bearers of ritual and sacred power—all generated profound changes in the identities of the indigenous peoples of Sonora and Chiquitos and in the landscapes they built and inhabited. Through repeated cycles of conquest by Catholic ecclesiastical hierarchies, Iberian military forces, civilian colonists, and the incipient nation-states of Mexico and Bolivia, Sonoran and Chiquitano peasants and forest dwellers encountered the modern arena of market exchange, state formation, and the secularization of spiritual authority.

Environmental history has developed into an established field within the discipline over the last quarter century. A number of studies that incorporate, to a greater or lesser degree, environmental and cultural domains of inquiry have guided the research design for this project. Formative works from North American historiography include William Cronon, *Changes in the Land*, and his *Nature's Metropolis* as well as Richard White, *The Roots of Depen-*

Chiquitos landscape with Piraí River. Photograph by the author.

dency, and *The Middle Ground*, also by White.[42] White's former book achieves a three-part comparative history of the Choctaw, Pawnee, and Navajo peoples that integrates the central themes of environment, culture, and political economy. The latter book integrates a comparative interpretation of Algonquian, Iroquoian, French, British, and American conflicts and accommodations from the seventeenth to the nineteenth centuries. Both *dependency* and *middle ground* are widely influential terms, but neither is universally transferable to different historical settings. My discussion of mission economy, by way of contrast, hinges on the degrees of interdependency that bound colonized native peoples to colonial trade goods and, conversely, centered the colonial enterprise on both coerced and negotiated means of recruiting Indian labor.[43]

It is not accidental, by any means, that environmental history in the United States has been concerned largely, but not exclusively, with the contested advance of the western frontier and the confrontations among Native Americans, European and Anglo-American settlers, and African American slaves and free persons. Robert Williams advanced an alternative view of multicultural frontier societies that countered the precepts of U.S. constitutional law and treaty obligations with Native American nations through Iroquoian and Algonquian artistic and performative images of negotiation on the basis of consent and reciprocity.[44] Not all histories of conquest led to the containment of indigenous polities. Recent scholarship has pointed to the expansion of Native American territories during the colonial experience, as occurred with the Apache and Comanche bands of North America and the Araucanian and Guaraní nations of South America.[45]

Mexican historiography has taken a somewhat different path, honing its sensitivity to environmental issues through Mesoamerican traditions of anthropology. Teresa Rojas Rabiela, Bernardo García Martínez, Elinor Melville, and Arij Ouweneel exemplify the intersection of these disciplines and themes, centered on peasant ecologies in the villages and agrarian landscapes of central Mexico.[46] Alfred Siemens and Andrew Sluyter have contributed carefully honed historical and geographical studies of the wetlands and highlands of Veracruz, and Jonathan Amith has constructed a richly diverse cultural geography of central Guerrero.[47] Anthropologists and ecologists who work in the arid environment of northwestern Mexico have opened valuable interpretive windows on the seminomadic, farming-hunting-gathering cultures of this region on the periphery of Mesoamerica.[48]

Turning to South America, the notion of ecological niches commonly employed to explain the logic of peasant survival in the Andes is significant for the material and cultural ties between the highlands and the lowlands in both preconquest and colonial times. Its dramatic imagery reinforces the distances and contrasts between these two worlds, yet recent scholarship has questioned the spatial configuration of discrete niches in Andean geography and pointed to migrations and material exchanges across the presumed Andean-Amazonian divide.[49] Warren Dean's study of the Brazilian Atlantic rain forest over five centuries rearranges the familiar historical and anthropological themes of Indian and African enslavement, commercial agricultural cycles of sugar and coffee, and the growth of coastal urban centers around the forest, the principal subject of his history.[50] Three recent studies of late nineteenth-century developments in Argentina and Brazil bring together environment, economy, and the cattle frontier.[51] All of these works take up different narratives of destruction and persistence in the physical environments and cultural lifeways of the regions they represent.

The foundational questions that guide this study focus on how partially colonized peoples advance their claims for physical and cultural survival, how they reconfigure their territories in the face of colonial imperatives for the displacement of people and the appropriation of resources, and how their identities evolve historically. Recent powerfully molded histories of ethnic struggles for survival and dignity, set in both the colonial and national periods, address these issues of racialized social distinctions, discourses of exclusion, and the conflictive construction of the nation-state.[52] While these histories have, in large measure, set the terms for further debate on hegemony, agency, and alternative nationalisms, they have not made the environment or the social construction of space a central theme of their inquiry. This is, precisely, the contribution of the present work: to bring together nature and culture in a comparative study of colonialism and its historical outcomes viewed from the perspective of colonial frontiers that became ecological, ethnic, and political borderlands. Let us thus turn now to the humanly crafted landscapes of Sonora and Chiquitos, seen through the lens of Alvar Núñez Cabeza de Vaca before Europeans had named these provinces or established colonial boundaries through their deserts and savannas.

CHAPTER 1

Ecological and Cultural Frontiers in Sonora and Chiquitos

> After two days had passed that we were there, we decided to go in search of the maize [corn]. And we did not want to follow the road of the cows [bison] because it is toward the north, and this was for us a very great detour, because we always held it for certain that going the route of the setting sun we would find what we desired. And thus we followed our course and traversed the entire land until coming out at the South Sea.—ALVAR NÚÑEZ CABEZA DE VACA, *Relación*, 1542

The year was 1536 by the Christian calendar: Alvar Núñez Cabeza de Vaca, his two European companions, Captain Andrés Dorantes and Captain Alonso del Castillo Maldonado, and Estevanico, a Christian, Arabic-speaking slave from northern Africa, entered the final phase of their eight-year odyssey.[1] These four survivors of the failed Pánfilo de Narváez expedition to "La Florida" were near their goal to reunite with "Christians," Spaniards who had extended the dominion of the Castilian crown northwestward from central Mexico. The Cabeza de Vaca company at this point traveled on the Río Grande river upstream from its confluence with the Río Conchos. Here they turned southwest along the so-called trail of maize, traversing the Sierra Madre Occidental and approaching what later would become the province of Sonora. Their itinerary, constructed through the journey, itself illustrates the ways in which Sonora was connected to the greater Mexican north—geographically through the Colorado and Grande river systems and culturally through complex trade networks among different nomadic and sedentary peoples.

Mountains and Deserts of Sonora

The four sojourners did not travel alone. Dependent on the native peoples for their physical survival, they were accompanied by indigenous guides and followers—hunters, gatherers, fishers, and cultivators—who supplied the strangers generously with the fruits of their lands, including piñon nuts, mesquite flour, bison and deer meat, hides, cotton cloth, and maize. Men, women, and children of unspecified numbers danced, wept, and sang, drawn to the strangers by their reputation as healers. Their territorial and cultural boundaries are inferred, if dimly, in Cabeza de Vaca's recollection of the journey, by the approach and retreat of different groups of people along the way and, in some instances, by his references to native "lords" who distributed food and stolen weapons, tools, and clothing among their followers.[2] Different bands "robbed" or seized one another's possessions as they passed from one village to the next, in what may be interpreted as ritual forms of pillage and exchange closely associated with the strangers' healing powers.[3] When their current guides feared to continue into enemy territory, as occurred in northern Coahuila and Texas, they sent women as messengers, "because women can mediate even when there is war."[4]

Cabeza de Vaca's *Relación* and its summary inclusion in Gonzalo Fernández de Oviedo's *Historia general y natural de las Indias* have inspired numerous accounts of the expedition featuring sketches of the lands and peoples along the route.[5] These texts provide themes and images that serve as signs for narrations of conquest and cultural encounters in this frontier and in many colonial theaters: the combination of Europeans and Africans in the Iberian expeditions to the Americas; the key participation of indigenous peoples as interlocutors, guides, and followers; practices of captivity and warfare; disease and shamanistic healing rituals. As the following passage illustrates, the Cabeza de Vaca entourage, appearing more like a pilgrimage than an expedition, precipitated far-reaching mobilizations of different peoples and provided the occasion for the invention of new cultural patterns of encounter and exchange.

> We went through so many types of people and such diverse languages that memory is insufficient to be able to recount them. And the ones always sacked the others, and thus those who lost, like those who gained, were very content. We carried so great a company that in no manner could we make use of them. . . . And those who carried bows did not go before us. Rather, they spread out over the sierra to hunt deer. And at night when they returned, they brought for each one of us five or six, and many birds and quail and other game. Everything, finally, that the people killed, they put before us without daring to take one single thing without our first making the sign of the cross over it, even though they might be dying of hunger, because thus they had it as a custom since traveling with us.[6]

Turning southwestward, then, from the Río Grande, Cabeza de Vaca and his companions followed the trail of maize, passing through distinct climatic zones and cultural domains. They crossed the arid plains of the northern Chihuahua desert for over a month before reaching the pine-and-oak forests of the Sierra Madre Occidental, its slopes, valley floors, and steep *barrancas* (raised earthworks) carved by deep-cutting streams. As the party advanced "toward the setting sun," the highland cordilleras gave way to the *zona serrana* (highlands zone), the western foothills of the sierra marked by roughly parallel alluvial valleys separated by ranges and plateaus that reveal a complex geological history of sedimentation and volcanic flows. After traveling more than a hundred leagues (approximately 270 miles) through the *zona serrana*, they turned southward, skirting the eastern edge of the Sonoran Desert to cross the Yaqui and Mayo rivers upstream from their wide alluvial valleys.

In this piedmont area of scrub forests and cultivable streambeds, Cabeza de Vaca saw ample evidence of settled villages and, finally, "signs of Christians," evident in the artifacts they had left behind, such as a metal buckle and nail that one man wore around his neck, perhaps as a kind of talisman. More ominously, Cabeza de Vaca observed the abandonment of village sites and heard stories of strangers "who had come from the sky" and of Indian enslavement. For, as the party advanced through the river valleys of Petatlán, Sinaloa, and Mocorito, approaching Culiacán, it entered the northern reaches of Nueva Galicia, a Spanish province created by Beltrán Nuño de Guzmán and Diego de Guzmán through the violent defeat of the Tarascan kingdom and the ruthless subjugation of native peoples along the western corridor of Mesoamerica extending from Compostela to San Miguel de Culiacán (1529–31).[7]

Cabeza de Vaca and his companions were strengthened by the abundant food they received in this phase of their journey but troubled by the wake of destruction that the "Christians" had left behind them. Yet they did not fully grasp the cultural significance of their itinerary along the trail of maize. The mountainous regions of Sonora through which they passed marked the northwestern boundary of Mesoamerican landscapes, heralded by densely populated villages of permanent houses, built of earth and reed mats; linguistic affinities with Nahuatl; cultigens of maize, squash, beans, and cotton; and communally constructed irrigation systems. Indeed, the gifts presented to the travelers became more varied and exotic; in addition to harvested grains, deer meat, and cotton robes of fine quality, Cabeza de Vaca remarked on beads of coral and turquoise that the people had acquired by trade with "villages of many people and very large houses" to the north.[8]

Cabeza de Vaca did not visit the pueblos of New Mexico, nor did he see the great urban center of Casas Grandes (Paquimé) in western Chihuahua, yet his native guides led him along trade routes that connected Sonoran villages with key sites of Mesoamerican influence. The earliest archaeological remains for Casas Grandes date from the eighth century C.E., showing a cluster of village settlements that later developed into an urban trading and ceremonial center (1200–1490). Sonoran villagers exchanged food surpluses, cotton cloth, tanned deer hides, coral, and human captives in return for turquoise, copper ornaments, and ceramic trade ware produced in Casas Grandes. In addition, some villages in the Altar valley of western Sonora became sites of craft specialization, working shells gathered from the Gulf Coast into adornments destined for Casas Grandes.[9]

The northern Mesoamerican frontier was unstable and contested even before the crises that followed European invasion. Contrasting and parallel forces of population concentration and dispersion created fluctuating human landscapes of settlement and land use, detected archaeologically from the thirteenth to the fifteenth centuries and attributed in part to climatic changes, in part to cultural innovations and struggles for territory and power. Even as the Aztecs extended their tributary empire from the Valley of Mexico to the shores of Veracruz, from approximately 1428 to 1521, in the northern Chichimec frontier Casas Grandes fell into decline, was partially abandoned, and burned. It is probable that some of the villages through which Cabeza de Vaca passed in the Sonoran sierra were either satellite communities or migrant colonies of Casas Grandes peoples.[10]

Villagers of the Sonoran piedmont adapted Mesoamerican techniques to their environment, creating distinctive regional cultural patterns. The human geographies they created developed historically from different kinds of material and social interactions, which we may group as follows: (1) means of production and modes of land use combining horticulture, hunting, fishing, and gathering; (2) demographic movements over time and space that resulted in shifting village settlements and oscillating population densities in particular areas; and (3) trade, political alliances, and warfare among different chiefdoms. The internal histories of conflict, innovation, and adaptation of the Sonoran *serrano* peoples, coming from the highlands or piedmont, to their environment—shaped in part by their long-distance trading networks—gave rise to the cultural landscapes that the Cabeza de Vaca party and subsequent sixteenth-century Iberian expeditions encountered in greater northwestern New Spain. Three major expeditions followed portions of the Cabeza de Vaca route: Fray Marcos de Nizza, 1539; Francisco Vázquez de Coronado, 1540–42; and Francisco de Ibarra, 1564.[11] The chronicles that each of these expeditions produced referenced similar landmarks, citing in turn Cabeza de Vaca's observations concerning prominent house structures, agricultural valleys, and arid stretches of *despoblado*, areas sparsely populated or abandoned. These explorers, who sought mineral wealth, land, and people to conquer, were looking for permanent towns with surplus products that could be turned into tribute commodities. Thus they focused on the built landscapes of northern Mexico, exemplified by the village of Corazones ("hearts"), where Cabeza de Vaca and his companions feasted on hundreds of deer hearts, and again by Chilchilticale, a construction of puddled adobe that signaled a frontier between the *despoblado* lying northeast of Sonora and the westernmost pueblos of New Mexico.[12]

Our reading of these sources in the light of the archaeological record and contemporary landscapes distinguishes three major cultural and environmental regions in the province of Sonora, moving from east to west: highland cordilleras that merged with the Sierra Madre Occidental; the *zona serrana* of alternating valleys and ranges; and the desert coastal plains. Passage from one region to another was gradual, but perceptible in the physical markers of elevation, precipitation, and vegetation that signaled all of these ecological frontiers. Within each region, different spatial patterns left their imprint on the land. Some of these territorial configurations were seasonal, referring to sites occupied intermittently, abandoned, and revisited; shell middens on the northern Sonoran Gulf Coast, for example, may well be linked to village sites farther inland on the Concepción-Altar-Magdalena river drainages.[13] Others present a historical sequence of distinct phases of construction, expansion, and partial destruction or retreat, as is observed in the village settlements of the Sonora river valley described below. These built environments corresponded to the material artifacts of production and shelter, the village polities, and the ceremonial and aesthetic domains of native cultures.

The following description of the outstanding features of these cultural spaces and ecological zones is meant to emphasize the necessary connection between nature and culture, without implying an exact correspondence between particular historical peoples and the landscapes attributed to them. It is well to remember that the major tribal groups whose names and identities endured in the colonial record spanned at least two of the regions, the highlands and the piedmont, and they traveled through the desert to trade, forage, and conduct

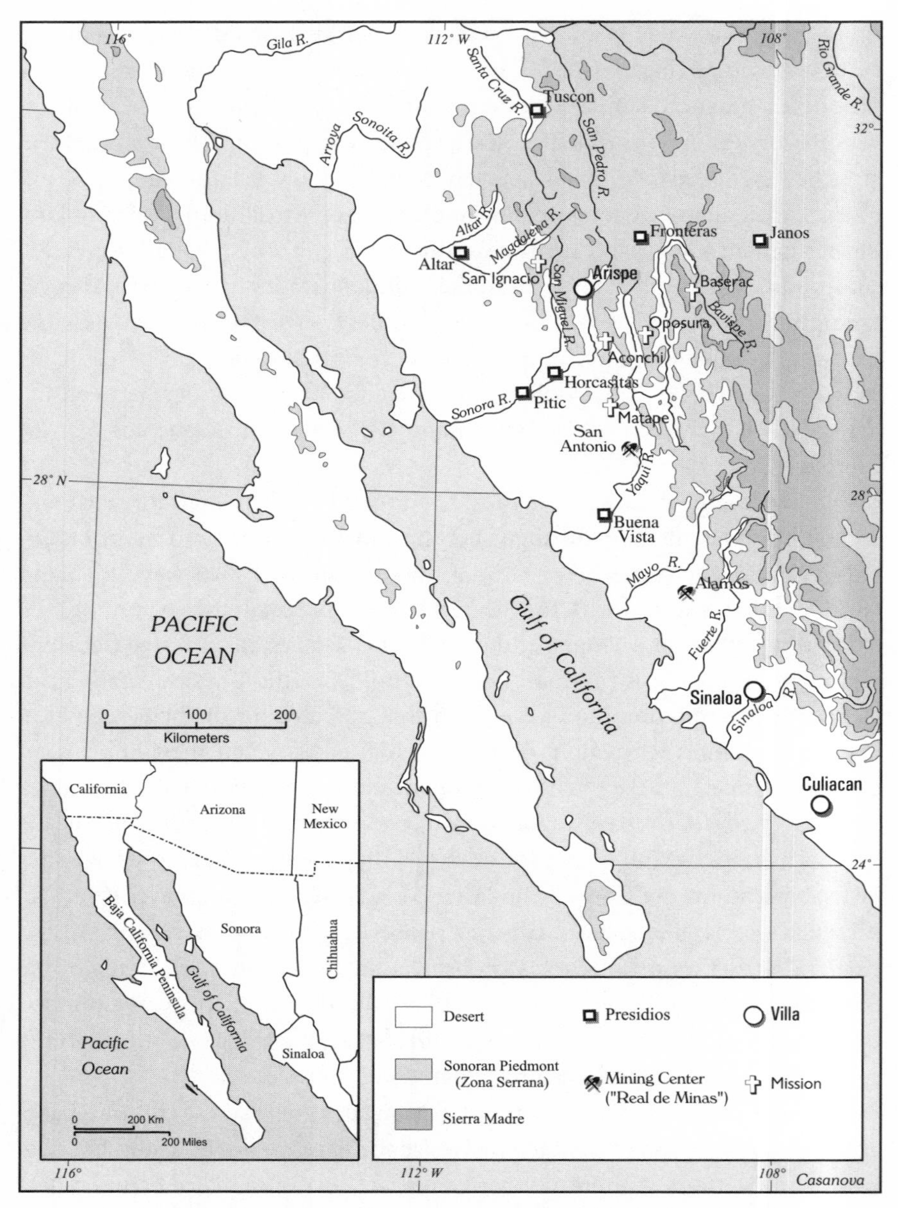

MAP 1 Sonora, place map drawn by Jason Casanova under the direction of Jane Domier, University of Illinois at Urbana-Champaign.

religious pilgrimages. This holds especially true for the Pima-Tepehuán, Yaqui-Mayo, and Opata-Eudeve peoples who dominated the mission communities of the seventeenth and eighteenth centuries, but whose territories were contested both before and after Spanish colonization. Ethnographic maps (including those reproduced in this book), although based on careful examination of both textual and archaeological information, offer but snapshots of probable territorial boundaries not fixed in time but changing through the course of historical events.

Cordillera

The Sonoran highlands extend westward from the escarpment of the Sierra Madre Occidental, forming a narrow band of approximately 150 km (90 miles) east-west and of 1,250 km (750 miles) north-south. The highlands are distinguished from the piedmont and coastal plains by their elevation, ranging from three thousand to six thousand or more feet above sea level, their relatively higher annual rainfall, registering from six hundred to eight hundred millimeters, and their relatively lower temperatures (average annual range of 14–16°C) than at lower elevations. Encompassing portions of the Bavispe, Oposura, Papigochic, and Saguaripa river drainages, and the ranges that surround them, the cordillera provides the climatic conditions for cultivation dependent on seasonal rainfall, using swidden techniques of field clearance and fallow.[14] Its narrow valleys and canyons supported permanent villages interspersed with encampments occupied for ephemeral plantings, hunting, and gathering. Building materials for houses and terraces comprised stone and sun-dried mud, reed mats woven from palm leaves and agave fiber, and light timber procured from saguaro ribs—*ocotillo* (*Fouquieria splendens* and *F. macdougali*), *torote* (*Bursera* sp.), and *encino* (*Quercus* sp.). Early Spanish chroniclers, beginning with Cabeza de Vaca, noted the *casas de tierra* or *casas de asiento*—permanent earthen dwellings—in both the cordillera and the *zona serrana* of Sonora.[15]

Highland villagers built their landscapes in a mixed setting of semideciduous forests, grasslands and, in the southernmost part of the province, tropical woods. Slopes higher than six thousand feet above sea level in northeastern Sonora were covered with various species of pine, juniper, and oak forests, while southern highland trees included *amapa* (*Tebebuia palmeri*), *pochote* (*Ceba acumminate*), and varieties of *torote* (*Bursera grandifolia*, *B. fragilis*).

At lower elevations different kinds of grasses, known generically as *zacate* (*Bouteloua* sp.), provided ground cover under the mixed growth of trees and bushes. In these transition zones between the cordillera and the *zona serrana*, the pine-oak forests gave way to varieties of mesquite (*Prosopis juliflora*) *torote*, *ocotillo*, and *palo verde* (*Cercidium microphyllum*).

In none of these vegetational zones did single species occur alone; rather, the extensive knowledge of plant husbandry that Sonoran peoples exhibited derived from the ecological diversity of their habitat. The savannas and forests of the highlands supported over five hundred known species and subspecies of animal life, ranging from insects and birds to rodents, reptiles, and mammals that included two kinds of deer (*Odocoileus hemidnus*, *O. virginianus*), coyote (*Canus latrans*), mountain lion (*Puma concolor*), bear (*Ursus americanus*), and wild boar (*Tayassu tajacu*). Many of these animals ranged widely through the province, but three species common to the desert lowlands included the iguana (*Ctenosaurus*), the desert tortoise (*Gopherus agasssizi*), and the wild great horned sheep (*Ovis canadensis mexicana*).[16] Sonoran men and women hunted and trapped many kinds of animals for meat, hides, antlers, horns, hoofs, bones, and feathers to feed, clothe, shelter, and adorn themselves, as well as to manufacture tools. We recall, here, that Cabeza de Vaca's party received gifts of dried deer hearts and finely cured animal skins at Corazones, probably in the heart of the Sonoran highlands. People constructed myths and stories about the animals' prowess, speed, and cleverness, referring especially to the deer and coyote, and drew pictures of different species of birds and mammals in their petroglyphs and rock paintings.[17]

Agricultural landscapes of the cordillera combined narrow alluvial floodplains, small garden plots watered by hand, and hillside plantings dependent on seasonal rainfall. Highland horticulturalists prepared swidden fields over two seasons of alternating wet and dry periods between the winter rains (*equipatas*) and the heavier rainfall of early summer. Fields were cleared and burned in two firings, requiring several weeks for the brush and trees to dry out between burnings. Farmers double-cropped with different varieties of maize, but they rotated planted fields and left individual plots to lie fallow. Several species of squash, gourd, and pumpkin complemented both winter and spring plantings of maize, along with tepary and pinto beans.[18] Fallow fields did not merely lie at rest, for they attracted small game and yielded amaranth (*quelites*; section *Amaranthotypus* Dumort) herbs, and other useful plants.[19] Beyond their cultivated fields and gardens, villagers of the sierra found a great array of resources in the forests and savannas that supplied them with food, fibers, medicines, soap, and building materials for shelter and tools.

Constructed spaces in the cordillera were distinguished by ribbons of narrow irrigated fields in the floodplains with small gardens and shallow wells at their edges; swidden plots in different stages of seeding, harvesting, and fallowing; and stone terraces built in canyons and close to caves and natural springs that provided planting surfaces, water, and protection against predators. Village settlements favored low hills or terraces overlooking rivers and arroyos since they depended on permanent sources of water and the floodplain itself was reserved for planting. Highlanders further altered their mountainous terrain by clearing trails and roads leading westward to the foothills and desert plains and eastward, through openings in the Sierra Madre, to Casas Grandes trade routes and to the Tarahumara and Tepehuán villages of Chihuahua.

Zona serrana

At once the heartland of Sonora and a transitional zone between the sierra and the coastal desert, the piedmont describes a wide territory in the northern part of the province that embraces the San Pedro, Santa Cruz, and Altar-Magdalena river drainages. Moving south and east, the *zona serrana* encompasses portions of the Oposura drainage and the Sonora and San Miguel valleys; south of Cumuripa, on the middle Yaqui river, it narrows to a slim band between the sierra and the alluvial deltas of the Yaqui, Mayo, Fuerte, and Sinaloa rivers. The piedmont is distinguished by a series of roughly parallel mountain ranges and alluvial valleys, with elevations ranging from 1,800 to 3,000 feet above sea level, and average annual temperatures of 14–20°C. Swidden is used rarely in this region since the average annual rainfall of four hundred to seven hundred millimeters, concentrated in the summer months of July and August, provides only scant possibilities for dry farming. Relatively wide floodplains in each of the valleys comprising the *zona serrana*, combined with sufficient gravitational flow, support irrigated plantings, created and maintained with *acequias* (irrigation ditches) and derivative canals, earthen weirs, and living fence rows. Native techniques of water and soil management have supported these cultivated landscapes in the piedmont, which integrate elements of visual and material culture, over many centuries of Sonoran history.

Serrano peoples created agrarian spaces characterized by irregularly shaped *milpas*, carved from the silt that flowed through the fence rows of cottonwood

Irrigation *acequia* in the Moctezuma River valley. Photograph by the author, 1979.

Newly planted fence rows at Mazocahui, Sonora River Valley, Sonora. Photograph by the author, 1990.

and willow saplings planted at the river's edge and intertwined with acacia brush. Repeated lines of planted trees at different angles to one another and the river channel, as well as the packed earthen weirs built across the streams were (and are) signs of a living agricultural architecture observed through archaeological remains, documentary descriptions, and contemporary practices in all the river valleys of the Sonoran piedmont. It bespeaks both the collective labor of peasant farmers and the ecological dynamic of human endeavor to maintain irrigated fields in the face of seasonal flooding, soil erosion, and the changing morphology of the rivers on which their livelihood depends.[20]

Village settlements in the *zona serrana*—as in the cordillera—favored the terraces and low hills of alluvial sediments and volcanic flows that overlook the rivers between the valley floors and the mountain ranges. Archaeological surveys and excavations in the San Miguel and Sonora river valleys suggest a history of domestic architecture and village distribution that, in turn, is related to the ebb and flow of population movements throughout the region. The earliest dated dwellings begin around 1000 C.E., located on the lower terraces; these houses in pits, semisubterranean shelters, were grouped in small hamlets. Two to three hundred years later, villages grew in size and number, combining houses in pits with surface structures. Approximately two centuries before Spanish contact, houses in pits had nearly disappeared; surface dwellings predominated in town sites that were less numerous than in the previous phase but that concentrated more population in each locality. Archaeological remains of fourteenth- and fifteenth-century dwellings in the Sonora valley conform to the *casas de asiento* that Cabeza de Vaca described, built of puddled adobe and stone. During this phase of village development, two large towns containing public architecture dominated key vantage points in the middle portion of the valley.[21]

This same period of pre-Hispanic history in the *zona serrana* represents a time of intense territorial rivalry within and between different river valleys. Population growth and increased social complexity in the Sonoran piedmont and cordillera may have fueled the intraregional wars that complicated indigenous responses to early Spanish invasions.[22] In addition to the agrarian and residential spaces carved into the floodplains and terraces of the Sonoran piedmont, selected hilltop sites were fortified with rock walls and embankments called *trincheras*. While archaeologists generally debate whether *trincheras* constituted defensive fortifications or ceremonial sites overlooking natural corridors between successive ranges and valleys, it is thought that some of

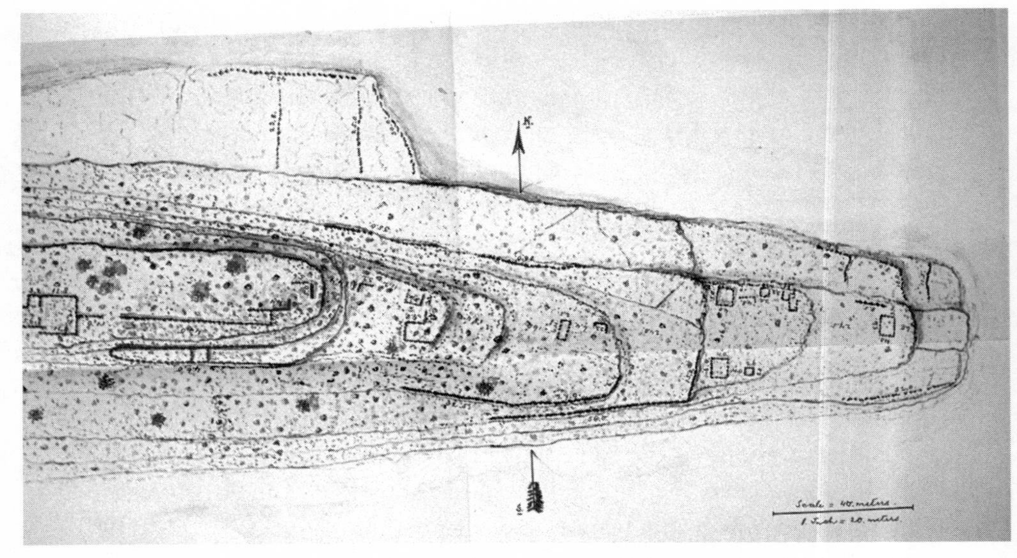

Pre-Hispanic *trinchera* (walled terrace). From Ernest J. Burrus, S. J., *A History of the Southwest* (Jesuit Historical Institute, 1969), plate 27. Fortified Town at Los Metates, Sonora. Catalogue number 500. Courtesy of Biblioteca Apostolica Vaticana.

the numerous *trincheras* sites mapped on lower slopes in northern Sonora and the central portion of the *zona serrana* served as agricultural terraces.[23]

Interspersed within this settled landscape of villages, towns, and fortified terraces, smaller hamlets signaled the movements of seminomadic peoples and seasonal displacements of *serrano* villagers to encampments for hunting and gathering in the scrub forest. While the Opata, Eudeve, and Pima (Nebome) were the principal ethnic groups associated with permanent villages and flood-plain farming, colonial documents ascribed names and linguistic affiliations to other groups, like the Jova and the Guarijío, reputed to move through the piedmont and sierra following game and the seasonal availability of plant foods and fibers. Mixed grasses, shrubs, trees, and various species of agave and cacti provided fruit, seeds, and beans; a storehouse of medicines; fuel; and the raw materials for thatch, cordage, building timbers, and baskets. Horticulture and gathering were closely related in the seasonal plantings dependent on summer rains and arroyo runoff (*temporales*) scattered in the larger arroyo beds away from the main river channels of the *zona serrana*.[24]

Sonoran Desert

The climate and vegetation most closely associated with the arid conditions of the desert extends westward from the Río Zanjón and the tributaries of the San Miguel and Altar-Magdalena rivers to the Gulf of California. The plains descend from a height of 2,100 feet to sea level, and the coastal lowlands are punctuated by numerous outcroppings of low hills and ancient fault lines. Average annual temperatures range from 20° to 25°C, with summer highs doubling these figures. Rainfall is significantly lower than in the *zona serrana* or the cordillera, at eighty-eight millimeters on the coast, but the distribution of annual precipitation over two seasons supports a varied and often beautiful vegetation of bushes, cacti, and agaves. Predominant species in this zone include acacia (*Acacia spp.*), iron wood (*olneya tesota*), *palo verde* (*Cercidium microphyllum*), *ocotillo* (*Fouquieria splendens*), *torote* (*Bursera microphylla*), mesquite (*Prosopis velutina*), and yucca (*Yucca arizonica*), which occur in the lower piedmont elevations as well. Plant life particularly identified with the Sonoran Desert includes jojoba (*Simmondsia chinensis*), *gobernadora* (*Larrea tridentata*), and veritable groves of distinctive cacti, including the organpipe cactus (*Stenocereus thurberi*), saguaro (*Carnegiea gigantea*), *pitahaya* (*Lemaireocereus thurberi*), and chollas and nopals (*Opuntia* sp.).

The Altar Desert holds the most arid lands of Sonora, in the extreme northwestern part of the province bordering on the northern coast of the Gulf of California and the delta of the Colorado River. The arid expanse of the Altar Desert is broken by several intermittent arroyos that form the Sonoita river drainage and by the dramatic beauty of the volcanic ranges and craters in the Sierra Pinacate. The lava fields and desert pavements of the Pinacate have yielded prehistoric archaeological artifacts, but since the late Holocene era, approximately eight thousand years before the present, extreme temperatures and aridity have prevented continuous human occupation of the area. Nevertheless, historic native peoples of Sonora have traveled through the Pinacate on pilgrimages to obtain lithic building materials and to fulfill religious needs—to dream new songs.[25]

The central Sonoran desert and gulf coast were home to numerous bands of Areñenos, "sand peoples" who gathered the fruit of the saguaro, *pitahaya*, nopal, and cholla buds, mesquite pods, sand roots, wild greens, and seeds of the *palo verde* and ironwood trees. These nomads developed stores of knowledge about their world and adapted their technologies to its constraints. Hunters, gatherers, fishers, and craftspersons, they exploited fully the seasonal cycles of the desert plains and the aquatic resources of the Gulf of California, its islands and estuaries. At the time of Spanish contact, two main groups lived in the desert lowlands: the Hokan-speaking Cunca'ac, known historically as the Seri, and the O'odham, related to the Akimel Piman peoples of the piedmont and highlands. The Cunca'ac, whose territory ranged from Guaymas in the south to Puerto Lobos in the north, subdivided further into several bands and subunits (*ihízitim*), distinguished by name and territory. Spaniards differentiated among the Guaymas, Salineros, Tepocas, and Tiburones, distinctions that corresponded roughly to overlapping territories but that did not express the complex kinship linkages of the Seri peoples.[26] The O'odham, in turn, were differentiated between the Hiach-ed or S-ohbmakam, nomads of the Altar Desert, and the Tohono or Papawi Ko'odham, seminomadic gatherers and seasonal cultivators: the "bean-eaters" whom the Spaniards would call Pápagos. Hiach-ed and Tohono O'odham defined their landscapes in part by the occurrence of *tinajas*, natural rock reservoirs and fissure springs where travelers could quench their thirst and small-scale plantings could be watered.[27]

Alvar Núñez Cabeza de Vaca heard stories of the desert people from his hosts in the village of Corazones, which he inserted into his report.

> [Corazones] . . . is the entrance to many provinces that lie toward the South Sea. And if those who should go searching for it do not pass through here, they will perish, because the coast has no maize, and they eat powders of *bledos* [wild greens] and of grass and of fish that they take in the sea with rafts because they do not have canoes. The women cover their shameful parts with grasses and straw. They are a people timid and sad.[28]

Cabeza de Vaca's characterization of the coastal tribes as "timid and sad," implying both material and spiritual poverty, reveals his European biases and preferences for agricultural lifeways and permanent settlements. The information he received from his fellow travelers, however, is strikingly accurate: the Cunca'ac harvested sea grass meadows, especially the seeds of the eelgrass (*Zostera marina*). Different species of littoral and mangrove scrub plants provided weaving fibers and sheltered oysters, crabs, and clams that, in turn, supplied them with nourishment.[29]

The architecture of desert peoples rested lightly on the land. Plant fibers and branches supplied their building materials; houses were assembled and disassembled, campsites revisited from season to season as nomadic O'odham and Cunca'ac moved frequently in search of game, ripening seeds and fruits, tubers, and the rich marine life of the littoral. Tohono O'odham occupied three different settings during the year: the wells (*wahia*), the fields (*oidag*), and saguaro groves. They hunted and gathered wild foods near permanent sources of water in the hills during the winter months. Tohono O'odham women harvested the fruit of the saguaro in the desert flatlands under the summer heat before the onset of rains; men then cleared and sowed their fields at the mouths of desert washes during the brief rainy season following the summer solstice.[30]

These three major regions of Sonoran landscapes comprised a complex mosaic of ecological and cultural frontiers across the desert, piedmont, and sierra. The mountain ranges, canyons, alluvial valleys, and desert plains that distinguished each of these regions served less as barriers than as corridors for travel and exchange within the province and beyond its borders. On the eve of Spanish contact, the Sonoran peoples had developed long histories of trade and warfare, population growth and dispersion, and of technological innovation. Their societies ranged from nonhierarchical bands to chiefdoms with possible tributary linkages to centers of Mesoamerican cultural influence. The religious and aesthetic expressions of both nomadic and farming peoples seem to have been closely linked to shamanism, health, and the natural environment that structured their world.

Our guide to northwestern Mexico, Alvar Núñez Cabeza de Vaca, returned to Spain in 1537, soliciting royal approval to seek wealth and extend the crown's dominion on yet another American frontier, but this time south of the equator. Cabeza de Vaca entered Asunción de Paraguay in 1541 with the title of *adelantado* (frontier governor) for the province of Río de la Plata, at a time when Spanish imperial objectives sought an inland route from the Atlantic to the mineral wealth of the Andes. Cabeza de Vaca's military expedition of 1543 northward along the Río Paraguay toward the Gran Chaco and Chiquitos plains pursued these objectives among diverse groups of allies and adversaries, both native and European. As we follow his route, we will explore the natural contours and historic peoples of the eastern lowlands of Bolivia.

BOX 1 Tohono O'odham Seasonal Calendar

Note: The O'odham annual cycle is arranged in twelve "months," interpreted to coincide with the solar months of the Gregorian calendar. The basis of O'odham seasons, however, is not exactly solar or lunar; rather, it rests on phenological observations of changes in vegetation, climate, and animal behavior. Source: Frank S. Crosswhite, "The Sonoran Desert," in *Reference Handbook on the Deserts of North America*, ed. Gordon L Bender, 254–55 (Westport: Greenwood Press, 1982).

Hahshani mashad: Saguaro harvest month, when families harvest, process, and eat the fruit of the saguaro (June)

Jakiabig mashad: Rainy month, time for the *nawait* ceremony to bring the rains and to plant bean, maize, and squash seeds (July)

Shopol eshabig mashad: Short planting month, last opportunity to plant crops; harvesting of early crops and gathering of mesquite pods; additional *nawait* drinking rituals to continue the rains (August)

Washai gak mashad: Dry grass month, after the summer rains have ceased (September)

Wi'ihanig mashad: Month of persisting vegetation, harvest of food plants that tolerate drought and frost, "when cold touches mildly" (October)

Kegh S-hehpijig mashad: Fair cold month, good for hunting (November)

Eda wa'ugad mashad: Inner bone month, of great cold when leaves fall (December)

Gi'ihodag or *Uhwalig mashad*: When animals have lost their fat, go into heat, and mate (January to February). Tohono O'odham might work for the Akimel

Kohmagi mashad: Gray month, when the landscape is bleak; *equipatas*, light winter rains (February)

Chehdagi mashad: Green month, new herbs and grasses, new leaves on the cottonwood and mesquite trees; saguaro ritual (March)

Oam mashad: Yellow-orange month, with flowers of desert poppy, brittlebush; stored food is running low (April)

Kai chukalig mashad: Painful month, or when the saguaro seeds turn black; desert food supplies are low, but the wheat harvests among the Akimel provides opportunities for work and food sharing (May)

Savannas and Forests of the Chiquitanía

The province of Chiquitos of eastern Bolivia is subtropical, located approximately within fifteen and nineteen degrees latitude (north-south) and fifty-eight and sixty-four degrees longitude (east-west) between the tributaries of the Amazonian and Paraguayan river basins and the Andean highlands (see map 2).[31] Comprising 370,621 square kilometers (143,098 square miles), its geography centers on the western portion of the Brazilian Precambrian shield, forming low outcroppings of mesas, buttes, and ranges that extend from southeast to northwest. The Chiquitano shield, in turn, is surrounded by alluvial plains and valleys, the humid llanos of Moxos to the north and the semiarid Chaco Boreal to the south. It is bordered by the Serranía de Santiago and Serranía de Chiquitos on the south, to the west by the Río Guapay and the alluvial plains of Santa Cruz, and to the east by the Paraguay River. A number of the rivers and streams that flow through the Chiquitanía lose their channels in swamps or bogs called *pantanales* or *bañados*, such as the Bañados de Izózog fed by the Parapetí River in the central southern part of the province and at the confluence of the Otuquis, San Rafael, and Tucabara drainages that straddle the present-day boundaries of Bolivia, Paraguay, and Brazil.[32]

Relief contrasts in eastern Bolivia form a less discernible pattern than in northwestern Mexico, so that it is difficult to speak of specific subregions like the highlands, piedmont, and desert described above for Sonora in Chiquitos. Local knowledge discerns different natural landscapes according to vegetation and the abundance or scarcity of water. The entire region presents a high diversity of forest, savanna, and wetland vegetation communities, distinguished as *bosque* (uncultivated forestland), *pampa arbolada* (wooded flatlands), *campo cerrado* (forested savanna), and *campo húmedo* (humid savanna). Their distribution corresponds to gradual changes in topography, soil structure, and precipitation. Average annual rainfall ranges between 1,000 and 1,600 millimeters; the average annual temperature registers at 24°C, but exhibits extremes of cold and heat reaching 3°C and 38°C, respectively. The soils corresponding to the Chiquitano Precambrian shield have low mineral content and are poorly adapted to agriculture, but the Chaco soils are fertile when sufficient water is available for continuous cropping cycles, as occurs in the Monte Grande and in the floodplain of the Río Guapay. Notwithstanding the relative poverty of these

MAP 2 Bolivia, place map drawn by Jane Domier, University of Illinois at Urbana-Champaign.

tropical soils, the *bosque alto* (high forest) supports scattered clearings for swidden plantings of seasonal crops of manioc (*Manihot esculenta* Crantz; known as *yuca* in the region), maize (*Zea mays*; or *oseórr*), peanuts (*Sterculia apetala*; called *maní* or *nankishiósh*), and cotton (*Gossypium barbadense* L.; or *nabósh*). Plantains and rice, two staples of Chiquitano cultivation at the present time, were no doubt brought to the region during the colonial period along with sugarcane, a commercial crop in the western part of the province.

The *pampa arbolada* and *campo cerrado* provide timber for fuel and construction, medicinal plants, game, and many different kinds of roots, bark, fruits, and seeds for gathering. People of the lowlands exploit all ecological canopies of their forests and savannas for herbs and grasses, bushes and trees, and they create multiple uses for many plants. The *almendra* (*Dipteryx alata*; or *nókümonísh*) provides firewood and its flowers and fruits are sources of food and medicines; similarly, the flower and fruit of the *chisojo* (*Terminalia argentea* C.; or *nánsübásh*), a tall savanna tree, yields medicine and dye, while *cuta de la pampa* (*Astronium fraxinifolium*; or *nópütotosh*) is used for building, fuel, medicines, and dyes. *Cuchi* (*Astronium urundeuva*; or *kükísh*) is commonly used for construction, and its flowers also provide medicinal benefits. *Bibosi* (*Ficus* sp.; or *nóbiosüsh*), observed frequently in village settlements and in the woods, has a medicinal fruit. Shorter trees and bushes that figure importantly in Chiquitano ethnobotany include the *chaáco* (*Curatella americana*; or *baküapósh*), found in the *campo cerrado*, and edible fruits of the *cusé de la pampa* and the *guapomó* (*Salacia elliptica*; or *nútachéns*). *Güembé* (*Philodendron undulatum*; or *tankósh*), a woody herb, is a common source of binding material used for construction, and its flowers are medicinal. Prominent species of palm trees that thrive in open savannas and in areas burnt for cultivation or for pasture include the *totaí* (*Acrocomia aculeate*; or *tutáish*), providing forage and edible fruit, the *motacú* (*Scheelea princeps*), used for construction and thatch, and *motacuchí* (*Allagoptera leucocalyx*; or *masúnunkutúsh*), with medicinal flowers and branches used as brooms (as indicated by its Chiquitano name). Thorny brush found commonly in the semiarid Chaco are the many species of *Mimosa* and three genera of cacti known in both Sonora and Chiquitos—*Bursera*, *Opuntia*, and *Cereus*.

Bosque, understood generally as uncultivated forestland for gathering or for clearing garden plots, is the locus of material existence and the source of symbolic meaning for the many different lowland peoples the Spaniards would call "Chiquitanos." Their cultures are marked by the rhythm of forest and garden in a counterpoint of natural power and domesticated space. Native cosmologies recognized diverse forms of power in nature to create, sustain, or destroy life, visualized as *jichis*. Distinct *jichis* are associated with water, wind, and different kinds of landforms—hills, woods, savannas—as well as with the human activities in nature such as hunting and cultivation. It is generally thought that *jichis* act as spirits or guardians to protect the forests, plains, and rivers, limiting human exploitation to satisfying basic needs. The Chiquitanos' spiritual world is complex—inhabited by *jichis* that watch over all animals and plants alongside many specific ones for particular species of fish, game, and plants. Most frequently invoked is the *jichi* of water, *tuúrr*, associated especially with the natural springs or openings in the earth with flowing fresh water.[33]

The human geography of Chiquitos followed the same general principles that we have outlined for Sonora: means of production and modes of land use combining horticulture, hunting, fishing, and gathering; demographic movements over time and space with shifting settlements and oscillating population densities; trade, political alliances, and warfare among different chiefdoms. Native peoples created cultural spaces in the *bosque* for scattered settlements and for their *chacos*, seasonal cultivated plots. *Chaco* may be related to the Quechua terms *chácara* and *chacú*, used widely throughout the Andean region.[34] *Chacú*, interestingly, refers to collective hunting, in which the hunters surround their prey in ever smaller circles. *Chaco* is unmistakably associated with agriculture, but it may figuratively allude to the image of encircling the forest by burning and clearing small portions of it through the combined labor of extended families.

Called *roza* or *roce y quema* in Spanish, the cycle of cutting, burning, clearing, planting, and harvesting in the *chacos* combines the rotation of crops within each plot with the rhythm of fallowing in which secondary growth partially reclaims the cultivated garden for the forest. Present-day crop cycles typically describe a sequence of three to five years, beginning with rice, peanuts, and maize—sown alternately for three seasons—followed by plantains, papaya, and manioc. When the *chaco* is left to fallow—in *barbecho* (Spanish) or *mocoé* (Chiquitano)—it is not unproductive; rather, the trees continue to bear fruit and much of the secondary growth has both edible and medicinal uses. *Camote* (*Ipomoea batata*; or *quibichorr*), a kind of sweet potato, is valued for its root and the curative properties of its leaves. Chiquitanos gather the fruit of the *Chirimoya* (*Annona* sp.; or *chirimoyarr*), a bush that grows in the forest near cleared patches, and plant the *Chirimoya cruceña* (domestic species) in

TABLE 2 Chiquitano Soil Classification Related to Crop Selection

Soils	*Cultigens*
Taturicürr (red)	Rice, maize, sugar, bananas
Cübusicürr (black)	Peanuts
Cüósosa (sandy)	Manioc, *camote*
Arricansa (rocky)	Peanuts
Taurr (if sandy)	Rice, manioc, *camote*, peanuts
Kanrr pacubeiza	Unfit for cultivation

Source: Schwarz, *Yabaicürr–Yabaitucürr–Ciyabaiturrüp. Estrategias neocoloniales de "desarrollo" versus territorialidad chiquitana* (La Paz: Ediciones Fondo Editorial FIA-SEMILLA-CEBIA, 1995), 106.

their gardens, using the leaves and seeds for medicine.[35] Arboreal cotton, usually planted at the edges of the *chaco* and in house gardens, is harvested regularly for its fiber, used for spinning and weaving.

Heavy cutting (*rozada*) begins at the beginning of winter, in the month of May, and burning takes place in the dry months of July and August. Clearing (*carpir*) of undergrowth following burning occurs before the onset of the rains in late September and October. Chiquitano native cultigens of maize and manioc (*yuca*) are found throughout North and South America, representing both seed- and root-propagated crops. Several varieties of maize are planted at different intervals from early October to November to account for their germination and maturation cycles in March and April: *maíz blando* (*oseomáto*), *maíz colorado* (*oseórr eturíki*), and *maíz duro* (*oseórr maskaárr*). *Maíz blando*, harvested as soon as it ripens, is used for making corn flour, bread, stews, and *chichi* (corn beer); *maíz duro*, used for *chicha* and animal fodder, can be left on the stalks for several months.[36] Similarly, manioc roots can be stored in the ground without spoiling, rendering the *chaco* a site for production and a reservoir of food and fiber from one season to the next. Rice is planted in early December, "before the feast of *la Vírgen purísima*," to allow the maturing stalks to absorb summer rains and produce their grain before the harvest four months later, at the beginning of the dry season.[37] Native horticultural techniques distinguish different types of soils according to their aptness for different crops, a criterion used for choosing where to clear the *chaco* and for selecting the crops.[38]

The *chaco* is basic to Chiquitano subsistence, but only part of the inhabitants' sustenance comes from harvested crops. Hunting, fishing, and gathering prove central to their physical survival and to their cultural expressions of sociability, aesthetics, and religiosity. I have noted above that the *bosque* and *pampa* supply a vast storehouse of edible and pharmaceutical raw materials, as well as of timber and fiber for building tools and dwellings. Hunting for mammals like the forest peccary (*Tayassu tajacu*; *taitetú* in Spanish; *nokichoriox* in Chiquitano), the tapir (*Tapirus terrestris*; *anta*; *nokitapakixh*), pampas deer (*Mazama gouazoubira*; *urina*; *noibox*), and the American deer (*Mazama americana*; *huaso*; *nunsiurixh*) provides meat, hides, and furs. Fishing for various freshwater species supplies a substantial part of the people's diet.[39] Rodents, especially two species of *jochi* (*Dasyprocta* sp., *nookixh* and *Agouti paca*, *nubaripirox*), as well as snakes, birds, and insects play an important role in Chiquitano hunting and gathering. For example, the people distinguish between multiple species of bees whose hives provide honey and wax. Honey is gathered during the wet season to assure a sufficient supply for human consumption.[40] Rhythms of foraging complement the planting and harvesting cycles of the *chaco*, and they serve as a social bond within and among families, involving whole communities in different phases of work and exchange.

The Jesuit Juan Patricio Fernández, one of the first missionaries to work among the Chiquitos peoples, described their daily routines and patterns of sociability as follows:

> At dawn they eat breakfast and together play musical instruments, similar to the flute, until the morning dew evaporates, because they consider it to be injurious to their health. Then they go to work, cultivating the land with wooden poles so hard that they compensate for the lack of plows and steel-tipped hoes. They work until the middle of the day, when they return [to their houses] to eat. They spend the rest of the day visiting one another, sharing food and drink, as a sign of amity and friendship. They pass around a jug of *chicha* and practice games and team sports. Women have this same visiting ceremony, and they have plenty of time for it, because their domestic chores are reduced to providing their homes with water and firewood, and cooking a broth of maize, vegetables, and other things that they find in the forest. And they are accustomed to spinning and weaving just enough [cotton] to make their *tipoy* [women's dress] or to make a shirt or a hammock in which to sleep with their husbands.[41]

Padre Fernández noted the Indians' combination of work and conviviality, but he did not report women's labor in the *chacos*. While men did the heavy work of felling tree trunks, women and children helped to burn and clear the plots, plant the seeds and yucca stalks, guard the plants from insects, harvest the crops, and process the food. Women made *chicha*, both the fermented and nonalcoholic corn beverages that provided (and still provide) a major part of Chiquitano nourishment and served as the centerpiece of their social and cultural ceremonies. In addition to their daily chores of child care and gathering wood and water, women spun and wove all their families' bedding and clothes.

Chiquitano communities created a variety of settled landscapes with different spatial dimensions. Precontact *rancherías* generally constituted small, semipermanent settlements. Their longevity over more than one generation and their distribution in the forest depended on patterns of mobility that, in turn, were linked to swidden cultivation and the shifting location of *chacos*, to the seasonal cycles of hunting and gathering, and to the consequences of warfare and changing political alliances. *Rancherías* comprised domestic communities of sibs, or extended families, among whom the harvests and spoils of hunting, fishing, and gathering were shared and work was exchanged through a system of reciprocal obligations known as *minca* (Quechua, *minc'ay*; Chiquitano, *metósh*).[42] *Minca* working parties came together for tasks like building and roofing houses, felling trees, and harvesting the *chacos*. To some extent the size of the community depended on the limits of these working relations and the forests and pampas through which its members moved to clear *chacos*, hunt, fish, and gather.

Chiquitanos carved their domestic spaces from the forests and the earth, using timber, reeds, vines, palms, and grasses of different species. Household structures observed in contemporary communities begin with *horcones*, timber columns sunk into the ground, cut from trees like the *cuchi*, *soto* (*Schinopsis brasiliensis*), or *cuta de la pampa*. The *horcones* support roof beams covered with light branches, such as the *cari-cari* (*Acacia polypilla*; *choborés*) and woven palm leaves of *motacú*. Horizontal strips cut from savanna trees like the *tajibo* (*Tabebuia ochracea*; *nónense*) and anchored to the *horcones* form the basic walls that are filled in with puddled adobe or woven plant fibers; the connecting beams are tied together with strong *güembé* vines. These same timbers in the interior of the houses provide surfaces for storage lofts (*chapapas*) and for hanging implements. Homes are spacious, furnished only with hammocks and low wooden benches. The enclosed spaces of Chiquitano houses, used

Roof repair in San Xavier, Ñuflo de Chávez, Bolivia. Photograph by the author, 1997.

principally for sleeping, are linked to outdoor patios and house gardens. Patios serve as kitchens where food is prepared and cooked, with fire pits and wooden implements—*batán* and *tacú*—for grinding rice, maize, and other seeds.[43]

The domestic architecture of these woodland peoples responded to the need to protect themselves from insects, a concern that emerges repeatedly in historical documents associated with the missions and in folkloric tales, while maintaining access to water and forested lands that could be cleared for planting. Thus native *rancherías* avoided the *bañados* and the dense forest, preferring raised ground and open savanna. Access to water was important, however, through *paurús* (waterholes) that were excavated and enlarged for bathing and fishing within proximate distances from settlements.[44] Houses were generally built without windows and had low-cut doors, and a small fire burned in the center of the house, its smoke coating the interior roofing to protect against insects and rodents.

The Chiquitanos' cultural and social life centered on their *rancherías*, even as the physical spaces created by these settlements linked them integrally with the forest. As we have seen, their skills for building, fashioning tools, procuring food, healing, and conserving the biota of their world grew out of their intimate knowledge of the *bosque* and the *pampas*. Just as forcefully, the spiritual guardians (*jichis*) who inhabited their cosmos derived their form and meaning from the natural environment of forests and streams in the Bolivian lowlands. The landscapes created by lowland peoples, illustrated by their *chacos* and dwellings, reflected concurrent patterns of settlement and nomadism that were grounded in their material subsistence and in bonds of kinship. Social and political allegiances were rooted in local communities, such that the ethnic identities that early Spanish chroniclers began to label as *naciones* and *generaciones* most likely signified temporary alliances among particular chiefdoms.

The paucity of archaeological studies for this region makes it difficult to construct in-depth cultural histories of the Chiquitano peoples. Nevertheless, ethnohistorical accounts and archaeological findings for greater Amazonia, the eastern highlands of the central Andes, and the humid llanos of Moxos suggest the contours of complex material and societal developments. The rhythms of subsistence described above provide important threads of continuity, but they do not constitute an unchanging culture. Patterns of settlement changed historically and created new landscapes through demographic concentration and dispersal, innovations in aquatic foraging and horticulture, and lithic and ceramic craft production. Cycles of conquest prior to the European invasions of the Americas, although only dimly understood in eastern Bolivia, point to territorial rivalries among riverine chiefdoms and on the frontiers of Andean imperial networks linked to the accumulation of different kinds of wealth and the quest for power.

The ecological diversity and abundant riverine resources of the South American lowland humid tropics created a promising scenario for cultural innovation and the development of complex communal and segmentary societies. From late Pleistocene nomadic hunter-gatherers, the Holocene brought significant adaptations to the warm tropical environment with seasonal flooding that established fertile floodplains for root cropping, fishing, and gathering, with initial ceramic remains perhaps as early as eight thousand years ago.

Dated archaeological findings suggest that between four thousand and two thousand years before the present, lowland populations grew with a proliferation of village settlements and the production of pottery; mound building may have begun in the Bolivian llanos during this era, where large habitation areas existed alongside small, dry-season fishing encampments. It is hypothesized that during the first millennium C.E., this early phase of horticultural and fishing villages was superseded by intensive seed cropping with the spread of maize cultivation. Continued population growth, craft innovation in technique and aesthetics—evident in polychrome ceramics—and long-distance trade gave rise during the period of 1000–1500 C.E. to complex societies and rival chiefdoms. The characteristic earthworks and raised fields of the llanos of Moxos belong to this late pre-Hispanic period. Complex societies throughout the lowlands created different landscapes, altering the topography, soil quality, and vegetation of their localities. Human industry changed the composition of forests and savannas through cultivation, nurturance, and adaptation of selected species.[45]

It is reasonable to suppose that the peoples of Chiquitos participated in the development of complex village cultures through their material subsistence base and their social and political lifeways, notwithstanding the lack of positive archaeological evidence for this region. Their proximity to Moxos and overland contacts with both the Andean foothills and the Paraguayan river network strongly suggests that Chiquitos constituted an important crossroads of cultural contact and exchange. Early colonial accounts report that some of the Chiquitano bands, most notably the Manazicas, had consolidated an ethnic territory of as many as twenty-two *rancherías* under the authority of a principal chief, extending northward toward the humid plains of Moxos.[46]

On the eve of Iberian contact, the Chiquitanía comprised a large inland frontier between the riverine cultures of the greater Río de la Plata and Amazonian drainages and the eastern periphery of the Andean world. It formed both a north-south and an east-west zone of transition, from savanna and forest to the dry brush of the Chaco. Chiquitano cultural contacts point to the Guaraní villagers of the Paraguayan tributaries and the Guaycurú hunter-gatherers of the Gran Chaco, as well as to the Arawak chiefdoms of Moxos. The central Brazilian cultures of Mato Grosso influenced Chiquitano lifeways, and the pre-Hispanic fortress of Zamaipata, in the foothills of the Andes, marked an important site of exchange and confrontation with Chiriguano and Quechua/Aymara peoples on the western boundary of the province. Zamaipata stands as a testimony to the limits of Inca dominion outside the central Andean cultures.

In lowland Bolivia, diverse cultural and ecological frontiers intersected the foothills and tropical plains extending eastward from the Andean escarpment. The provinces now comprising Valle Grande and Cordillera marked a contested border, traversed by the Chiriguano, Guarayo, and diverse forest peoples who would become the colonial Chiquitano.[47]

Europeans first sought access to the savannas and forests of these interior provinces from the Atlantic through the La Plata and Paraguay-Paraná river systems. The riverine port of Asunción de Paraguay, the first permanent Spanish settlement of the province of Río de la Plata, served as an important nexus between the Atlantic coast and the internal lowlands of South America. It was the point of departure for the early sixteenth-century expeditions that reached the eastern margins of Chiquitos. Pedro de Mendoza, Juan de Ayolas, and Domingo Martínez de Irala explored the Guaraní territories in the 1530s—after the first failed attempt to establish Buenos Aires at the base of the La Plata estuary and the founding of Asunción upstream at the confluence of the Pilcomayo and Paraguay rivers. When news of the deaths of Pedro de Mendoza (on his return voyage to Spain) and Juan de Ayolas (at the hands of the Guaycurúan nation of Payaguaes) reached the court in 1540, Charles I commissioned Alvar Núñez Cabeza de Vaca, conferring on him the titles of *adelantado* and governor of Río de la Plata, to return to the province to strengthen the frontier outpost of Asunción and end its isolation by founding new settlements and converting the Indians to Christianity and making them subjects of the king.[48] Cabeza de Vaca invested eight thousand ducats in the enterprise and, in return, he should have received the effective governance and dominion of the territories he claimed for the crown, as well as one-twelfth of the riches obtained from the lands and their peoples.[49] He served as governor from 1541 to 1544; from his base in Asunción, Cabeza de Vaca led and commissioned a number of voyages and overland explorations through the riverine network of Paraguay and portions of the Chaco Boreal as far north as the *bañados* of Otuquis near the present-day border between Chiquitos and Mato Grosso. His governorship ended abruptly, however, with his arrest and forced return to Spain, where he remained a prisoner of the court until 1547, seeking written and verbal testimonies to vindicate his service and reputation before the Council of the Indies.[50]

The narrative of his Río de la Plata experiences, entitled *Comentarios* and written by the chronicler Pedro de Hernández, constructs a studied image of Cabeza de Vaca as a just and capable governor, yet it reveals multiple conflicts among Spaniards, Portuguese, and different indigenous nations during this precarious stage of Iberian conquests in the Americas. The persona created for Cabeza de Vaca through the *Comentarios* does not bespeak a swashbuckling tale of daring conquests, as might have become Francisco de Pizarro or Hernán Cortés; rather, it emphasizes the correctness and orderliness of his actions and, above all, his scrupulous attention to the needs and sensibilities of the Indians. In Hernández's narrative, Cabeza de Vaca was careful to gather information about the land and its people, foreseeing possible dangers, before undertaking new expeditions. The governor paid for all the food, labor, and supplies that he received from natives at every stage of his explorations and generously bestowed additional gifts of trade goods on the chiefs—*principales*—with whom he conversed and negotiated to obtain guides and interpreters. A detailed account of his dealings with Guaraní villagers occurred early in his travels, inland along the Iguazú river toward Asunción, in December of 1541.

> Proceeding by land through the province the governor and his people arrived at a Guaraní village, where the principal lord of this pueblo, with all his people, came out to the road joyously to meet him, bringing honey, ducks, chickens, flour [ground piñon nuts], and maize. Through the interpreters [Cabeza de Vaca] arranged to speak with them and reassure them, thanking them for coming and paying them for what they brought, for which they were very happy. Beyond this, he ordered gifts of scissors and knives and other things to be given to the *principal* of this pueblo, whose name was Pupebaje; from there, they continued their journey, leaving the Indians of this village so happy that they danced and sang for pleasure in all the pueblo.[51]

Furthermore, Cabeza de Vaca forbade the expeditionaries traveling in his party to traffic with the Indians on their own account, insisting always on the prodigious distribution of gifts. It would seem that he brought to Río de la Plata two cardinal lessons from his previous adventures in northern New Spain: the importance of communicating (through interpreters) in the Indians' languages and the necessity of establishing good relations with the native peoples on whom the Spaniards depended for information and their very survival. Cabeza de Vaca articulated his policy explicitly during one of his first forays north of Asunción in 1542—at the behest of the Guaraní—to make war on the Guaycurús, avenging their destruction of the Guaraní town of Caguazú. The governor embarked on military action only after conferring with the clerics in his company and ordering the pronouncement of the *re-*

querimiento (an initial declaration of Spanish dominion) to the Guaycurús, thus following legal procedure no matter how meaningless the act was for the native warriors. He was solicitous of retaining the Guaraní as allies, because, "if we were to break with the Indians, and it were not remedied, all the Spaniards who were in the province could not sustain themselves nor live there, and they would be forced to abandon it."[52]

Hernández's portrayal of Cabeza de Vaca countered his accusers, who claimed that he had restricted trade with the Indians to his own closed coterie, mistreated the Indians, and aborted an expedition in search of gold in the region of the upper Paraguay river against the advice of the royal officials who accompanied him. These policies seemed to bring Cabeza de Vaca greater favor with the Indians than with the Spaniards, both those who accompanied him and those who were already settled in Asunción, restless to seek new conquests and to acquire *encomiendas*, grants of Indians held in service to them.[53] He would later pay dearly for his standards of conduct. The events surrounding Cabeza de Vaca's governorship of Río de la Plata, including pointedly Juan de Ayolas's earlier failed exploration and Cabeza de Vaca's arrest at the hands of royal officials ostensibly under his authority, occurred when Spaniards and Portuguese fervently hoped to find mineral wealth in the inland expanses between Asunción and the Inca treasures of the Andes. The importance of this second Cabeza de Vaca narrative for our story lies in its descriptive approaches to the Chiquitanía, with references to the environment and indigenous material culture, and in its composed portrayals of Spanish logistics, native chiefdoms, and the practices of warfare.

The overall impression gleaned from the *Comentarios* on the natural environment praises the bounteousness of the land. Rivers, springs, and arroyos flowed with clear water through meadows, savannas, and woodlands, which Hernández judged apt for farming and livestock.[54] Spaniards and Indians hunted deer, otters, and forest peccary, and the Indians' gifts of food combined the harvests of maize and manioc from their swidden plots with gathered tubers, fruit, honey, and dried fish. These landscapes of abundance were found in riverine floodplains and the territories settled by native villagers, largely the work of indigenous cultivation and selective gathering. When Cabeza de Vaca or his lieutenants diverted their route through *despoblados*, the narrative turned to stories of hunger, thorny brush, and swampy reeds that obstructed their path.[55]

In the lands of the Mataraes and Guaxarapos, along the tributaries of the upper Paraguay river, the Cabeza de Vaca party observed seasonal fishing camps at low water. At this time "of good life," the natives "dance and sing all day and all night, like those who are sure of having enough to eat." When the rainy season beginning in January brought flooding, the people withdrew to *barrancas* and, at the height of the floodwaters, lived in their canoes, fishing and hunting in the remnants of highland forests. As the floodwaters receded, in March, the people navigated toward their villages on the *barrancas*, but remained in their canoes until the stench of dead fish had passed, left behind on the newly exposed fields. Indians and Spaniards alike took sick from the intense heat and the smell of rotting fish. At the end of the tropical winter, in October, the expedition passed through deep, swiftly flowing water laden with golden fish (*peces dorados*) that supplied the entire party with food and oil.

Hernández characterized the fishing peoples of this aquatic world as "savages, moving around from one place to another." They were "a frontier people" without principal leaders, in contrast to the agricultural nations downstream. During this part of the journey Cabeza de Vaca ordered that none of the Spaniards or Guaraní who accompanied him should leave the ships to roam on land, nor should they mix with the local people, but leave them in peace.[56] The expedition had, indeed, crossed both an ecological and a cultural frontier. A little distance upstream from the golden fish, the river channel split into numerous shallow streams and lagoons, so many that even the Indians who navigated in canoes at times lost their way.

Cabeza de Vaca left the Paraguay river and entered another stream that local Indians called Iguatu, probably at the southern margins of the *pantanales* that extended northward into Mato Grosso. He led the company through several more lagoons, entering the country of Sacocies, Saquexes, and Chanés, where the governor waited to proceed until he was assured of a peaceful reception. Finally, it was necessary to portage the brigantines and canoes through low water until the party reached a navigable depth and, in November of 1543, they arrived at the Puerto de los Reyes, founded and named earlier that year (6 January, Epiphany) by Domingo Martínez de Irala. Irala's smaller expedition had explored four days' journey westward of the river port into the eastern plains of Chiquitos. Cabeza de Vaca took possession of the Puerto de los Reyes, as well as the lagoons and lands surrounding it, in the name of the king; he ordered a cross to be built and placed on the shores of the lagoon, and two settlements to be built for Spaniards and Guaraní apart from the Sacocies' villages. Once again, he enjoined the allies under his command not to enter the local villagers' houses, nor to disturb them. It was his intention to use the port as a base for further explorations inland, counting on the abundant resources and the Indians' knowledge of the territories and peoples that lay beyond.[57]

Cabeza de Vaca's careful questioning of the Chanés *principales* elicited the kind of information that encouraged him to undertake new expeditions. All the tribal peoples of the interior provinces, he was told, cultivated maize, manioc, sweet potatoes (*batatas*), and *mandubíes* (perhaps similar to *chirimoyas*), at times gathering two harvests a year, and raised domesticated ducks and hens. Different groups, like the Chimeneos, Carcaraes, Gorgotoquies, Payzuñoes, Estarapecocies, Candirees, and Xarayes—each comprising numerous villages—recognized the authority of singular chiefs and were continually at war with one another. Their weapons were bows and arrows, which served also as trade items along with cloth (*mantas*) and women; and, yes, some of these village-dwelling peoples had gold and silver. Notwithstanding these optimistic reports, and the governor's initial attempts to select capable guides and find a passable route to the land of the Xarayes, his preparations were delayed by news of the Guaxarapos' attack on the second part of the expedition led by Gonzalo de Mendoza; the Spaniards had considered that nation trustworthy allies.[58]

In late November 1543, Cabeza de Vaca led a company of three hundred armed Spaniards and twenty guides, which cut a path through thick brush without advancing to the villages and riches they had hoped to find. Eight days later, with dangerously dwindling supplies and fearful for the safety of the colony he had left behind, Cabeza de Vaca returned to Puerto de los Reyes. Confronted with the news that the Guaxarapos, allied with the local groups, had conspired to kill the Spaniards and drive them from their land, and facing the specter of hunger during the approaching flood season, Cabeza de Vaca sent emissaries by land and by water to search for food supplies in return for trade goods of beads, knives, iron ax heads, and metal hooks. Two expeditions led by Gonzalo de Mendoza and Francisco de Ribera, in December 1543 and January 1544, met with overt hostility; their guides and interpreters began to falter, and they returned empty-handed. Ribera, traveling west for twenty-one days, encountered Indians harvesting maize, and whose language none of his party could understand. Their *principal* led him by the hand to a large house built of wood and straw, where he offered them fermented maize *chicha*. Even as the Spaniards observed large clay vessels with stored maize and women occupied with emptying the house of cotton cloth and silver plates, bracelets, and hatchets, they were surrounded by armed warriors plumed and painted for battle. Amid war cries and a cascade of arrows, Ribera and his small band escaped, wounded, to return to Puerto de los Reyes. Cabeza de Vaca now confronted the situation he had most feared: without the support of indigenous nations, in unfamiliar territory, afflicted with sickness and insufficient food, it was necessary to abandon Reyes and return to Asunción.[59]

Cabeza de Vaca lost his governorship and failed to complete his ambition to discover new lands. Despite his entreaties for peace, fair trade, and Christian vassalage, the *adelantado* learned that warfare governed the indigenous polities of the greater Río de la Plata provinces and constituted one of the defining principles of their culture. In the context of early Indian-Iberian encounters, warfare provided an irreplaceable measure of male status through combat, ritual killing, and the capture of enslaved persons. Equally important, war operated as a political exercise to test alliances, forge new intertribal networks, and create reciprocal sets of obligations across lineages and bands. Native chiefdoms played a pivotal role in both the peaceful, negotiated and the violent, destructive phases of war, as is illustrated by the prominent appearances of indigenous *principales* in Cabeza de Vaca's narratives and in other early conquest chronicles. Nevertheless, the chiefdoms of the greater Paraguayan-Chaco-Chiquitano theaters were not as centralized or as hierarchical as the Spaniards imagined. If, indeed, the *principales* exhibited the insignia of power and held sway over their domestic societies, the ethnic identities they claimed were splintered among numerous bands and settlements and did not support unitary political authority.

Warfare was intimately entwined with the principles of gender and family that underwrote their social foundations. Women (and children) were taken as captives and hostages in the frequent skirmishes that marked the rhythm of Native American political cultures. Women were offered in marital alliances through rituals of both violence and peaceful negotiation; the exchange of women signified a palpable recognition of political alliances that crossed ethnic lines, including the Indian-European divide.[60] No less important, women in marriage or war captivity brought with them dual access to labor: their own productive work as agriculturalists, craftspersons, and food processors, and the labor of their male relatives. Women's reproductive labor raised new generations of dependents who would soon work for their masters. Spaniards eagerly accepted the sexual services and work of the women they received, but contributed only partially—through expedient wartime alliances—to the social networks of reciprocal obligations so important to the Indians' cultural expectations.

In this setting, the numerous references in the *Comentarios* to indigenous chiefs who came out to the road from their villages to greet the approaching expedition *with their women and children* point to a deliberate sign of their peaceful approach. It may well have been their intention to offer women to the Spanish soldiers and Guaraní warriors together with the copious supplies of food they exchanged for trade goods. Cabeza de Vaca's chronicler refers only tangentially to the women who, in fact, were given to expeditionaries along their journey. When the governor made the decision to leave Puerto de los Reyes and return to Asunción, he admonished the captains who had received as many as a hundred young girls (*muchachas*) from their fathers not to bring them with them, but to leave them with their native families in their own land.[61] This single act, more than any other, according to Pedro de Hernández, kindled the Spaniards' wrath against Cabeza de Vaca—a sign, perhaps, that they had begun to absorb Guaraní cultural values without fully understanding their meanings.

The privileged alliance that this first generation of conquistadores, including Cabeza de Vaca, forged with the Guaraní may signal another significant misreading of indigenous political culture. Unquestionably the material support of different Guaraní chiefs first gained Spaniards entry into the Río de la Plata, but their military alliance soon proved to be more limiting than empowering, since the Spaniards' highly visible association with the Guaraní earned them the enmity of many Guaycurúan tribal peoples. The spectacular Spanish-Guaraní defeat of the Guaycurús at Caguazú undoubtedly had repercussions across the Gran Chaco. We shall recall that the Guaxarapos attacked the Spaniards and Guaraní who were camped at Puerto de los Reyes; the inland tribe of Tarapecocies had denied Francisco de Rivera food and driven him and his small force from their lands. When Cabeza de Vaca later questioned some of their tribesmen in Puerto de los Reyes as to why they had attacked the Spaniards who came to them in peace, they admitted that they were not at war with the Christians, but that they had turned on them because they had come with the Guaraní, who were their enemies.[62]

Different chiefs and tribal bands accumulated experiences with the separate European invading forces. The Sacocies, Chanés, and Tarapecocies who met with Cabeza de Vaca at Puerto de los Reyes had earlier encountered Domingo de Irala and recalled their first meeting with Alejo García, a survivor of the

Juan Díaz de Solís expedition, who guided Juan de Ayolas into the Gran Chaco.[63] Native *principales* sought out their counterparts among Spanish and Portuguese expeditionaries as temporary allies in pursuing their own objectives of defense or aggrandizement in war. At the same time, indigenous leaders tested Spanish resistance. As related in the *Comentarios*, the trial expeditions led by Cabeza de Vaca, Gonzalo de Mendoza, Francisco de Rivera, and Hernando de Rivera northwest of Puerto de los Reyes—during the 1543–44 season of rain and flooding, with diminished numbers and limited supplies—emboldened local tribes to refuse the Spaniards' overtures for gifts and guides. Observing that the Spaniards were forced to beg for food and that their navigational technologies failed them in the *bañados* and plains of Chiquitos, the Indians saw them as vulnerable intruders who could be routed. Domingo de Irala again crossed portions of the Chaco Boreal and Chiquitanía in 1548, but Spaniards did not establish permanent settlements north of Guaraní territory for another two decades.

Uneven Conquests and Shifting Colonial Frontiers

Together, these two narratives of the landscapes built by the peoples of Sonora and Chiquitos, their multiple ecological and cultural frontiers, and their first encounters with Iberian invaders, provide a comparative picture of the tentative character of European conquests in both northwestern New Spain and the greater Río de la Plata. Spanish colonialism only gradually incorporated Sonora and Chiquitos into the northern and southern frontiers of its American empire, for reasons integral to the natural environments and complex cultural mosaics of both provinces. The Spaniards' need for reliable sources of food and forage met formidable obstacles in the contrasts of desert and highlands in Sonora and in the savannas and tropical forests of the greater Paraguayan basin. Just as pointedly, their expectations of wealth and labor in servitude provoked armed resistance on numerous occasions by the decentralized chiefdoms in both of these frontiers.

Sixteenth-century conquest expeditions to Sonora proceeded north from Compostela—following the westernmost trade routes from Mesoamerica in search of mythic cities, precious metals, and Indian laborers—and westward from the silver mining *reales* of Zacatecas and Durango. Neither the ruthless conquests of Nuño and Diego de Guzmán, nor the epic explorations associated with fray Marcos de Nizza and Francisco Vázquez de Coronado, succeeded in creating enduring Spanish settlements or in establishing *encomiendas* among the indigenous nations north of San Miguel de Culiacán. During this same period, the Chiquitanía faced three avenues of approach. As we have seen, numerous expeditions comprised of a few hundred Spaniards and several thousand Guaraní auxiliaries used Asunción de Paraguay as their base of departure to explore routes of access to the mineral wealth of Alto Perú. Portuguese explorers and slave hunters, *bandeirantes*, invaded Chiquitos from Mato Grosso; looking westward, the development of the Andean mining economy created linkages between the lowland plains and La Plata and Potosí. At about the same time that Francisco de Ibarra crossed the Sierra Madre Occidental into Sonora, in 1563, Ñuflo de Chávez founded Santa Cruz de la Sierra, a precarious villa of adobe dwellings at the base of the serranía of San José on the southern edge of the Chiquitanía. The settlement subsequently retreated westward to a site known as San Lorenzo de la Barranca, above the floodplain of the Río Guapay and within view of the foothills of the Andes. Under both names, the villa remained the focal point of contact between Spanish *encomenderos* (recipients of an *encomienda*) and the indigenous peoples of Chiquitos for over a century.[64]

Early colonial contacts in the Chiquitanía grew out of the conflictive demands of *encomienda* service, intermittent raiding and warfare, and the trade for iron tools. As was illustrated by Cabeza de Vaca's attempts to forge alliances, native peoples throughout the greater Río de la Plata accepted the Spaniards' terms of gift exchange largely to procure metal implements. The introduction of iron ax heads, hooks, steel knives, and scissors into the lowlands initiated a veritable technological revolution by altering methods for clearing the forest, hunting, fishing, and construction. Access to metal tools most likely intensified intertribal warfare, and their continued supply proved central to negotiations between native peoples, Spanish colonists, and missionaries. Metal tools circulated in the Sonoran frontier, as well, although their importance is less consistently documented there than in Chiquitos. Eighteenth-century Jesuits reported that northern Pimas received knives and iron implements from Spanish colonists and, in turn, traded them to the "inland" *rancherías* of the Colorado and Gila river valleys.[65]

Sixteenth-century Spanish conquests in both provinces remained inconclusive, yet they brought lasting changes to the human ecologies of Sonora and Chiquitos. The vulnerability of frontier outposts like San Miguel de Culiacán and Santa Cruz de la Sierra demonstrated the need for missionary orders to "reduce" native communities to supervised towns in fixed locations, a process that marked the transition from conquest as private enterprise to state imperialism. Mission culture had a profound impact on indigenous lifeways within the contexts of a transformed disease environment, technological innovation, and the development of the colonial political economy. The ecological and cultural frontiers of northwestern Mexico and eastern Bolivia endured, but their dimensions and boundaries as zones of contact were altered significantly as the colonial order gradually imposed its dominion on the territories and peoples of the desert and forest. These themes provide the organizing framework for the following chapter on political economy, community, and mission culture in the Jesuit provinces of Sonora and Chiquitos.

CHAPTER 2

Political Economy: Communities, Missions, and Colonial Markets

This chapter narrates the historical experiences of missionaries and native peoples who built economic and political networks in Sonora and Chiquitos, based on the productive technologies described in chapter 1. It portrays reconstructed communities within the missions, underscoring the contrasts between forest and desert and the mission towns (*reducciones*) intended to enforce permanent settlement in both of these nomadic frontiers. Beginning with the early stages of missionary *entradas*, referring to the overtures—both peaceful and coercive—made to native *rancherías* to accept mission life in the uncertain climate created by the fearful sequence of epidemic diseases and by Spanish demands for forced labor under the *encomienda*, the chapter then turns to the political economy of the established missions. *Political economy* is used here in its fullest sense: the polis, or polity of internal governance, representing the institutionalized indigenous community, and the *oikos* (house), the managed economies of production and distribution that sustained the ecological and economic foundations of the missions.

The imagery of anvils and looms, scythes and grinding stones illustrates gendered divisions of work and technology in the day-to-day life of the missions. These symbols of material culture are related to the mission-as-enterprise through trade linkages to the mining economies of New Spain and Upper Peru and the development of market economies that forged divergent trading networks in each region after the decline of the Jesuit administration. This chapter analyzes the different kinds of marketing circuits that operated in Sonora and Chiquitos in terms of the native peoples' experience with barter, trade goods, monetary values, work, and contradictory notions of wealth. It concludes with the initiatives of Sonoran and Chiquitano mission residents to bend the colonial market system to their own purposes and their defense of the communal assets of their mission towns.

The contrasting natural environments and different social settings of each region illustrate the complexities of trade and market exchange under the constraints of colonialism. Moving necessarily beyond the mission communities themselves, our comparison shows the different means by which ambitious colonial settlers accrued wealth and political sway over their respective frontier regions. The Jesuit administrative system provides a common institutional reference, even as the contrasting geographical parameters of Sonora, in northwestern Mexico, and Chiquitanía, in eastern Bolivia, enrich the dual histories. The economic life of these two mission districts, and the stories of which they form a part, outlived the Jesuit administration of both Sonora and Chiquitos and extended beyond the formal limits of Spanish colonial rule into the early national period. For comparative purposes, however, the present chapter is centered on the period of 1720–1810. During this time, the expulsion of the Jesuit order from all Spanish dominions, occurring in 1767–68, provides a fulcrum point of before and after to assess comparative processes of change that are both temporal and spatial.

As we saw in chapter 1, Sonora constitutes the Sonoran Desert and the semiarid highlands, bounded by the Gulf of California and the Sierra Madre Occidental. Five major river valleys flow southwestward from the foothills of the Sierra Madre to the gulf, creating successive basin and range formations that support irrigated cultivation in the lowlands (see map 1). The entire mission enterprise in Sonora rested on the agricultural and livestock grazing resources of the valleys and grasslands of the cordillera and *zona serrana*. While hunting and gathering in the *monte* of scrub forest remained important for the Sonoran peoples, their individual cultivation plots were watered by the irrigation canals (*acequias*) that served the mission as a whole. By way of contrast, the fluvial arteries and distinct vegetation communities in the humid tropical province of Chiquitos, lying between the Andean highlands and the tributaries of the Amazonian and Paraguayan river basins, supported colonial missions in different ways (see map 2). The labor of indigenous men and women was as essential to the economic life of the Chiquitos missions as in Sonora, but the agrarian subsistence base of these settled communities was less integrally tied to the market economy of the Jesuit reductions or to their spatial location in the forest. Livestock formed an essential part of the missions' patrimony, bred and herded in separate *estancias*. Indigenous families met their basic needs through their swidden plots, hunting, fishing, and gathering, and turned to the missions for meat rations, tools, and trade goods that became a central part of their material culture and their standards of aesthetic beauty and social prestige.

Sonora and Chiquitos both constituted marginal areas of the Spanish empire on the fringes of the mining centers of northern New Spain and Upper Peru. Each of these frontier provinces had relatively sparse European settlement; in both areas agriculturalists and seminomadic hunters and gatherers submitted to Spanish rule through the institution of the religious mission administered by the Jesuit and Franciscan orders. Notwithstanding these broad similarities, the cultural processes of subdivision and commingling among different peoples brought about by the colonial missions led to different kinds of ethnic mixtures and identities in the two regions. By the mid-eighteenth century, the numerous ethnic and linguistic groups of highland Sonora had coalesced into a few polities or nations, identified in colonial nomenclature as the Opata, Pima, and Eudeve, and these would form the core of the Spanish-speaking mestizo peasantry of the region by 1900. A different picture emerges in eastern Bolivia, where the larger, more urban mission centers of Chiquitos continued to recognize distinct language and kin groups known as *parcialidades* that retained ethnic and cultural significance. The peoples known as the Chiquitanos, while differentiated among themselves, remained separate from the non-Indian population of the province, an ethnic divide that remains even to the present day.[1]

Seasonal migratory patterns made for a defining feature of the cultural systems of Sonoras and Chiquitanos.[2] Their physical existence depended on the multiple resources of game, wild plants, and cultigens that, in turn, required access to different ecological niches within widely defined territories. Equally significant, the kinship systems and lines of ethnic differentiation that supported their rationale for conjugal coupling and societal groupings implied a spatial distribution of shifting communities, as well as larger nucleated villages. The territorial mobility on which the Sonoran and Chiquitano peoples relied for their economic production and their social reproduction clashed repeatedly with the colonial policy of concentrated settlement of mission neophytes in fixed settlements. In both of these frontier provinces, the missionaries eventually reached a compromise with the cultural and ecological imperatives of the indigenous environment. Their demands for the Indians' labor took into account the fruits of gathering—especially honey and beeswax—which they put to commercial purposes, and they learned to accommodate mission discipline to alternating seasons of hunting and cultivation.

Conflicting Conquests: Mission *Entradas*, Mines, and *Encomiendas*

As we begin both stories, we shall keep in mind the following questions that guide our discussion of missions as institutions, communities, and economic networks: What were the degrees of coercion and choice that led the Sonoras and Chiquitanos to enter the missions and recreate distinctive ethnic communities there? What were the systems of production and redistribution that underlay the religious and political life of the missions? What different relations of accommodation and conflict developed between the mission, a European transplant and a vehicle of conquest, and the indigenous peoples?

Chiquitos

The Society of Jesus began operating in San Lorenzo in the 1580s, their contacts with Chiquitanos limited to the families assigned in servitude to the *encomenderos*, the recipients of *encomiendas*. Only a century later, in 1691, did the Jesuits begin their program of mission *reducciones* in the Chiquitanía proper, with viceregal authorization and the material support of the Jesuit province of Paraguay. By the mid-eighteenth century, the black robes had established seven stable mission compounds, a number that would reach ten by the time the Jesuit enterprise was forceably halted in 1767. Each of these mission towns was settled by a considerable Indian population that oscillated between one thousand and three thousand souls and comprised different *parcialidades*.[3] Formal civil government was not instituted in the Chiquitanía until after the expulsion of the Jesuits, at which time the province was erected into a governorship separate from that of Santa Cruz and the missions were placed under the diocesan authority of the bishop of Santa Cruz. Although curates in name, the Chiquitano missions continued to function under the corporate structures that the Jesuits had established; however, the political and economic life of these *reducciones* was increasingly dominated by the conflicting policies that emanated from ecclesiastical and civil authorities under Bourbon administration.

During the formative phase of the Chiquitos missions (approximately 1691–1730), as reported by Fathers Julián Knogler and Juan Patricio Fernández, Jesuit efforts were expended in the search for new converts through repeated *entradas* that brought different bands speaking distinct languages into the mission villages. The mission at this time was less a settled compound than an itinerant foray into the forests, a "spiritual hunt" that resembled indigenous practices of hunting, warfare, and taking captives. Missionaries depended on groups of Christianized Indians to accompany them for up to four months at a time to locate and surround the encampments of forest peoples. Some of the expeditions proved successful, bringing new groups of extended families into the missions, but often these Christians on trial returned to the forest.[4] The mission enterprise was contingent on their persistent migratory patterns and on the persuasive powers of the caciques, indigenous leaders of the diverse bands that came to reside in the Chiquitos missions.[5]

European contact both in and out of the missions transformed native cultures. Old World diseases reduced their numbers even as *encomenderos* and slave hunters, operating out of the villas of Santa Cruz de la Sierra and San Lorenzo, exploited the labor of diverse tribes and bands that, in turn, raided these Spanish settlements in search of iron tools. Throughout the colonial period and well into the nineteenth century, Chiquitos remained a porous frontier traversed by different groups of cultivators, hunter-gatherers, and warriors, Portuguese contrabandists (*mamelucos*), and Spanish colonists. Jesuits and their successors persisted in bringing additional forest dwellers into the mission towns, further complicating the diverse mosaic of *parcialidades*.[6]

Sonora

The Jesuit mission enterprise in northwestern Mexico followed the advance of the mining frontier during the late sixteenth and early seventeenth centuries. As mining *reales* multiplied in Nueva Vizcaya and military presidios were established along the western flank of the Sierra Madre, Spanish governors and commanders focused their attention on the Cáhita and Sonora village peoples. Responding to an appeal by the governor of Nueva Vizcaya, and counting on viceregal support, the Jesuits began their missionary *entradas* in 1591, working northward from the Colegio de San Felipe y Santiago in Sinaloa. The black robes advanced gradually through the river valleys, their progress halted at times by native resistance and by conflicts with Spanish colonists who vied with the Jesuits for access to the Indians' land and labor.[7] By the early decades of the eighteenth century, the Jesuit mission field of Sonora and Sinaloa had reached its maximum contours, but the northernmost missions of Pimería Alta remained an open frontier, where new *entradas* among the Pima and

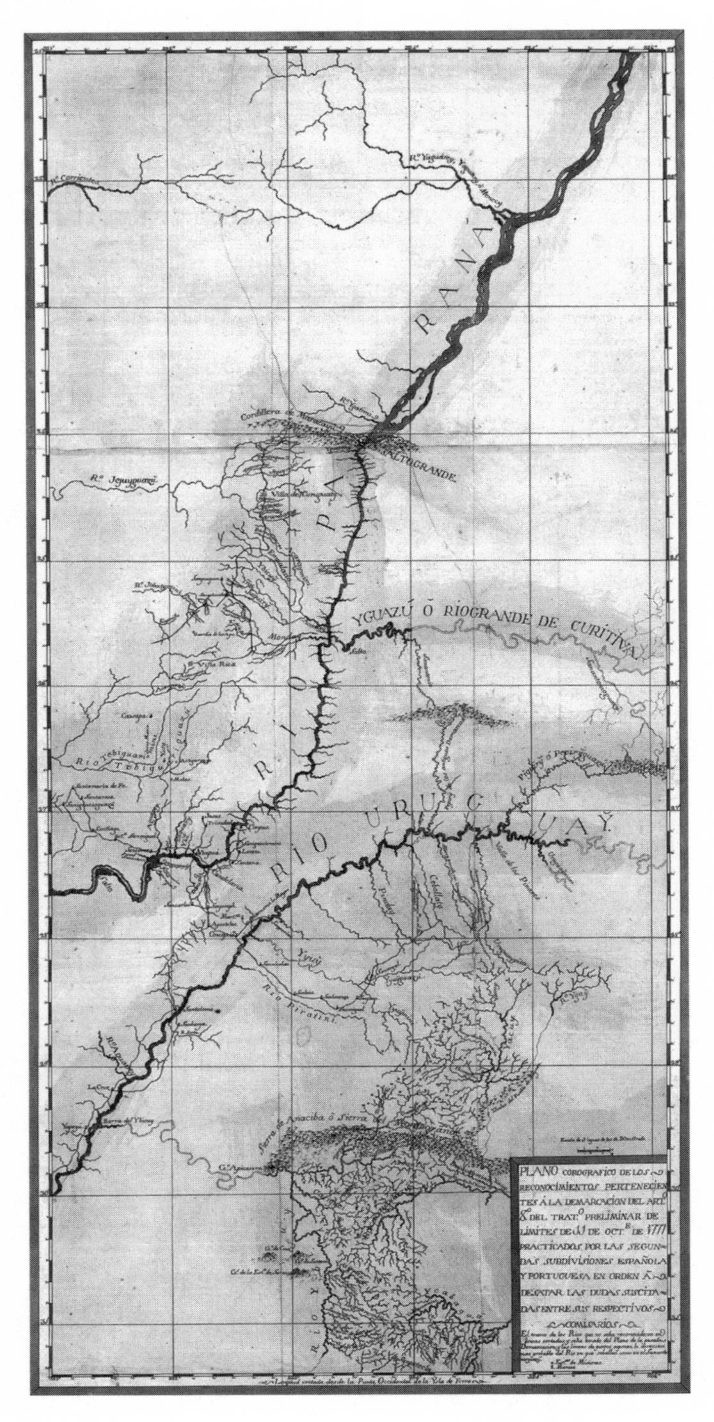

Plano corográfico de los reconocimientos pertenecientes a la demarcación del Artículo 8 del Tratado Preliminar de límites de 11 de Octubre de 1777. From Library of Congress, courtesy of Robert H. Jackson. Product of the boundary disputes between the Spanish and Portuguese territories of South America, this map illustrates the network of rivers in these vast lowland provinces.

Maricopa peoples of the Colorado and Gila valleys awaited the stalwart Sons of Loyola. Despite the concerted efforts of missionaries and Jesuit provincials, the order never accomplished its goal of uniting the mission provinces of Sonora and Baja California by land.

During the first century of evangelization, the Jesuits represented the only consistent line of authority in direct contact with the Indian communities in Sinaloa and Sonora. In 1732, however, the provinces west of the Sierra Madre Occidental were separated from the large territorial unit of Nueva Vizcaya and placed in a new governorship. As the century wore on, the governors and their *tenientes* (local Spanish officials) and *alcaldes* (judges) constituted a political and administrative column distinct from both the Society of Jesus and the military officialdom linked to the presidios. Following the eclipse of the Jesuit enterprise in 1767, Bourbon policies militarized the frontier even further with the expansion of presidial troops and the establishment of the Commandancy General of the Internal Provinces, even as they fused together the lines of administrative accountability by converting the governorship into the Intendancy of Arizpe.[8] The mission pueblos of Sinaloa and the southern provinces became secular parishes under the care of diocesan priests; many of them, in fact, were ethnically mixed villages. Franciscan friars from the province of Xalisco and the Colegio de la Santa Cruz de Querétaro replaced the Jesuits in the Pima, Eudeve, and Opata missions of Sonora and Ostimuri.[9]

Several distinguishing features emerge from the ecological and cultural parameters of the Sonoran and Chiquitano mission frontiers to set the terms of their comparative histories. First, indigenous agrarian systems, responding to the environmental conditions of each of these regions, conditioned the character of the mission enterprise. Not only the size and composition of the *reducciones* but also the types of linkages that each district would establish with the colonial economy depended on native subsistence practices, technical skills, and cosmic beliefs. Sonoran mission compounds generally followed pre-Hispanic settlement patterns, notwithstanding the concentration of outlying *rancherías* into larger villages, while the Chiquitano *reducciones* represented a sharp contrast to indigenous encampments and villages. Second, the Jesuit missions established in Sonora and Chiquitanía belonged to specific hierarchical networks that developed under the aegis of royal patronage. Jesuit *rectorados* (administrative districts) in northwestern Mexico, numbering six in Sonora and Sinaloa, maintained regional surveillance and support for forty mission districts—comprising as many as 103 pueblos at the height of the missions—and, in turn, were accountable to the Jesuit provincial in Mexico City.[10] Chiquitos mission province, although geographically close to the Moxos missions of the tropical savannas of northern Bolivia, was tied administratively to the Jesuit provinces of Paraguay and Tucumán; following the expulsion of the Jesuits, the ecclesiastical and governing points of reference for the ten mission towns of Chiquitos centered on Santa Cruz de la Sierra (San Lorenzo) and La Plata, seat of the *audiencia* of Charcas.[11] While the history of Jesuit evangelization in northwestern Mexico covers nearly two centuries (1591–1767), the black robes' effective presence in the Chiquitanía was limited to three-quarters of a century (1691–1767). Third, in this latter province the *reducciones* competed with the agricultural and service *encomiendas* that colonists of Santa Cruz tenaciously defended. The *encomienda* and *rescate*, a ransom of war captives, forceably drew an undefined number of Indians away from the Mercedarian missions of Porongo and Buenavista and from the Jesuit towns of the Chiquitanía proper.[12] The *encomienda* did not endure beyond the sixteenth century in Sonora; however, the proximity of the missions and mines in the *zona serrana* brought the Indians into direct contact with regional colonial markets for labor and produce.

Reducciones and Constructed Communities

In 1628, Father Pedro Méndez spoke of his initial success among the Eudeve-speaking peoples of central Sonora:

> I have congregated [the Sisibotari in] three pueblos with Crosses and churches, and these are located in two very fertile valleys that produce maize and various legumes. They skillfully irrigate their fields with fresh and healthful waters drawn from streams. . . . Some have thought that these were mountain people because their land is surrounded by gentle hills and mountains, but this is not the case. Their pueblos and cultivated fields are situated in level valleys and all the inhabitants are very peaceable.[13]

Jesuit Arts of Persuasion and Coercion in Sonora

The above passage from Méndez's report seems to underscore the complementarity of piedmont agricultural villages with Jesuit norms of settled town life. We may recall the evidence culled from archaeological and early ethnohistorical sources, summarized in chapter 1, for late preconquest settlements located on terraces above their cultivated fields on the floodplains. Sonoran *reducciones* consolidated smaller *rancherías* in larger *aldeas* (fixed settlements), often located in preexisting village sites, thus providing a demonstrable continuity with indigenous spatial and social patterns of residence. Méndez recounted, further, that church construction proceeded apace in all three mission villages, using the Sonoras' building techniques of puddled adobe and dressed wooden beams. Yet the apparent harmony represented in the Jesuit's narrative masked the contradictions inherent in the missionaries' relocation of Christianized Indians and the contrasting landscapes that emerged from colonial *reducciones*.

Andrés Pérez de Ribas, a Jesuit provincial, missionary, and historian, admitted that drawing different tribal groups into mission villages required patient negotiation and compromise with local customs. His own efforts among the Ahome and coastal hunter-fisher-gatherers of Sinaloa, two decades earlier, illustrated the negotiated contingencies of evangelization. Pérez de Ribas found it necessary to deal with a number of different caciques, both individually and in groups, and he learned to respect their material cultures and political alignments. He acceded to the solution proposed by the Batucari caciques to establish their own pueblo some three leagues downstream from Ahome, near their accustomed fishing sites. The Batucari had agreed to join the Ahome mission because they were currently at peace with their agricultural neighbors and came to their villages to barter for maize at harvest time. Despite these carefully orchestrated negotiations, Pérez de Ribas acknowledged that coastal fishing people found it difficult to adapt to farming and fixed dwellings in the new pueblo. They abandoned the mission for their coastal lands, where they remained rooted in a familiar habitat and free to practice their own religious rites: "They were drawn by the *monte*, where at times they would celebrate their drunken revels. Because they were exiles in the Ahome pueblos, they sought their former solitude, where they would not be seen by Christians and could have freedom of conscience." Pérez de Ribas contrasted the *monte* of "rock-strewn hills and thickets" unfavorably with the "beautiful airy plain above the river" where the Batucari had been offered a village site. Yet the *monte* offered them freedom from Christian surveillance and the material security of fishing and gathering coastal resources that were so well known to them.[14]

Mission landscapes constituted a radical spatial reorientation with political, economic, and spiritual consequences. Even in those areas of central Sonora where the *reducciones* followed the contours of the land and the location of indigenous settlements, the Jesuits' designation of head villages (*cabeceras*) and visiting stations (*visitas*), dependent missions visited from time to time by the priest who lived in the head village, established a hierarchy of *aldeas* and their caciques that may have not existed in this way before evangelization. Consequently, the relative independence of different chieftains and their close identification with particular family lineages may have been compromised in the reshuffling of village locations. Furthermore, resettlement aroused fears of displacement by distinct ethnic bands from the territories in which they farmed, fished, hunted, and gathered. Again, it is well to remember that the fortified terraces found throughout the cordillera and *zona serrana* of Sonora during the late preconquest period indicate territorial rivalries that frequently erupted into warfare—motivated by the enslavement of captives, control over trade routes, and de facto possession of fertile lands and resource bases. Finally, the construction of new head villages centered around a sacred plaza marked by the church and the Christian cross may have altered significantly indigenous notions of social and religious space. Although neophytes like the Ahomes welcomed Pérez de Ribas with green boughs bent into arches and adorned crosses, and the first Jesuit chapels were simple armadas, or one-room adobe structures, the Christian association of height with godliness must have contradicted indigenous notions of sacral power located *in* the earth—in caves, canyons, and springs—and mediated by shamans. Later acts of apostasy and rebellion disputed both the locus and the representation of spiritual authority.[15]

Jesuit evangelizing strategies centered on the caciques who cooperated with them in the early stages of reducing numerous *rancherías* to mission towns. The missionaries baptized the caciques and their immediate families, conferring on them Christian names and rewarding them with gifts to be redistributed (glass beads, knives, horseshoes, hatchets, needles), as well as with special insignia of prestige for themselves, such as horses and cloth. Missionaries and military officers formalized their dependence on these leaders, awarding them titles of captain, fiscal, and even governor over their respective "nations." Missionaries held them responsible for assuring attendance at catechism and other religious exercises and for maintaining order in the pueblos. Early Jesuit evangelizing practices selected the male children of native caciques to be educated at the Colegio de San Felipe (in Sinaloa); they would return to their pueblos as literate catechists and teachers.[16] *Cacique*, then, took on a colonial meaning of ranked leadership that, over time, was institutionalized in a hierarchy of offices, symbolized by the silver-tipped canes Spanish authorities distributed among indigenous leaders. The significance of native officers for governing the missions will be discussed more fully in chapter 5, but we shall see later in this chapter that indigenous council members took an active role in the economic administration of the pueblos.

Jesuit Towns in the Forests of Chiquitos

The points of conflict and estrangement between indigenous landscapes and mission *reducciones* in Sonora are perhaps even more dramatic in Chiquitos. Jesuits carved ten town centers in the Chiquitano forests and savannas, comprising over one thousand inhabitants each and separated into residential sections for different *parcialidades*. The complex ethnic mosaics that characterized the Chiquitos *reducciones* originated in the multiple *rancherías* concentrated in the missions through numerous *entradas*. Father Julián Knogler underscored the central role that the *reducciones* played in the evangelization of the Chiquitanos. "To humanize these children and accustom them to living together in civilization," the Jesuits deemed it necessary to build towns with a central plaza and rows of houses separated by streets; the Indians' dwellings covered three sides of the plaza, the fourth remaining reserved for the church and the *colegio* where the missionaries lived and workshops and schools were established for the neophytes.[17] Despite the Jesuits' considerable work of building and maintaining the mission compounds, the shifting communities that lived within them remained subject to nomadic patterns of seasonal mobility. Like the coastal fishing peoples of Sinaloa who confounded Pérez de Ribas's efforts to join them in a mission together with the farming Ahome, many ethnic bands of Chiquitanos dispersed and returned to the forest after their arrival in the pueblos. Notwithstanding the formal urban grid that defined public and residential spaces, the central plaza, and processional pathways, the Chiquitos *reducciones* remained a contested frontier, exposed to the forest and dependent on the cultural rhythms of *chaco* cultivation, hunting, and gathering.

Disease, Warfare, and Survival in the Missions

Disease and curing emerge as recurrent themes of conquest narratives, beginning with Alvar Núñez Cabeza de Vaca's *Relación* of his odyssey across the deserts of northern Mexico. Cabeza de Vaca and his three companions acquired the aura of healers during their journey, associated with shamanic rituals using pierced gourds and the Christian blessing with the sign of the cross. His account of throngs of Indians pressing to be touched by the strangers juxtaposed sickness, debility, and death with life, health, and power. In his telling of these episodes, acts of healing were accompanied by the exchange of food, ritual pillaging, dancing, and feasting.[18] The palimpsest of meanings derived from the fear of disease and the curative power of spiritual healing resurface repeatedly in Jesuit narratives of conversion and *reducción*.

Pérez de Ribas, Pedro Méndez, and other coreligionists of the first generation of Jesuits to work in northwestern Mexico nearly a century after Cabeza de Vaca's journey openly addressed the issues of disease and native population decline. Their chronicles and letters reported the missionaries' observations of the occurrence of epidemics, rationalized with theological explanations of the moral significance of disease and the symbolic meaning of baptism, the Catholic sacrament that marked the Indians' conversion to Christianity. Pérez de Ribas's *Historia* provided ample evidence that different native peoples in Sinaloa and Sonora both sought out and feared the rite of baptism in connection with illness and natural disasters like earthquakes. The missionaries saw in the shamans their chief rivals as intermediaries for divine power, and they accused them of preaching to the people that baptism caused sickness and death.[19]

Pérez de Ribas addressed the debate over indigenous population decline current in the mid-seventeenth century with a mixture of rational observation and scholastic intellectual traditions. He attributed the decreasing numbers of Native Americans to disease, Spanish labor demands, indigenous migration, and divine intervention. While admitting that "the Spaniards' employment of these peoples for their farming, mining, and other work . . . has contributed to the decrease,"[20] Pérez de Ribas chose to emphasize disease as the principal cause and attribute it to God's will to both punish Indians who had practiced idolatry and sorcery and save their souls.

> One general [cause] is very well known throughout the Western Indies: the many illnesses that they call cocolitzles [Nahuatl: *cocoliztli*]. These are particular to the Indians and are like a plague to them. Through this kind of illness Our Lord, in His highest judgment, has willed to diminish the populations of almost all the nations discovered in the New World. Today there are probably not even half the Indians left of those countless numbers who existed when the Gospel arrived.[21]

Furthermore, Pérez de Ribas noted that the decline appeared to be greater than it actually was because Indians had migrated to Spanish settlements at cattle ranches, mines, and sugar estates, leaving their former pueblos diminished or abandoned. Referring to the Mayo *rancherías* of Ostimuri and Sinaloa, he stated that "they are remarkably fond of traveling and are curious about other lands. . . . They set out to see the world and to seek clothing with which to cover and adorn themselves—clothing which today they are never without. Because they are very good workers . . . there are a lot of Mayo scattered throughout nearly all the pueblos of New Spain."[22]

The venerable Jesuit's empirical observation of native population decline and his catalog of its causes touch on some of the major points in the modern debate over the demographic history of conquest and colonization. Scholars have combed the texts of early exploration chronicles, Jesuit histories, and periodic reports (*cartas anúas*) to compile sets of comparative figures that represent population estimates for particular localities and dates, as well as for recorded diseases. That Eurasian diseases spread through Native American populations in epidemic proportions is a point of common consensus; what remains debated is the magnitude of population decline, its duration (whether and when demographic recovery occurred), and the relative importance of its different causes. Repeated waves of smallpox, measles, typhus, influenza, malaria, and dysentery struck the mission towns of northwestern Mexico in epidemics documented from the end of the sixteenth century to the mid-nineteenth century, when cholera made its first appearance in the region.[23] The questions that remain are whether the observed population decline was biological, due mainly to pathogens and representing an absolute loss of people, or social, due in part to displacement, Spanish demands for labor and produce, and—as Pérez de Ribas claimed—voluntary migration of Indians to colonial settlements and enterprises. Even the attribution of divine will that runs so strongly in the contemporary accounts is historically relevant to the issues surrounding the struggle for power between missionaries and shamans and the probable motives that led significant numbers of Indians to accept baptism and the Jesuit *reducciones*.

Comparable data on specific disease episodes and population estimates are far less accessible for Chiquitos than for Sonora. This is so probably because systematic evangelization of the many different bands that lived and moved in the province of Chiquitos did not occur until the close of the seventeenth century, and because their nomadic movements were more pronounced than in Sonora. The formative institution in Chiquitos during the early period of colonial expansion was the *encomienda*, which supplied the frontier settlers of Santa Cruz de la Sierra with labor for their ranching and agricultural operations, centered on small-scale plantings of sugar cane. It served the interests of Santa Cruz *vecinos* (Hispanic landholding residents)[24] to avoid government surveillance and taxation in order to retain control over a private labor force; conversely, the Indians held in servitude sought opportunities to flee their bondage and return to the forest. For these reasons, the service *encomienda* of the Chiquitano frontier generated far less systematic counts of native population than did the *encomienda* and *repartimiento* arrangements established among the tributary Indian towns of the Andean highlands and in central Mexico that were located closer to the administrative seats of viceregal authority.[25]

These limitations in the data make it difficult to construct rates of population decline or design graphs showing convincing directions and degrees of demographic change. Eighteenth-century population counts generated by the Jesuits and the clerical priests who succeeded them do supply systematic estimates of native families who resided in the missions, but they do not offer reliable standards for calculating what proportion they represented of the total indigenous population of Chiquitos. The figures culled from different archival sources and summarized in the tables presented in chapter 4 show a relative stability in the demographic levels sustained in the missions, although with oscillations over time and from pueblo to pueblo. A significant, if temporary, drop in mission population occurred in 1767–68 following the forced removal of the Jesuits and simultaneous with a region-wide epidemic. My interpretation of the available information is that the apparent stability reflected in the figures rested not so much on the natural increase of a sedentary population in each of the missions, but rather on the mobility of different ethnic bands through the missions and between the *reducciones* and the forest.

Warfare, as much as disease, was a constant presence on both the Sonoran and Chiquitano frontiers. It set the rhythm for the *reducciónes*, in which distinct bands settled somewhat warily in the missions, and for their frequent flight back to the *monte* or the forest. War, often carried out as raiding and targeted homicides, accounted for periodic fluctuations (sudden peaks or dips) in mission population. Skirmishes within tribal groups and between bands constituted a strong motive for seeking out Spaniards as allies for taking vengeance or seeking protection from enemies. In 1700, Alférez Juan Bautista de Escalante, a presidial officer of Sonora, led a squad of soldiers through the desert lowlands of Seri and Pima territory, ostensibly to enforce peaceful settlement and curtail raiding that had occurred in the mission pueblos of the San Miguel valley and Pimería Alta. Escalante took formal depositions and administered exemplary punishments to individuals identified by native witnesses speaking under oath and through interpreters as culprits who had stolen livestock or murdered Indians that had settled in the missions. During the six months of Escalante's itinerant watch over the western perimeter of the settled colony, this small expeditionary force was drawn into the patterns of raiding and vengeance killing that underlay relations among the Salinero and Tepoca bands of Seris, as well as nomadic groups of Pimas. Native governors and justices, recently named in the mission pueblos, called on Escalante to administer justice and punish their enemies, using the language of kinship (*parientes*) and the craftsmanship of arrows to identify both their allies and their foes.[26]

I thus argue that the destructive impact of disease has been overplayed as the single explanatory factor for the destructuring of late precontact societies and the Indians' acceptance of life in the missions. While larger, consolidated villages were dispersed as their populations fell below a certain threshold, it is less clear that whole peoples died out. Disease undoubtedly continued as a strong biological and cultural factor in the history of mission communities from the seventeenth to the nineteenth centuries. In the *reducciones* Indians adapted their curative strategies and sought out new sources of spiritual power to cope with the recurrence of epidemics and high rates of morbidity and mortality. Nevertheless, warfare should be considered more closely, along with disease, in our interpretation of the early stages of mission *entradas* and *reducciones*. Patterns of warfare among the native peoples of Sonora and Chiquitos, as well as violent encounters between Indians and Europeans, created a matrix for defining power relations among the different contenders: the caciques and the people they presumed to lead; Spanish and Portuguese expeditionary forces; and the emerging hierarchy of ecclesiastical and civil officers. War and the taking of human captives, forced abandonment of campsites and villages, and the theft or destruction of stored possessions threatened the survival of indigenous communities and engendered an environment of fear perhaps as potent as that of disease in conditioning the cultural politics of mission towns.

Work, Production, and Exchange in the Missions

By the eighteenth century, population levels had stabilized sufficiently to develop economic systems of commodity production in both Sonora and Chiquitos, achieved through imposition of colonial policies and negotiated terms of exchange in the pueblos. The mission economies of both frontiers created commercial and symbolic wealth whose significance varied for the native peoples, the missionaries, and the merchants and governing officials of both provinces. Despite their physical distance from colonial administrative centers, these frontier pueblos became entwined with the mercantile networks of exchange and transport of goods that operated in northern New Spain and the southern Andes. The flow of material goods in and out of the mission districts, duly registered in the ledger books of the Jesuits and their successors, constituted tangible evidence of the Indians' productive labor channeled to the circulation of commodities.

Commercial Circuits in Sonora

Routine correspondence and accounting ledgers provide a richly detailed picture of the commercial transactions that supported mission economy. "The muleteers took all the wheat that the mules could carry, and we should not despair over what I still owe [you] of the 60 fanegas, because as things are with me, when I look for it, it is not to be found. I would like to send you a few fanegas soon after Easter, so that they can be ground into flour and stored there for me."[27] During March and April of 1766, Father Miguel Almela, the missionary of Opodepe and its *visita* of Nacameri sent thirty-two *fanegas* of wheat (48 bushels) and eleven *fanegas* of maize (16.5 bushels) to his superior, Father Andrés Michel, in Ures through the offices of Marcial de Sossa, a local merchant. The following year, in June 1767, Father Almela reported on the wheat harvest in progress: "I am writing to you from Nacameri, where I am watching over the cutting of wheat, it is necessary to keep a close eye on these my people [the Indians]. The crop was not very good this year, but I hope to raise a little more than last year, why without having sold as much as a fanega, I have almost none left [from the previous harvest]."[28]

Mission crops of wheat, maize, beans, and chickpeas—planted and harvested on the communal fields that were cleared and irrigated under the supervision of indigenous officials—supported the families that lived in the mission beyond their own individual plots (*milpas*) in return for their labor. In addition, the Jesuits channeled surplus grains into the provincial circuits that linked the mission pueblos to one another and to the mining *reales* that operated in Sonora and Sinaloa and across the mountains in Chihuahua. Mission produce served as payment for the semiannual shipments of merchandise that missionaries ordered by means of *memorias* (lists of purchasing requests), ostensibly from the Jesuit provincial in the viceregal capital, but in practice through the merchants established in the principal mining centers like San Antonio de la Huerta, Río Chico, Santísima Trinidad, and the provincial capital of San Miguel de Horcasitas. The debt that Father Almela acknowledged, in *fanegas* of wheat, was an integral part of this commercial circuit. The relative values of harvested crops and merchandise, rendered in their equivalent in silver *reales*, were calculated and tabulated for each transaction. Normally, these debts were allowed to accumulate and, over several years, would tend to even out. At times, however, rancorous correspondence among the missionaries, and between them and the merchants, did occur;[29] nevertheless, all parties were mainly concerned with keeping goods and produce circulating in a regional commercial system that bred interdependence.

The merchandise that flowed to the missions typically included different kinds of textiles, which varied from the simple *mantas* (cotton or woolen shawls or blankets) produced by the *obrajes* (workshops) of central Mexico to fine woolens and silks imported from Europe. In addition, missionaries regularly purchased tobacco, shoes, needles, stockings, ribbons, writing paper, and, less frequently, tools and implements such as plow heads and harnesses. Eagerly awaited mule trains brought these trade goods to the mission pueblos, mining camps, and presidios, where they were destined for redistribution to the mission Indians, to *vecinos* living in the missions or nearby settlements, and to the missionaries themselves who had purchased mirrors, paintings, and statuary to adorn the churches. In return, the missionaries dispatched shipments of grains, ground wheat flour and corn *pinole* (basic corn meal), dried chilies, chickpeas, brown sugar, soap, lard, and candle wax. Livestock was also a marketable commodity, as the missions supplied the mining *reales* and presidios with horses, mares, mules, and cattle for hides and meat. Not infrequently, missionaries received uncoined silver, stipulated in *marcos*, a measurement of silver equal to 8.5 pesos, as payment for their livestock and produce.

Jesuit ledgers for the mission of San Pedro de Aconchi in the Sonora River valley, including the head village of Aconchi and its *visita* downstream at Nuestra Señora de la Concepción de Baviácora, provide an important record of the production and sale of mission surpluses from 1720 to 1766.[30] From the patiently compiled accounts of sales and purchases, as well as the volume of grain harvested in the mission from 1749 to 1762, we can calculate the portions of mission produce stored or consumed in the pueblos, as well as those sold.

Mission income came from three main sources: grains, cattle, and silver payments. Expenditures comprised productive assets, that is, plows, iron tools, planting seed, and livestock; trade goods to distribute to the Indians; religious cult and church adornment; and wages, paid in money or in kind to nonmission laborers for specialized work often associated with church construction.

The ledgers of Aconchi, combined with the Jesuits' correspondence for these same years, reveal several important patterns in the economic administration of the missions.[31] Grain harvests varied significantly from one year to the next, dependent on the vagaries of the weather and on the available labor supply. Wheat harvests were consistently higher than maize, by a ratio of as much as 2.8 to 1, which may indicate that irrigated fields along the floodplain were reserved for wheat, while maize was planted in *milpas de temporal* dependent on rainfall. The relationship between production and sales for both crops, however, is inverted: only one-fifth of the wheat crop was destined for the market for the sample years included in this study, while over two-fifths of the maize harvests were sold outside the mission. While both of these grains served subsistence and commercial purposes, it appears that maize, cultivated during the summer, had a strong demand in the mining camps and presidios. The missionaries may have kept less maize than wheat in storage—perhaps mainly as seed grain—in view of their expectation that mission Indians would plant maize in their own *milpas*. Wheat, sown in the winter and harvested in the spring, supplied mission granaries for shipment to new *reducciones* in the Pimería Alta and Baja California and was sold throughout the year to Hispanic settlements in the presidios and *reales de minas*.[32]

Turning to the figures on revenues and expenditures, mission harvests constituted the most important source of income, representing over 80 percent of the total value in pesos accruing to the mission of San Pedro de Aconchi for the entire period of 1720–66. Livestock sales accounted for a little over 10 percent of total income, followed by remittances of silver at 7.6 percent. The sale of livestock declined significantly during the last decade of Jesuit administration at Aconchi, due possibly to the increased productive capacity of haciendas and ranches that had built up their own herds and competed with the missions for a greater share of the provincial market. The preponderance of grain sales indicates the dependence of the missions on the labor of the Indians to produce the surpluses that sustained the commercial exchange of merchandise and produce. Resident mission families were bound to serve the *común* (mission communal lands) in lieu of paying tribute to the crown.

As is reflected in their priorities for investment, the Jesuits considered it important to maintain a stable resident population. Over two-thirds of the missionaries' purchases comprised clothing and other gifts for the Indians, and these outlays continued even in years of lean harvests when expenditures exceeded income. Although the mission Indians did not receive a monetary wage for their work, it had proven necessary to institute a form of payment in kind through the semiannual distribution of trade goods. The second highest outlay of mission resources—nearly one-quarter of all expenditures—was destined for the construction and adornment of the churches. Jesuits, and the Franciscans who administered the missions after 1767, engaged in an almost continual round of church construction, amplification, and repair, requiring investment in skilled masons and carpenters, materials and implements, and the religious imagery that served for the celebration of ritual and the teaching of Christian doctrine. During this period the simple chapels made of puddled adobe, described earlier, were replaced by churches built of adobe, stone, and fired brick, their roofs constructed with long wooden beams and barrel-vaulted masonry. Statuary and religious paintings were imported from central New Spain, and musical instruments were either brought in or built in the missions.[33]

Overall production and sales declined significantly after 1750, a circumstance that suggests several possible explanations in reference to changing conditions in the wider provincial economy. Sonoran missions shipped produce to the *reales* of Chihuahua and Cosihuiriachic in Nueva Vizcaya as well, and these mining centers suffered decline in production and population during the middle decades of the eighteenth century, thus shrinking the potential market for mission goods.[34] During the seventeenth and early eighteen centuries, the missions had constituted the sole or the best source of foodstuffs for the mining camps and military outposts of Sonora, but during the late eighteenth century, haciendas and ranches produced increasing quantities of grains and livestock for local markets. These estates expanded their landholdings especially along the floodplains of the San Miguel and Sonora rivers, impinging on the missions' resources of arable land and drawing laborers away from the pueblos. Initially, small rancheros depended on the missions for the loan of a few *tapisques* (harvesters) during the peak seasons, but by the end of the century, a number of haciendas recorded resident families of Indian *peones* (laborers).[35] The increased flow of foodstuffs into the commercial economy from different sources affected not only the volume of grains that the missions could sell but also the prices they could command.

Generally, the Jesuits sold both wheat and maize at 3 pesos a *fanega*, although prices rose to 4.5 and 5 pesos a *fanega* in 1749, a year of relative scarcity; *cargas* (a measurement of volume; see glossary) of ground wheat flour brought 10–12 pesos each; and *pinole* usually sold for 6 pesos a *fanega*. By the waning years of Jesuit administration, these prices had fallen to 2 pesos a *fanega* for maize and wheat, 4 pesos a *fanega* for ground *pinole*, and 7 pesos for a *carga* of wheat flour.[36] The median prices at which wheat and maize were sold in Sonora, as listed in the mission ledgers, compare with grain prices in San Felipe el Real de Chihuahua during this same period, but they were higher than reported prices for Zacatecas, Guanajuato, and central New Spain during nonfamine years. An important contrast occurred during the latter third of the eighteenth century, when the cost of basic grains at first began falling in Sonora, but then rising and remaining high after 1785, the "year of hunger" following devastating droughts and frosts that caused crop failures in central New Spain.[37]

Following the expulsion of the Jesuits, the corporate mission economy declined even further in Sonora. No systematic ledger books for San Pedro de Aconchi survived beyond 1766, although scattered references to wheat and maize harvests for 1779 and 1780 indicate that the crops for those years were stored and consumed within the mission villages. Two accounts for the Pima Alta missions of San Ignacio de Cabúrica and Santiago de Cocóspera, in 1787–88, show that an abbreviated version of the Jesuit system of harvest sales and purchase of trade goods for redistribution continued to operate in these northern missions.[38] In contrast to the Jesuits before them, the Franciscan friars who administered Cabúrica and Cocóspera included their stipend together with the sale of mission produce in their calculations of annual income; expenses covered items for the church and religious cult, as well as clothing to be given to the Indians, following established practices. New expenditures, however, that indicate a shift in the internal economy of the pueblos, include the purchase of food (nine hundred pesos in Cabúrica) and wages for herdsmen, shepherds, gardeners, blacksmiths, carpenters, and "servants." The increased use of monetary wages and references to "the minister's charity" suggest that living standards had suffered impoverishment even as the missions developed a money economy. Furthermore, missionaries had to resort to direct payment to obtain skilled labor for construction and for the seasonal tasks of cultivation and herding.

The intersection of indigenous systems of livelihood, the economic administration of the network of mission *reducciones*, and the commercial economy of this colonial frontier over nearly two centuries worked, in the end, to undermine the communal economy of the mission pueblos. Falling economic returns for mission produce, as reflected in the pattern of declining prices, and the loss of communal assets brought the Sonoran peoples into the regional market under conditions of marked inequality. Indians sought out alternatives to the missions, in the traditional foraging reserves of the *monte* and in the wage economy of Spanish mines and estates, as the corporate management of labor, production, and redistribution of goods that the Jesuits had established faltered and weakened during the closing decades of the eighteenth century. The waning mission economy, observed by clerics and civil authorities alike, was not merely coincidental but was foreseen and even intended by royal visitors and intendants who implemented Bourbon policies designed to increase the crown's share of colonial wealth and enhance the power of the metropolis in the northern frontier of New Spain. During this later period several key missions of the Opatería and Pimería Baja districts were secularized, notably Aconchi, Banámichi, Ures, Mátape, and Ónavas.[39] Once converted into parishes, their mixed population of Indians and *vecinos* was expected to pay alms and tithes. These same policies initiated changes in the administration of the missions of Chiquitos, with far-reaching consequences for their political economy and communal culture.

Gathering, Weaving, and Trade in the Chiquitanía

Three years after the departure of the Jesuits from the Chiquitanía, the indigenous officers of San Xavier drew a marked contrast between the good years of plenty, when the Jesuits ran the missions, and the scarcities of the new administration.

> Father Thomas, we present ourselves to you, all the people and the justices [of this pueblo of San Xavier] so that this Governor [of Chiquitos] will go home and leave only the ministers here, as it was in the time of the Padres of San Ignacio. Then there will be plenty of wine for Mass, because the priests do not drink up the wine as the governor has done. . . . When we look into the mission storehouses, we find there is no wine and the flour has spoiled, because the governor is stingy [and] there is no cloth nor medallions, scissors, and other things for us. [But] when the Padres of San Ignacio were here there was plenty of cloth, knives, scissors, religious medallions, beads, and all things.[40]

"In the time of the Padres" became a familiar refrain, used to condemn the offenses and omissions committed by the civil and military governors of the province or to protest the abuses of the secular clerics sent to the pueblos.[41] The Chiquitanos' perceptions of change here have direct bearing on the economic system that the Jesuits had established for the Chiquitos missions and the alterations in that system during the post-Jesuit period. They express which practices the Indians interpreted as standards of good administration for their pueblos in both the material and spiritual realms of their experience.

Extant ledger books allow us to analyze the economy of the Chiquitos missions during three distinct periods representative of the Jesuit system of *reducciones* and of the Bourbon policies implemented after 1768. Jesuit records for a continuous eighteen-year sample of their administration from 1750 to 1768 provide solid evidence of the economic system established by the black robes during the mature mission period for seven of the Chiquitos pueblos.[42] Following the expulsion of the Jesuits, the Junta Provincial de Temporalidades undertook surveillance of the accounting system of revenues and funds disbursed to the Chiquitos missions, reporting in turn to the *audiencia* of Charcas, in the highland city of La Plata. The records kept by Juan Antonio Ruiz Tagle, the official in charge of the Junta de Temporalidades from 1771 to 1776, account for the investments made by the junta in the Chiquitos missions during those five years.[43] This sample, unfortunately, does not include the income from the products that the junta received from the missions, nor does it indicate the distribution of funds to the individual pueblos. It does, however, demonstrate how the costs of the Jesuits' removal and their replacement with secular clergy were charged against the missions. The final period for which quantitative data is available, including both income and expenses, covers the years 1783–94; these records are notably consistent for all ten of the missions, especially during the administration of the Junta de Temporalidades in La Plata by Joaquín de Artachu, from 1786 to 1794, who turned the office over to Jácobo Poppe Rendón in 1795.[44]

Indigenous technologies provided the basic materials from which the Jesuits developed a commercial nexus between the mission enterprise and the mining economy of the highlands. Local products that merited shipment to Andean colonial markets fell overwhelmingly into two categories: cotton cloth and wax. Chiquitana women spun and wove different lengths and thicknesses of cotton textiles that were used for clothing (*tipoi*) and bedding (hammocks). Under Jesuit tutelage women were employed in mission workshops, where they turned out varying sizes and qualities of *lienzos* (linen or canvas cloth) and *tocuyos* (plain cotton cloth) that were sold in the great mining center of Potosí and in the colonial capital of La Plata. Their utility ranged from cheap clothing for mineworkers and the urban poor (*lienzo ordinario*) to pieces of canvas for carrying burdens (*toldos*). Chiquitanos had traditionally gathered wax and honey from a number of species of wild bees native to the region. Beeswax readily became a commodity on the colonial market for lighting churches, homes, and mining shafts. The Jesuits turned this opportunity to commercial advantage by appropriating large quantities of wax gathered by the Chiquitano men, weighing it, classifying it for its color and quality, and packaging it in *marquetas* (crude cakes of wax) for shipment to the Andean cordillera.

In return, the Indians received recompense for their labor with meat from the livestock herds pastured in the *estancias* that belonged to the pueblos and with a steady flow of trade goods that came from the highlands. The Jesuits maintained an agent (*procurador*) in Potosí, who stored and sold the products received from the missions and dispensed the merchandise that the missionaries requested in their annual *memorias*. The diversity of trade goods shipped regularly to the Chiquitos frontier ranged from imported cloth and textiles produced in the highland *obrajes* to beads of different colors and sizes, religious medallions and crosses, metal tools, knives, scissors and needles, fishing hooks, dyestuffs, and pieces of iron, copper, and steel. In addition to the goods destined for the Indians, the missions imported sumptuous textiles for priestly vestments and church ornaments, silver chalices and other vessels for Mass, and glass, metal locks, and other building materials for church construction. Furthermore, the missionaries invested considerable sums in cattle to replenish mission herds. The cost of these commodities, added to the charges for freight and storage in Potosí, the missionaries' stipend (*sínodo*), a nominal tribute charged to each mission to be forwarded to the crown, and a monetary contribution to the Jesuit province together constituted the debit levied against the Chiquitos mission district each year.

The evidence gleaned from the ledger books corroborates the Jesuits' histories and official reports of life in the missions. In mission workshops the Chiquitanos practiced the arts of tanning, weaving, smelting, carpentry, lathe work, and cabinet making. Father Julián Knogler explained that the supply of steel and iron needed for this production came from Potosí, and that the women spun and wove cotton on the looms that had been brought to each village. Further, he described the process of gathering and cleaning the wax from wild beehives, sending it "to the nearest Peruvian city [La Plata]," from there to be sold in Potosí.[45] *Cabildo* (town council) officers and their families, who lived prominently in the first rows of houses that lined the central plaza, were favored in the distribution of trade goods—as measured in *varas* (yards) of cloth, strings of beads, and numbers of knives, scissors, medallions, and the like.[46] Significantly, then, the trade networks that linked the Chiquitos mission economy with the colonial markets of Upper Peru served, in turn, to reinforce the social rankings on which the political life of the missions depended.

This same commercial enterprise supported the religious life of the Chiquitos *reducciones*, as in Sonora, through the architectural splendor of the churches and the liturgical calendar. During the second third of the eighteenth century, from the early 1740s to the time of their expulsion, the Jesuits erected spacious chapels in the seven missions that were well established: San Xavier, Concepción, San Miguel, San Ignacio, San Rafael, San José, and Santa Ana.[47] Native artisans were employed to raise the walls and roofs out of local materials. To build these temples, enormous wooden trunks were cut into pillars and arches, the earth was molded into thick adobe walls, ceramic tiles were baked to cover the roofs, and hollow reeds (*guapá*) were woven together to line the ceilings. Indigenous carpenters carved and incised pulpits, chests to hold the sacraments and other liturgical implements, and some of the religious icons that gave each church its identity.[48]

Following the expulsion of the Jesuits, the formal outlines of the economy they had established remained in place. The bishop of Santa Cruz, Francisco Ramón de Hervoso, elaborated a lengthy set of instructions for the temporal and spiritual governance of the Chiquitos pueblos based on the liturgical rituals and the exchange of goods for labor that had become an acceptable standard among the Chiquitanos. In addition, the *audiencia* of Charcas instructed the newly installed governor of Chiquitos as well as the secular priests and administrators sent to the missions to maintain the commercial circuits linking the province with the highlands. Two priests and one administrator replaced the Jesuit missionaries in each pueblo. Their goals reflected Bourbon fiscal policies: to guarantee the subsistence of the *reducciones* and to enhance their production while minimizing the maintenance expenditures administered through the Junta de Temporalidades.[49]

A comparison of the income and expenses recorded for the Chiquitos missions during the Jesuit and post-Jesuit periods shows a marked contrast in the volume of inputs and outputs, expressed in their monetary value, and in the total cost of sustaining this mission province. Employing the terms established by the administrators of the period, *cargo* refers to the products shipped from the missions to La Plata and Potosí, and *data* comprises the trade goods sent to the missions and freight costs, as well as the clerical stipends and lay salaries paid to the administrators and priests established in each pueblo and to the governor of Chiquitos and his subordinate staff. In addition, beginning in the 1790s, the missions were expected to support the salaries assigned to the small detachment of soldiers stationed in the eastern frontier with Mato Grosso.[50]

The products of the Indians' labor that supported the mission regime continued to be wax and cloth, supplemented with the sale of carved wooden rosaries; however, it is noteworthy that the market value of wax increased dramatically over that of *lienzo* during the latter decades of mission administration. More important, the imbalances between *cargo* and *data* increased over the years. The meticulous records kept by Joaquín de Artachu show that after 1784, mission production had expanded in both volume and market value over the levels maintained in earlier years, but the costs charged against the missions (*data*) had risen far more than the income derived from the sale of native crafts. What, then, were the expenditures underwritten by the Junta de Temporalidades and the mission pueblos? After 1768, the debit side of mission ledgers comprised three main categories: commodities shipped to the missions for redistribution to the Indians, clerical stipends and local salaries, and administrative costs that extended beyond the region, including the governor's salary and merchant commissions. A striking shift in the proportion of trade goods and administrative costs occurred in the expenditures attributed to the missions in the post-Jesuit period. It was the added burden of salaries that determined the imbalance between *data* and *cargo* and that drove the overall expansion of mission costs under Bourbon administration.

The written petition of the justices of San Xavier cited at the beginning of this section indicates that the Indians were aware of economic dysfunction even during the early years of the new regime. Although the justices had targeted their anger at the governor of Chiquitos, their complaints centered on the scarcity of trade goods and the spoilage of eucharistic wine and flour in the mission storehouses. Their voices were echoed by the first governor of Chiquitos, Diego Antonio Martínez de la Torre, who confirmed the Indians' perception of hardship. In 1768, he warned that the grains, wax, and *lienzos* stored in the missions would rot if not removed before the next rainy season, and that supplies were critically low in the pueblos since the expulsion of the Jesuits had stymied the flow of commerce to and from the province for two years.[51] In addition to the trade goods that Indians had been accustomed to receiving punctually, regular distribution of meat rations constituted a central feature of their material well-being. Without a dependable supply of beef, to which they felt entitled, the Chiquitanos left the missions to hunt and gather in the rain forest. Priests and governors alike emphasized the need to replenish mission herds with new livestock and repeated a litany of poverty in the mission pueblos.[52]

Martínez de la Torre held the priests responsible for the pending crisis. He denounced to the *audiencia* that the clerics sent to replace the Jesuits did not perform their religious duties and were sacking mission warehouses of all supplies—even violins—and selling them clandestinely. "They came here without a priest's frock or cassock, and they have no intention of staying, most of them are only waiting until the rains cease to leave. If mendicant friars do not come [to the missions] all will be lost in less than two years."[53] Successive governors repeated similar accusations of clerical misbehavior, reporting that the distribution of goods in the pueblos had taken on the character of commerce between the Indians and their priests. The priests, for their part, appealed to the bishop of Santa Cruz to expedite the collection of their annual stipends. Lengthy correspondence among the clergy in the missions, the provincial governors, the bishop, and the *audiencia* concerning the payment of clerical stipends indicates that, in fact, the *sínodos* were in arrears for years. Moreover, the priests were required to appear in person in Santa Cruz to receive the *sínodo* or entrust its collection to an authorized representative. These *apoderados*, typically local merchants, charged a commission for their services. In the absence of regularly paid salaries, priests and the local administrators assigned to each mission supported themselves by engaging directly in the regional trade in cloth and wax.[54]

Priests, administrators, and *apoderados* often came from the same small circle of landowners and merchants based in Santa Cruz; through the collection of stipends and the exchange of mission products for trade goods, these families generated commercial wealth from the productive resources of the pueblos. Cloth became the currency of the province as unauthorized trade routes began to operate between the missions and Santa Cruz de la Sierra. The approved flow of goods passed through the official *receptoría* (receiving office) established in San Xavier, the oldest of the missions and the closest to Santa

Cruz, where the merchandise should be taxed and then remitted to La Plata. The revenues so collected would have been destined to the support of salaries and stipends in the province. In practice, however, local merchants, as well as long-distance contrabandists operating from as far away as Mato Grosso, bypassed the *receptoría* to move their goods through the missions to Spanish markets in Santa Cruz and the highlands.[55] Contraband flourished due to the natural barriers for surveillance in a land in which most roads were shifting foot paths and seasonal flooding hindered the movement of people and goods for weeks or months at a time, and in which the Bourbon administration had promoted only a limited development of legitimate commerce.[56]

Even as crown officials and local interests disputed the returns from mission produce and trade, contemporary observers agreed that the province depended on the economic resources that it generated internally. Jesuit missions in the neighboring province of Moxos had enjoyed income from several haciendas in the eastern foothills of the Andes, managed by the Jesuit order to benefit the missions. Following the expulsion of the Jesuits and the confiscation of the order's properties, these rents were redirected to the general fund for diocesan tithes and ceased to benefit the Mojeño *reducciones* exclusively. In contrast to these missions, however, the Chiquitos *reducciones* had no endowment beyond the fruit of the Indians' labor.[57] It was the intention of Bourbon authorities to place both provinces on a self-supporting basis by increasing production and intensifying the commercialization of the commodities raised in the missions. Despite repeated ordinances that conditioned the payment of clerical stipends from the Real Hacienda on reimbursement from the sale of mission products, the income derived from the surpluses generated by the Chiquitos pueblos did not reach royal coffers in sufficient amounts to cover either tribute or the priests' *sínodos*.[58] The Bourbons sustained the corporate mission economy developed by the Jesuits only in its outward form; the wealth it generated was increasingly channeled to the emerging provincial elite of Santa Cruz.

Chiquitanos, for their part, protested the withering of their communal economy in the hands of Cruceño priests and merchants. Governor Juan Barthelemí de Verdugo received numerous complaints from indigenous justices and militias concerning the clergy's sale of mission produce and cattle. During his periodic visits to the eastern pueblos of the Chiquitanía, in 1779–81, Verdugo met with formal delegations from the mission *cabildos*, often with an interpreter, who denounced mistreatment and flagrant privateering with mission property. Opportunities for abuses surfaced with the issuing of licenses to extract salt from the area and remit it for sale in the *receptoría* of San Xavier, the need to supply cattle to the troops stationed in the province, and the porous frontier with the Portuguese in Mato Grosso.[59] The commanding officer of the small military force stationed in the pueblo of Santiago testified that the priest, Joseph Jacinto Herrera, had bragged in front of all the soldiers that "the King gave him a 50 peso stipend each month and, with another 50 pesos that he stole, he could end the year with 1,200 pesos, and that all the parish priests stole as much as they could behind the governor's back." Father Herrera forced the Indians to work to serve his own pecuniary ends, threatening them with lashes and irons, and paying the women for their woven *lienzos* with mere trinkets.[60]

The entire *cabildo* of San Juan confronted Governor Verdugo to denounce the illicit sale of cattle from the *estancia* of San Joseph where the Indians maintained the mission's herds. On various occasions the priest assigned to San Juan, Father Yldefonzo Bargas y Urbina, had received "Portuguese" agents known to him and sent them to the *estancia* to round up livestock and drive it to Mato Grosso. When the Indian *mayordomo* (foreman) and herdsmen resisted turning over five hundred head of cattle to these unwelcome visitors, they were threatened with punishment and flogging. Sometime later, when native *vaqueros* (cowboys) returned to the estancia of San Joseph—now abandoned without a resident *mayordomo*—to bring a few head back to feed the pueblo, they reported seeing "many Portuguese" who were armed and rounding up the livestock. For fear of the intruders' weapons, the Indians of San Juan returned to their pueblo and left them to rob at will.[61]

The Chiquitanos reserved their most bitter testimony for the hated memory of Domingo Muñoz, a former priest of San Juan whose moral crimes had reached scandalous proportions, violating both his spiritual and temporal duties. The justices of San Juan testified before the governor of Chiquitos and the vicar of Santa Cruz that Muñoz celebrated Mass only on Sundays and festive days and refused to administer the eucharistic sacrament to the neophytes; furthermore, he had ordered the flour reserved for making communion hosts ground into bread for him to eat. Muñoz had denied the sacraments of confession and last rites to the ill and dying, ignoring the repeated warnings of the *cruceros de día* (minor Indian officials), whose job it was to notify the priests of such emergencies, threatening even to whip them. Moreover, Father Domingo terrorized the entire community with vicious floggings, causing the death of two Indians from their wounds and the flight of a married woman who was stripped and beaten in the presence of the *cabildo* and "all the people who could be found in the pueblo." Muñoz cheated the Indians by refusing to weigh the wax and pay for it by the pound. Finally, the justices of San Juan presented Governor Verdugo with a long list of the mission property that Muñoz had stolen or degraded to sell on his own account. He ordered native blacksmiths to melt down door locks and hinges and the bellows used in the workshops, and to remove the iron nails that held the frames and pictures in the church—even the large nail that supported the cross of Jesus of Nazareth—and replace them with wooden joints. He took ornamental cloth and the pictures of saints that the Jesuits had placed in the convent.[62]

The native officers of San Juan vehemently protested Muñoz's excessive violence *and* his negligence as a priest. They showed that they understood the equivalent monetary value of their produce: in effect, they demanded a just price, in goods, for their wax and cloth. Furthermore, in their denunciation of Muñoz's theft of mission property, the *cabildo* justices particularized their valorization of material possessions in the church and *colegio* (school, meeting place for the cabildo) that were endowed with religious symbolism. In their view, the very caretaker charged with the Indians' spiritual ministry and material well-being had violated the sacred space of the church and seized the collective assets of the mission, thereby trampling on the Indians' most fundamental rights.

Conclusions

Our comparative analysis of established mission communities in Sonora and Chiquitos, in their distinct cultural and natural settings, has highlighted the conditions under which early mission *entradas* took place, the spatial dislocations involved in the mission *reducciones*, the importance of disease and warfare in their early histories, and the secular development of their political economies. We return now to the three questions posed at the beginning of this chapter concerning the degrees of coercion and choice that led Sonoras and Chiquitanos to enter the missions, the systems of production and redistribution that sustained the *reducciones*, and the opposing signs of accommodation and conflict that characterized their long-term history. It is undeniable that violence underlay the tense accommodation and the negotiations, both tacit and explicit, through which native peoples settled in the mission pueblos. Violence emerges in the historical record of both provinces in repeated episodes of intertribal warfare that, compounded by the forced labor of *encomienda*, led Indians to submit to the *reducciones*. Nevertheless, coercion alone cannot explain, nor did it accomplish, the physical and political structuring of the mission pueblos. Native caciques and commoners conditioned their labor and obedience to ecclesiastical and civil authorities on both economic and political terms.

The productivity of indigenous technologies and labor, and the landscapes they created, supported the political economy of the missions even as the tensions underlying their negotiations limited the economic networks that connected these frontier pueblos to the wider colony. Jesuit commercial circuits converted mission produce into commodities, thus redirecting a significant portion of the work of native men and women from direct subsistence to surplus. It can be argued that in the Sonoran highlands, mission commerce reconstituted, in part, the long-distance trade routes that had established important preconquest linkages between this region and the urban centers of Mesoamerica, as described in chapter 1. Colonial commerce introduced radical innovations in the political economy of the Sonoras, however, through the monetization of the terms of exchange and the extraction of surplus value to benefit the Jesuit order, the crown, and private colonists. In Chiquitos, the crude terms of barter that characterized early colonial relations between Indians seeking metal tools and *encomenderos* seeking bonded labor were only gradually supplanted by the systematic trade of labor for goods in the Jesuit *reducciones*. Despite the porous frontier between the mission towns and the forest, and the itinerant seasonality of *chaco* cultivation and gathering, the longevity of the corporately managed economy remained an impressive feature of the Chiquitano missions.

The eighteenth-century transition from Jesuit to Bourbon administration points significantly to the ways that the mission-as-enterprise contradicted the ecological and cultural foundations of the native communities. In both regions, the pecuniary demands of market exchange overtook the bases of indigenous economies. Post-Jesuit developments exhibited different trajectories in each of these areas, and their structural differences arise from the ways in which the mission districts developed commercial ties with the broader colonial societies to which they belonged. Chiquitos mission economies maintained their corporate structure and expanded, as measured in the volume and monetary value of their production and expenditures, but the administrative overhead and commercial advantage gained by the Cruceño elite took wealth out of the pueblos. Chiquitanos did not have direct access to the colonial market; rather, they were forced to rely on priests and lay administrators for the clothing, tools, and religious medallions that they had come to prize. Neither Jesuits nor their successors directly controlled the production of foodstuffs in the missions, although *chacras* worked by the Indians for the benefit of the missions produced cotton and food for the priests and administrators. Native swidden plots continued to yield yuca, maize, rice, and plantains (the latter two crops were introduced under the colonial regime), and mission herds provided meat rations on feast days.

Conversely, the agrarian economies of the Sonoran missions declined as the former *reducciones* lost control over labor and productive resources. The market economy of northwestern New Spain drew the Indians directly into its sphere as petty producers and laborers, increasingly so after 1767. This was facilitated by the proximity of the mines to mission villages, so that the Sonoras moved between the two worlds of their own communities and Spanish estates and towns. Jesuits worked directly with local merchants, as evidenced in their correspondence and account books, notwithstanding periodic conflicts between missionaries and colonists over access to the land and labor of the Indian villages.[63] The missions ceased to feature as the centerpiece of the provincial economy during the last quarter of the eighteenth century. Refined silver bars and minted coins provided a regional currency that underwrote a partially monied economy in which a few miners and merchants flourished in the wake of new gold placers and silver strikes. Royal treasuries (*cajas reales*) established in Rosario and Alamos deposited regular annual payments of silver *reales* in the exchequer (*pagaduría*) of Arizpe, destined to pay presidial troops and missionary stipends. Despite crown investment in the province designed to increase the flow of specie and the imposition of the *alcabala* (sales tax), public revenues did not meet Bourbon expectations. The ultimate beneficiaries of administrative reform were the landed and mercantile families of Sonora who gained access to the communal lands of the missions and garnered the wealth generated by the mines.[64] Private greed overran mission properties in the wake of the Jesuits' exodus as lay commissioners charged with conserving the pueblos' lands, herds, and granaries pillaged them, but their theft was partially curtailed when economic administration of the missions was restored to the Franciscans who replaced the Sons of Loyola.[65] Provincial elites in both cases gained control of the productive assets of their respective regions, but through different modes of commercial exchange and the exploitation of indigenous labor.

The missions left an indelible imprint on the indigenous cultures of both regions. Undeniably the evangelization project centered on the *reducción* constituted a vehicle of conquest and a foreign transplant in Sonora and Chiquitos. Nevertheless, the native peoples of these frontier provinces made of the missions a viable way of life, developing strategies of negotiation that tempered the full impact of Spanish domination. The organization of space in compact villages, the exchange of labor for commodities, and the development of a political hierarchy in the pueblos—while serving imperial interests—became integral parts of Sonoran and Chiquitano cultural identity. No less significant, the cultivated lands and cattle ranches nominally included in mission holdings provided a legal rationale for defending ethnic territory. Indian militias and justices—*gobernadores* (governors) and *alcaldes* of different levels of authority, also called *jueces* (judges)—so distinguished by their canes of office, assumed their role as spokespersons and guardians of the communal property that they understood to be the product of the labor of their people. Male indigenous leaders of Sonora protested overwork in their assignments to presidial patrols, missionaries' arbitrary hoarding or distribution of mission harvests, and the *vecinos*' encroachments on mission lands—issues that will be analyzed in subsequent chapters on land tenure and political culture. Sonoran complaints concerning the internal regimen of the missions were less numerous and detailed, however, than was the case for Chiquitos, especially in the post-Jesuit period.

Indigenous communities endured under both the Jesuit and Bourbon regimes, but they underwent significant modifications in their cultural and political formation and in their relations of production. The gendered division of labor under missionary tutelage seems to have been more pronounced in the Chiquitano *reducciones* than in the mission villages of Sonora. Surplus products from Chiquitos traded on the Andean colonial market comprised wax, gathered by men from wild bee colonies, and cloth, woven by women in the missions. Freedom of movement was probably more restricted for Chiquitano women than for men in light of the supervised labor required to spin thread and weave cloth. Resident families in the mission towns of Chiquitos, perhaps composed largely of women and children, lived in the compacted urban spaces of the *parcialidades*. Herding and gathering beeswax, by way of contrast, took the men farther afield to the cattle *estancias* and into the surrounding forests and savannas. Agricultural labor in Sonora occupied both men and women, although gender specialization was apparent in food gathering and processing, as well as in craft production. Hunting and herding were male occupations, while women predominated in gathering wild seeds, roots, and fruit, and grinding grains and seeds on the *metate* (grinding stone). Women specialized in the indigenous crafts of pottery and weaving, while mission construction favored men as masons, carpenters, and metalworkers.

What were the alternatives to mission life for the Indians in these two colonial frontiers? Village peasants of Sonora gravitated to wage labor or peonage in the mining camps and haciendas of the province or, conversely, dispersed into scattered *rancherías* where they subsisted by hunting, gathering, and seasonal cultivation. Military service became an important outlet for Opata and Pima men, who were recruited as auxiliary troops for the presidios and punitive expeditions against the Apaches or rebel bands within the province. Honored with titles and remunerated with rations and a modest wage, these indigenous militias acquired prestige among their peers and recognition of their rank within the colonial order. At the same time, their repeated assignment to military expeditions occasioned their absence from the pueblos, giving rise to complaints of overwork and the neglect of their families from both Indians and missionaries.[66] Chiquitanos were reported to have fled the mission pueblos periodically—due to shortages, disease, or mistreatment. Their alternatives, however, remained limited. As late as the mid-eighteenth century they were vulnerable to forced servitude in an attenuated form of *encomienda* and ran the risk of capture by Portuguese/*mameluco* slave hunters who raided the province.[67] Nevertheless, many Indians did alternate between the mission pueblos and seminomadic life in the *bosques* and pampas.

We are left with two important interlocking questions: How did the distinct natural and cultural environments of Sonora and Chiquitos affect the historical outcomes of the mission economies that developed in both regions, and how did the colonial missions impact the ecosystems of these two frontier provinces? In each area, the introduction of European livestock, the expansion of agriculture, and the consolidation of nucleated villages created new landscapes. The Sonoran missions were essentially agrarian villages, using indigenous technology for irrigation but producing surplus harvests for commercial sale by introducing a two-crop regime (wheat and maize) and planting orchards and gardens with European fruits and vegetables. Moreover, the missions of eighteenth-century northwestern New Spain competed with private haciendas and ranchos for land, water, and resident laborers. Late eighteenth-century mission documents registered contemporary perceptions that water sources were diminishing under the pressures of grazing and cultivation.

The *reducciones* of Chiquitos became centers of protoindustrial production in textiles, carpentry, and metallurgy, in addition to gathering and cleaning beeswax. The grazing and breeding of livestock in numerous *estancias* increased pressures on pampas grasslands and savanna scrub forests. Agriculture, however, remained a subsistence endeavor, and the cyclical rhythms of *chaco* and fallow encouraged a counter-rhythm of dispersal away from the villages to the *bosque*. Jesuit and Bourbon political economies appeared to dominate the pace and direction of change, but the efficacy of imperial policies and the contours of commercial networks of exchange were conditioned in both regions by the ecological constraints of tropical and desert environments and by cultural patterns of resistance.

CHAPTER 3

Territory: Community and Conflicting Claims to Property

The bonds between human communities and the land that lie at the heart of all historical landscapes constitute a central theme of this history of colonial frontiers in Sonora and Chiquitos. Chapter 2 illustrated the connections between polity and economics in the mission towns and the different modes of production and distribution that supported both subsistence and marketable surpluses in each of these provinces. It emphasized as well the conflicting communal and entrepreneurial claims to the wealth generated by the mission communities. This chapter focuses on the historical construction of social ecologies that brought culture and technology together with the physical elements of land, water, vegetation, livestock, and game to show how the environment conditioned the significance of land tenure in both regions. Land tenure in Sonora is laden with notions of water scarcity, land competition, and the defense of the commons; in Chiquitos, it connotes the expansiveness of grazing and shifting cultivation in the renewing forest.

The concept of social ecology employed here is closely linked to territory, space, and humanly crafted landscapes to explain the different rhythms and stakes of conflict over material resources in mixed environments of foraging (hunting, gathering, fishing), cultivation, and pastoralism.[1] It addresses (1) the ways in which different kinds of land use gave rise to distinct notions of territoriality, (2) the processes through which village polities of both regions lost control of vital resources, and (3) the strategies they used to rebuild their communities. In exploring these issues, I have learned that territory cannot be defined simply as dominion over a particular extension of land. The contrasting histories of land tenure in Sonora and Chiquitos show us that territorial boundaries change through the collision of competing claims and modes of land use. Moreover, the notion of *territory* evokes different cultural values for distinct social actors in these two colonial provinces. The following two episodes illustrate the material and spatial dimensions of territory and community in eighteenth-century northwestern Mexico and eastern Bolivia.

Ocuca in the Mission and Mining Frontiers of Pimería Alta

Ocuca (Oacuc in the Pima language) came into the documentary record of colonial Sonora in 1770, when Prudencio de Salazar registered his petition to measure and gain formal title to grazing lands occupied by livestock belonging to him and his brothers. The Salazar brothers lay claim to *tierras realengas*, nominally vacant lands not yet reduced to private ownership under the terms of Spanish law, extending for over 3 leagues (approximately 12 kilometers or 7.2 miles) between the small settlements of Santa Ana and Ocuca, a *sitio* owned by Miguel Antonio Velasco on the eastern slope of a low range of hills that separated the Altar and Magdalena valleys.[2] O'odham and Akimel Piman villagers sustained stable mission communities in both of these riverine valleys. Santa Ana, the Salazars' home base, was a site for mineral prospecting and cattle ranching located south of Magdalena pueblo and directly east of Ocuca. The *realengas* (public lands) that Don Prudencio petitioned to reduce to *sitios* surrounded two smaller properties named Aguaje (a watering hole) and Carrizalito, the latter registered by his father Alejo de Salazar in 1746 as a *sitio* to raise mixed herds of cattle, horses, and mules.

> In the royal name of his Majesty (may God keep him) I register for me and in the voice of my brothers the realengas that are comprised [of the vacant lands] from the property boundaries measured for the Puesto of Santa Ana [southward] to the limits of the sitios of [Santa Rosa] Arituaba, belonging to militia captain Don Joseph Ignacio Salazar, and [westward] to the limits of the sitios of Don Miguel Antonio Velasco known as El Ocuca, and also to the measured boundaries of the sitios of the deceased Don Juan Cosío, that are found in the place known as El Aguaje. It should be pointed out that this area includes the sitio of El Carrizalito, already registered; we have established our possession in it, and in all this land, by virtue of occupying it with our livestock and building corrals that we use to brand and mark our cattle. And concerning the fact that until now we have possessed this land without legitimate title, it is because we have not been able to register it, due to the unceasing warfare of the enemies.[3]

Salazar's claim was measured publicly in August of the following year to assure that none of the adjoining neighbors suffered encroachment on their land and to assess the correct payment that the Salazar brothers would owe the royal treasury in order to establish title to their new property. Over the better part of a week the appointed officials measured linear extensions "to the four winds" to then enclose the quadrant, thus establishing the perimeter of the Salazars' enlarged holdings in thirty-four and three-quarters *sitios*. Three *vecinos* who accompanied the land commissioner concurred in setting the value of the land at four pesos and four *reales* per *sitio*, considering that it contained good pasturage but lacked permanent running water and lay in the path of Apache incursions into the province. The Salazars' payment to the royal treasury totaled 156 pesos and 3 *tomines*, plus the *mediannata* tax of 16 pesos and 1 tomín.[4]

The apparently uncontested appropriation of the enlarged Salazar holdings, and the absence of native voices in the proceedings, masked an accelerating process of land enclosures that changed the boundaries of private and communal properties and altered the terms of usufruct and resource use for Indians and Hispanized *vecinos* in the entire Sonoran province. Ocuca, watered by a small stream that flowed westward from Santa Ana during the summer rains, would have provided the O'odham with fields (*oidag*) for a short planting season. The nearby *wahia* (hills with natural pools) and the very grasslands that made the site attractive to the Salazars for their cattle herds supplied seeds and foliage for gathering and game for hunting. Jesuit Ignaz Pfefferkorn described Ocuca as a "pleasant" *ranchería*, abandoned, he said, due to native population decline in the disease environment that followed European contact and resettlement in the missions.[5]

Several major events occurring during the quarter century between Alejo de Salazar's settlement of Santa Ana, his entitlement of Carrizalito in 1746, and the younger generation's enlargement of the family's holdings in 1770 constituted a turning point in the formation of social landscapes in Sonora. The northern Pima uprising of 1751–52 and its military repression brought destruction and dislocation to mission villages and caused *vecinos* to flee (if temporarily) their ranches and mines. Jesuits Jacobo Sedelmayr and Juan Nentvig, stationed at the missions of Tubutama and Sáric, fled to Santa Ana, where they took refuge and Sedelmayr recovered from his arrow wounds.[6] Thus the "enemies" Don Prudencio referred to in his petition of 1770 were probably Apaches and Seris, as well as dispersed bands of rebellious Pimas who rejected mission life and defied Spanish authorities. One of the Spanish settlements burned by Pimas and Seris during the uprising was San Lorenzo, a ranch (*poblado*) located between Santa Ana and the mission of Magdalena. Following the rebellion, two permanent garrisons were established at San Ignacio de Tubac (1752), on the middle Santa Cruz river, and Santa Gertrudis del Altar (1755), on the broad floodplain formed by the confluence of the Magdalena and Altar rivers. The presidio of Altar was first stationed in Ocuca, and then subsequently moved to the western bank of the Altar river.[7] The imperial order of 1767 to expel the Jesuits from all Spanish dominions forced sudden and far-reaching changes in mission administration, leading to their secularization and the enclosure of their communal lands and water.[8]

A generation later, *vecinos* from the presidio of Altar and Santa Ana played an important corroborative role in the commercial sale of Ocuca. Miguel Antonio Velasco, named above in the Salazars' land claim, and his sister María Guadalupe Velasco had each inherited a portion of the Ocuca grant from their father Lorenzo Velasco. María Guadalupe, widowed, initiated proceedings in 1795 to sell her portion to Francisco Xavier Redondo, a merchant and *vecino* of Altar. Her land title had burned in the Pima and Seri attack on San Lorenzo during the 1751 revolt, which obliged María Guadalupe to request the current presidial commander José Sáenz Rico to call witnesses who could attest to her legitimate holding. Their testimony confirmed her right to a portion of the property and revealed that Miguel Antonio had been obliged to cede his portion to one Manuel Buingoechea (son-in-law of the former presidial commander Bernardo de Urrea) in payment of a debt. Buingoechea had profited from the sale of Velasco's cattle, without replenishing the herds, until they were depleted. For this reason, at the time of Redondo's purchase of Ocuca, it was described as "an ancient ranch, empty of livestock." When the transaction was completed, Redondo paid María Guadalupe 200 pesos (in silver *reales*) to acquire twelve *sitios*; the land was assessed at a value of 8 pesos per *sitio* to determine the taxes and fees that Redondo deposited in the royal treasury in the provincial capital of Arizpe—an additional 105 pesos and 6 *reales*.[9]

This short history of Ocuca and the Salazar landholding in northwestern Sonora exemplifies the early stages of a land market in the enclosure of *realengas*. By the close of the eighteenth century, divided inheritances, indebtedness, and the land sales converted range land into pieces of property that circulated among the Hispanic population of the province. The commercial value of the land was derived from the resources of vegetation and water that sustained livestock, itself a moveable good that could be traded, and reinforced by the power of the colonial state to impose a schedule of fees for land titles. Territory, understood as geographic space and the material resources in it needed to support life and livelihood, became property through the legally sanctioned practices of denunciation (staking a claim to particular tracts of land), measurement and demarcation, and entitlement.

Privatization of land, as institutionalized by colonial law, developed more slowly in the humid tropical frontier of eastern Bolivia, notwithstanding commercial traffic between the lowlands and the Andean cordillera, the Hispanic nucleus in the villa of Santa Cruz, and the growth of livestock in the *estancias* established for the missions and by *vecinos*.[10] The explanation for this contrasting profile of land tenure points to the natural environment and rhythms of swidden cultivation, sparse Spanish settlement, and the strength of the mission economy.

Santo Corazón on the Eastern Perimeter of Spanish Chiquitanía

On 18 April 1779, six officers, "at the head of the other justices and the people of this pueblo of Santo Corazón de Jesús," signed a petition that they presented in person to the visiting governor of the province of Chiquitos. Their plea to move the pueblo to a more favorable location occupied ecclesiastical and civil authorities from the province of Chiquitos to the *audiencia* of Charcas.

> It is true, sir, that in this our pueblo we suffer calamities of disease and hunger caused by our pueblo's location in a place that is very low, with no drainage at all, so that [the water] is consumed in the same pueblo, giving rise to the corruption of the land . . . [for this reason] there are hardly any people in our pueblo due to the many accidents [deaths] that result from its location. . . . For these reasons we appeal to your [sense of] integral justification to give us a resolution that will favor our advancement (all the more, because this is the principal port to all the missions) by allowing us to move our pueblo and houses to a less watery and more comfortable place. If not, the land will consume us, and we shall all be finished in this territory.
>
> Benito Putarés, corregidor [highest male officer] and commander of the militias
> Macario Yabacürr, lieutenant corregidor and sergeant major of the militias
> Macario Torubes, *alferez real* [second lieutenant] and adjutant officer of the militias
> Miguel Xores, first magistrate [*alcalde de primer voto*]
> Paulino Torubes, second magistrate [*alcalde de segundo voto*]
> Simón Pooris, provisional magistrate[11]

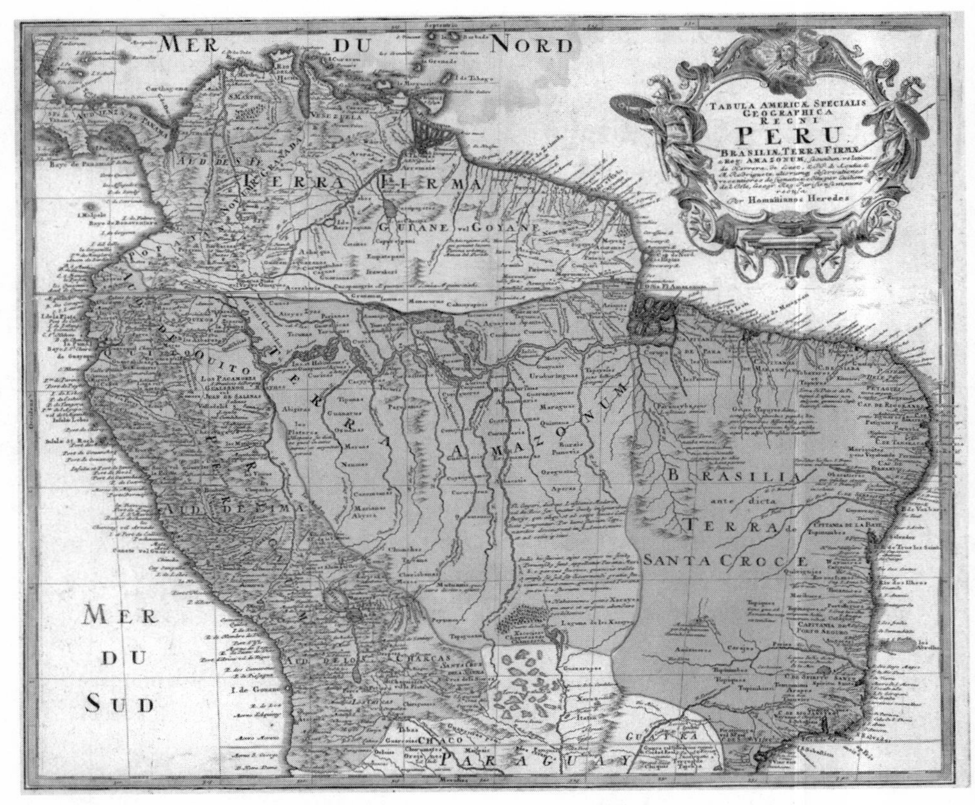

Tabula America specialis geographica regni Peru, Braziliae, Terra Firme, Amazonum, 1737. Courtesy of Map and Geography Library, University of Illinois at Urbana-Champaign. Santa Cruz de la Sierra, the Laguna de los Xarayes, and the "terra pantanosa" are shown in the lower central part of this map, north of Paraguay and the Chaco.

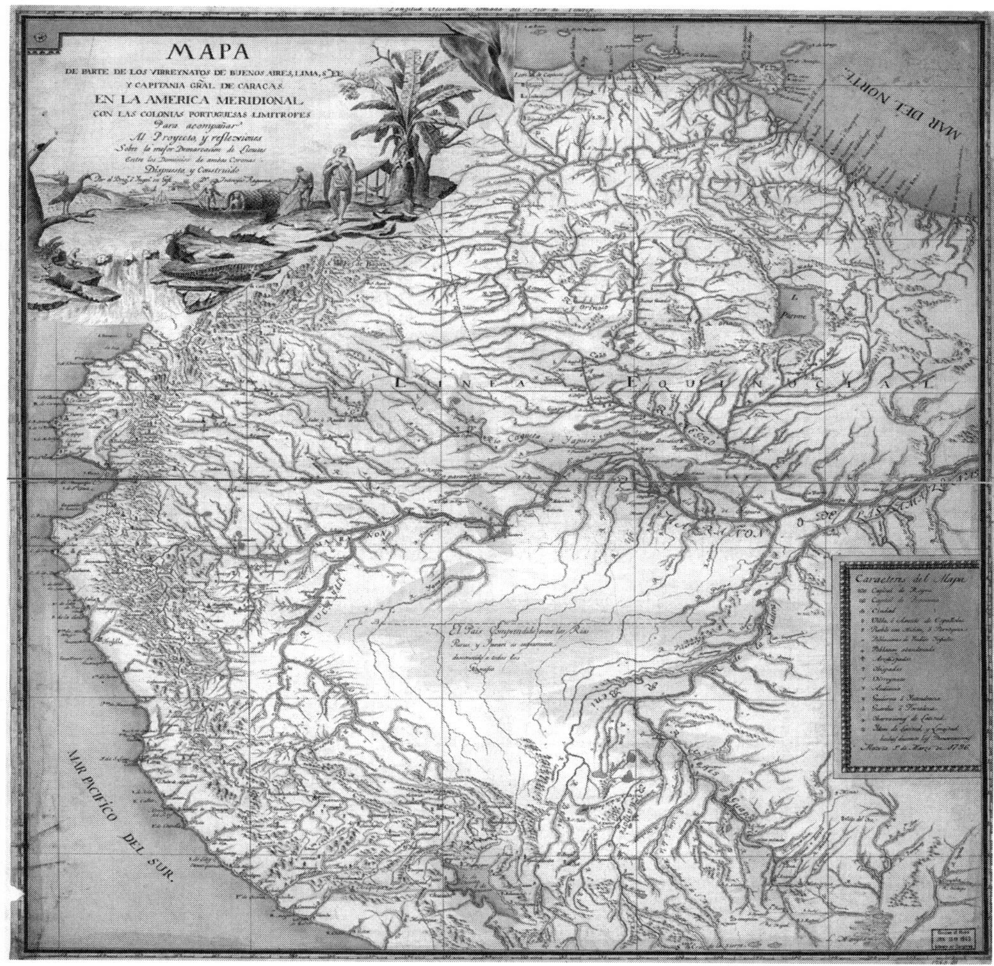

Mapa de parte de los virreynatos de Buenos Aires, Lima, Santa Fé, y Capitanía General de Caracas en la América Meridional con las colonias Portuguesas limítrofes. From Library of Congress, courtesy of Robert H. Jackson. This map produced in Madrid in 1796, is centered on the Amazon and its many tributaries. Santa Cruz de la Sierra and the northern Chiquitos missions appear at the very bottom of the map.

Jesuits had founded Santo Corazón as early as 1717, with the objective of establishing a port that would communicate with the Paraguay river and the mission provinces of Tucumán and Asunción. Located in the vicinity of swamps and shallow streams where Alvar Núñez Cabeza de Vaca and his auxiliary expeditions had failed to find a passable route in 1543, the mission had occupied several different sites in search of a more favorable setting. The Indians' petition of 1779, strongly seconded by the resident priest, Augustinian friar Tomás Duque de Estrada, complained of fevers and fetid odors, which they attributed to the lack of good drainage from the pueblo, resulting in the corruption of the soil. Duque de Estrada claimed that in four years, 1776–79, as many as 1,700 persons had died from diseases resembling malaria (*calenturas tersianas*) that seemed to hang in the environment itself. Livestock as well as humans perished in these humid lowlands, where Duque de Estrada had seen pigs, hens, doves, and other breeding animals die without benefiting the mission. The Indian officers and their priest asked to relocate the pueblo some twenty leagues (eighty kilometers) north to the banks of the Boturiquis river, a place they claimed that the Jesuits had selected for the permanent settlement of Santo Corazón for its clear waters, fertile land in which to plant *chacras*, and healthful surroundings. Their petition was forwarded to the *audiencia* by the governor of Chiquitos, Berthelemí Verdugo, and approved that same year.

There is no documentary evidence that describes the transfer of the pueblo, but the account provided by French naturalist Alcides d'Orbigny, who visited Santo Corazón in 1830, seems to confirm that the move did take place. He described the approach to the village through the woodland valley and Bokis river (a local term referring to reedlike vegetation) and related that the mission had been moved northward from an earlier site, although he (erroneously) attributed its relocation to the Jesuit administration.[12] Built on a low plateau overlooking several small streams, the village supported planted fields of cotton, maize, plantains, and manioc. At the time of his visit, the population of Santo Corazón numbered over eight hundred persons, representing four different ethnic groups: Chiquitos, Zamucos, Curavés, and Otukés. The latter three "nations" (corresponding to the *parcialidades* or bands that settled in all the missions) were identified as having raided the mission and fled, or, conversely, sought refuge in the village from enemy Guaycurúan nomads that roamed northward from the arid Chaco plains to the south.

In these references to alien raiders, d'Orbigny echoed the perception voiced by the authors of the 1779 petition that Santo Corazón represented an important outpost on the frontier between Spanish and Portuguese dominions and, equally significant, between Indians settled in mission towns and nomads of the forests and savannas. The frontier character of Santo Corazón was further reinforced by the titles that the indigenous signatories of the 1779 document held: *corregidor*, *alférez real*, captain, lieutenant, and sergeant major of the militias. Indeed, their successful petition was based on the warning that should the pueblo not receive healthful and fertile lands in a new location, it would soon be abandoned. The prospect of the disbandonment of this, the easternmost of the Chiquitos missions, did not prove inconsequential to Spanish colonial authorities whose vision of territory comprehended the geopolitical sphere of imperial boundaries. The Indians used the language of their colonial overlords in their description of the unhealthful nature of their environment in order to gain new ground for their territorial objectives that were tied intimately to their notion of community. Their words and actions foreshadow present-day political movements among highland Andean peasants, recent periurban migrants, and lowland indigenous federations to claim land and local representation.[13]

These two historical narratives of eighteenth-century Sonora and Chiquitos illustrate changing territorial boundaries in colonial Spanish America. The different sets of social actors in each province created distinct kinds of frontier landscapes. From the documented texts we learn that territory is a cultural construct, but one derived from the spatial dimensions and material resources of the natural environment. Hispanic *vecinos* of Sonora rendered territory as *terreno*, that is, titled properties with fixed boundaries marked visibly with wooden and masonry landmarks (*mojoneras*) and traced on paper in maps and spatial diagrams. As we have seen, *vecinos* guarded their boundaries to enforce their exclusive use of the resources contained within them, at the same time that they sought opportunities to expand their properties, most frequently through the growth and migratory grazing practices of their livestock. Boundary disputes among different landed families, and between *vecinos* and native communities, maintained a lively interest in the adjudication of property titles that filled much of the province's documented history.[14] In the arid environment of Sonora, both village-dwelling, farming Indians and private landowners faced the limitations of irrigated cultivation. Yet village horticulturalists were also hunter-gatherers, like the O'odham of Pimería Alta, whose subsistence strategies required territorial mobility and access to dispersed resources in different localities and seasons.

Forests and waterways marked the principal contours of territory for Indians and Europeans alike in the open topography of the Chiquitanía. Native bands sought access to the high forest (*bosque alto*) to create their *chacos* through the seasonal labor of clearing, burning, planting, and harvesting; two to three years of cropping were followed by long fallowing periods of twenty years. Cultivated plots were dwarfed, however, by the spaces in the forests and savannas reserved for hunting and gathering that provided wood for construction, fibers, hides, tools, food, and medicines. Streams, lagoons, and swamps—as much as the land—defined Chiquitano territory and supplied indispensable resources for fishing, sustaining human communities, and watering herds of livestock.

Spaniards who settled in Santa Cruz, first as *encomenderos* and later simply as *vecinos*, vied for Indian labor to open irregular sugar plantations and cattle *estancias*. Their cultivation methods resembled enlarged versions of indigenous *chacos* that produced marketable surpluses. Livestock, mainly bovine cattle and horses and mules, constituted a centerpiece of communal patrimony in the mission towns and became the property of *vecinos*—the bulwark of their capital, family inheritance, and social prestige. Even more markedly than in Sonora, *territory* in the Chiquitanía did not refer to fixed portions of land claimed as individual or family property through purchase, royal grant, or adjudication.[15] Here, *territory* described an expansive and moveable spatial concept, tested and disputed through conflicts among different groups of Indians and Europeans. Not until the latter nineteenth and twentieth centuries, as we shall see below, did the privatization of herds and land become a major issue undermining the resource base of the pueblos and setting in motion the conversion of territory into property. In contemporary lowland Bolivia, population growth and the notable expansion of commercial markets for cattle and agricultural produce have intensified territorial disputes and politicized conflicting claims to land and productive resources.[16] The following sections of this chapter narrate outstanding episodes of conflict over land in Sonora and Chiquitos in order to show distinctive patterns of change in the social landscapes generated in each area and the normative values associated with territory.

Sonora: *El Común* and Bounded Territories

Eighteenth-century Sonoran landscapes bore the indelible imprint of numerous mining *reales* and encampments of varying size and duration that, beginning in the 1630s, spread through the eastern piedmont and, by the 1770s, reached the Altar desert. Silver shaft mining, as well as gold placers, altered the terrain of Sonoran hillsides and streams and required timber for fuel and construction.[17] No less significant were the herds of livestock that bred and multiplied in ranches established by the Jesuits to support the agrarian economy of the missions and by private settlers, as illustrated above in Ocuca and Santa Ana. Cattle ranching developed as an industry and way of life that supplied meat, hides, and track animals to the colonial markets that centered on the mines and presidios of Sonora and Nueva Vizcaya.

By the mid-eighteenth century, the colonial enterprises that linked together mining *reales*, missions, and military garrisons had altered the natural environment of the Sonoran Desert and upland savannas and woodlands through deforestation, soil erosion, and surface water runoff. At the same time, population growth of humans and livestock put additional pressure on the irrigable cropland and scrub forest vegetation of the *monte*. The Eudeve villages of Cucurpe, Tuape, and Opodepe on the San Miguel drainage, for example, had experienced a marked reduction in the extent of their irrigated fields, resulting in diminished harvests, due to the heavy silt that flowed in the summer rainy season and floodwaters that carried precious topsoil downstream.[18] Changing conditions in the natural environment, coupled with the institutions of colonial society, gave rise to several distinct patterns of land tenure that reflected the tensions between communal and private territorial claims. Sonora's changing landscapes involved the settlement of *vecinos* in the mission pueblos, the efforts of native communities to protect and circumscribe their holdings, the enclosure of grasslands and *monte*, and the parallel processes of amalgamation and fragmentation of landholdings.

The number of *vecinos* who lived in mission pueblos or in their vicinity, in proportion to the resident native families, grew appreciably during the second and third quarters of the eighteenth century. In the central Sonoran valleys, where missions and mining *reales* were located in close proximity to one another, putatively Indian villages registered mixed populations of Opatas, Eudeves, Pimas, Spaniards, Hispanized Indians, mestizos (*coyotes*), and mulattoes. Arizpe and Ures, two pivotal villages in the Sonora River valley, had relatively evenly divided populations among *vecinos* and Indians, but successive Jesuit and diocesan reports from the 1740s to the 1780s showed that the Spanish and mixed populations grew more rapidly than did the Indians. Similarly, the Opata head villages of San Miguel de Oposura and Señora de la Asunción de Cumpas were surrounded by seven mining and ranching settlements of Spaniards and Indian laborers who had left the missions.[19] These mixed populations of peasants and ranchers increasingly came into conflict over floodplain cropland and the open spaces of the *monte* for grazing and hunting. Pima Indians of San Miguel de los Ures and Santa Rosalía, in the lower Sonora valley, nominally held title to mission holdings extending nineteen leagues east and west of the head village of Ures. They protested, however, that the *vecinos* based in the nearby Real de San Josef de Gracia and neighboring ranches were squatting on mission cropland and letting their cattle overrun the Indians' planted *milpas*.[20]

In the face of these pressures, a number of mission pueblos took measures to protect their communal holdings through the same juridical instruments used by the *vecinos* to secure title to their properties. In 1716, Pima *alcaldes*, council officers from the mission of Cumuripa in the middle Yaqui valley, petitioned Thomás de Esquivel, the Spanish magistrate (*teniente de justicia mayor*) at the Real de San Miguel Arcángel in the name of the *común* of indigenous villagers to restore to them the lands of Buenavista that Captain Antonio Ancheta had violently wrested from them and converted into a ranch for raising horses and cattle. Following Ancheta's death, leaving Buenavista *despoblado* (abandoned), the Pimas secured approval from Esquivel and the *audiencia* of Guadalajara to reclaim the site for their own use, "so that we may found a pueblo there for the good of our souls and the protection of our lands."[21] In the following decade the pueblos of Santa María de Mobas and Santa Ana de Nuri, lying east of Cumuripa in the province of Ostimuri, each paid the entitlement fees and taxes to secure their tenure over village commons and additional *sitios* for grazing mission herds.[22] During these same years the Jesuit Juan Echagoyan paid ninety pesos and two good mules for official measurement and title to the mission lands adjudicated to the pueblos of San Pedro de Aconchi and Baviácora in the middle Sonora valley.[23] Nearly a century later, in 1806–7, the Pima governor and four council officers of San José de Tumacácori, in the middle Santa Cruz valley, with the encouragement of their missionary Narciso Gutiérrez, secured title to a *fundo legal* (minimal allotment), an *estancia* of two *sitios*, and additional lands south of the pueblo that, according to local witnesses, had been purchased for the mission by the Jesuits.[24] Such precautions, seen in the light of mounting incidents of overt conflict as seen in Cumuripa and Ures, suggest that during the seventeenth-century development of both missions and mining *reales*, when the reduced native population and relatively sparse settlements of *vecinos* provided a scenario of land abundance, population growth (of humans and livestock) during the eighteenth century brought with it a contrasting awareness of land scarcity.

The *denuncia* (new land claim) and enclosure of grazing land became the most common procedure used to accumulate individual and family landholdings. In the second half of the century, the number of land claims filed before colonial authorities by Spaniards for private grants, through the established procedures of *composición* (paying for legal title to lands already occupied) and *denuncia*, increased markedly. Salazar's successful claim to *tierras realengas* west of Santa Ana and the history of private ownership of Ocuca, summarized above, exemplify the process of land privatization. The eastern piedmont region of savannas, forested slopes, and irrigated valleys lying between the Sonora and Bavispe valleys was the scene of increasingly numerous *denuncias* and enlarged holdings among the Hispanic *vecinos* that comprised Sonora's emerging landed elite. Here, as in the Altar valley, the presidial garrisons of Fronteras, Terrenate, Horcasitas, Pitic, Tubac, and Buenavista—founded in 1742 on the site reclaimed by Pima villagers in 1716—constituted focal points of commercial exchange that supported capital accumulation among interrelated networks of notable Sonoran families.

An analysis of sixty-seven land titles from 1788–1859 points to the conditions for a competitive land market on the Sonoran frontier. The size of properties placed in public auction increased, and the prices paid for rangeland rose appreciably during this transitional period from colonial to national rule.[25] Private holdings resulting from the public auction of grazing lands in north-central Sonora typically measured from one to four *sitios*. Ephemeral streams, augmented with wells (*norias*), enhanced the value of the land for intermittent cropping and herding livestock. By the second quarter of the nineteenth century, the number of *sitios* claimed in each transaction and the prices paid at auction rose, especially in the northeastern portion of the province. *Denuncias* of sites measuring from four to eight *sitios* in Teuricachi, Mututicachi, and Cuchuta, formerly Opata villages, were valued at from thirty to sixty pesos a *sitio* in view of their rich grasslands, springs, and natural reservoirs.[26] By these standards, the Salazars' claim to over thirty-four *sitios* and Redondo's purchase of twelve *sitios* in Ocuca comprised large properties, but they were located in the arid setting of the western Pimería Alta.

Enclosure of village commons and public lands led to the fragmentation as well as the amalgamation of private holdings. As we have seen above, Miguel Antonio and María Guadalupe Velasco, siblings, had each inherited a portion of their father's claim to Ocuca. For the different motives of indebtedness and widowhood, each of their portions passed into the hands of different proprietors, the Buingoechea and Redondo families associated with the presidio of

Altar. Buingoechea, who had demanded Miguel Antonio's land and livestock in payment of a debt, seems to have managed the property poorly and passed from view among the provincial landholders of Sonora. In contrast, Redondo and his descendants established a lasting presence in the Pimería Alta as landowners, merchants, and political officers. Santiago Redondo, as *alcalde* of Altar, was appointed a visitor of the northern Pima missions in charge of overseeing their economy following the expulsion of Spanish priests in 1828, and thus was well positioned to dispose of mission lands and assets.[27]

Several land claims and disputes involving the mission villages of Mátape and Batuc give further evidence of the widespread use of the *denuncia* and the subdivision of communal lands in central Sonora. During the final decade of colonial rule, the Indian pueblo of Mátape and Hispanic *vecinos* of the same village forwarded parallel land claims to contiguous plots of grazing and cropland. Three extended families represented by sets of brothers denounced and bid successfully on a ranch called Las Animas that measured 1.5 *sitios* and 1 *caballería* (2,663.5 hectares or 6,579 acres), including both cropland and pastures. Two Indian officers from Mátape who witnessed the measurement of Las Animas and signed the documented proceedings "for me and all the *común*" did not dispute the alienation of this piece of *realengo*.[28]

This case exemplifies the social and spatial divisions within peasant communities that comprised networks of patrilineal, extended families whose agropastoral economy had common roots in the colonial mission. The prevailing distinction between the Indian and non-Indian sectors of the community turned on their differential access to land: the former held title by virtue of the *común*, a recognized ethnic and social unit, while the latter—the *vecinos*—resorted to the *denuncia* and formal purchase. These different routes to land entitlement erupted in conflict in 1831, when Indian commoners of Mátape and the neighboring village of Nácori protested a *denuncia* filed by Rosalía Sánchez and her son-in-law to three *sitios* (5,241 hectares or 12,948 acres) that infringed on the communal property of both villages. Indian leaders and their representatives defended their communities' landholdings on the basis of effective occupation and colonial legal antecedents that forbade measuring private property adjacent to Indian pueblos. They appealed their case to state authorities, who upheld the priority of their possession: the Sanchez family *denuncia* was canceled and the community's competing claim was admitted as a *composición*, in the absence of prior land titles.[29]

The history of land distribution at the presidio of Pitic illustrates further these same themes. San Pedro de la Conquista de Pitic was founded at the confluence of the Sonora and San Miguel rivers in 1741. Four decades later, the settlement comprised a mixed population of *vecinos,* Cunca'ac (Seris and Guaymas), and O'odham (Pimas) in the combined presidio, villa, and mission of Nuestra Señora de Guadalupe. Land grants distributed among the *vecinos* and indigenous villagers in 1785 combined individual allotments with communal grants measured not in *sitios* but in *suertes*, small plots of farmland equivalent to 10.5 hectares or 26.3 acres. Eight *suertes* were reserved as municipal lands (*propios*) for the Villa de Pitic; twenty-seven *suertes* were assigned to the Seri mission, five for the community and twenty-two as individual plots; twenty-five *suertes* were designated for the Pima Indians living in Pitic, five for the community and twenty for individual households; and a total of forty-two *suertes* were divided unevenly among the eighteen *vecino* households that comprised the villa.[30] The parish census compiled in 1796 showed a combined population of 775 residents in the presidio and villa of Pitic, among soldiers and their families, *vecinos*, and Pima and Guayma Indians. An additional 30 Spaniards and 204 Seris lived at the mission. Although some of the 36 Pima male heads of household living in Pitic were presidial soldiers and thus entitled to an allotment of land, subsequent documentation reveals little trace of Pima, Guayma, or Seri smallholders.[31] Among the *vecinos*, however, notarial records including wills, inventories, rental contracts, and land titles show that the planned distribution of urban lots and cropland in Pitic gave way to parallel processes of accumulation and fragmentation of property over the next two generations.[32]

The irregular size of landholdings and the transfer of property rights among the *vecinos* of the lower Sonora and San Miguel river valleys correspond to gender, ethnic, and incipient class divisions in late colonial Sonoran society. Not surprisingly, the descendants of the two Pitic smallholders who received the largest shares in 1785, Juan José Valencia and Manuel de Monteagudo—eight *suertes* each—had amassed additional holdings of arable land and livestock herds in the floodplain between the Villa de Pitic and San Miguel de Horcasitas. Widows and children of smaller recipients sold or further divided their inheritances; purchasers who increased their holdings by gaining ownership of numerous scattered plots were frequently merchants or landowners who marketed surpluses and supplemented their own production through sharecropping and land rentals. José María Vidal, who had received two *suertes* of farmland, exemplifies the common stratum of smallholders. Vidal gave his daughter María Josefa approximately one-fifth of a *suerte* when she married José María Fernández; her modest dowry included a vineyard and a two-room adobe house. When Vidal's widow, María Juana Espinoza Bernal, drew up her will and testament in 1809, she left two small separated plots of irrigated land to the rest of her children.[33] This is not to suggest that women invariably or even routinely emerged as the losers in transactions involving the sale or transfer of land. It does appear, however, that women in this median range of peasant smallholders were vulnerable to the subdivision of land and loss of assets, especially when faced with indebtedness, drought, and life-course events like illness or death.

Legal changes in the institutions governing land ownership accelerated the privatization of property during the transition from colonial to national rule, responding to the growing complexity of this frontier society. Beginning in the mid-eighteenth century, Bourbon policies established administrative procedures that favored, and even directed, the privatization of communal land tenure associated with the missions and Indian pueblos. Royal visitors José Rafael Rodríguez Gallardo and José de Gálvez—who oversaw the expulsion of the Jesuits—undertook to raise productivity and promote commerce, thus bringing Sonora fully into the imperial orbit. These objectives were served, in their minds, by dividing mission lands into *suertes* to be distributed among native households and *vecinos* living in the pueblos. Gálvez's decree of 1769 was reiterated by successive intendants and commanders of the interior provinces: Pedro de Corbalán (1772) directed his attention especially to the presidio and the Seri mission of Pitic; Pedro Galindo Navarro (1785) defended the property rights of *vecinos* established in Indian pueblos; and Pedro de Nava (1794) ordered the division of mission lands and their distribution to resident families.[34]

In anticipation of Nava's order, his subdelegate Hugo Ortiz Cortés carried out surveys of Opata pueblo lands in the Bavispe and Oposura valleys of the upper Yaqui river drainage in northeastern Sonora; his descriptions and measurements reveal an intricate pattern of communal fields (*labores*) and *milpas* that shared the same irrigation canals and floodplain soils. The mission of Santa María Bacerác and the presidio of San Miguel de Bavispe, established at the mission *visita* with a garrison of Opata troops in 1781, were located within ten miles (four leagues) of each other; Indians, soldiers, and *vecinos* depended on the gravitational flow of the Bavispe stream and farmed contiguous portions of the valley. Here, as in Pitic and in the Eudeve pueblos of Mátape and Batuc cited above, the formal division of floodplain land would disrupt Sonoran floodplain cultivation and patterns of communal labor to maintain canals, living fence rows, and cropland.[35]

Following Mexican independence, the state legislatures of Occidente and, after 1831, of Sonora passed laws that defined citizenship in terms of landed property and, once again, enforced the division of communal lands still held in usufruct by Indian pueblos. In 1828, the deputies of the State of Occidente approved Decree 89, centered on the corporate holdings, understood as the *fundo legal* that in theory comprised the public domain of the mission communities. Citing selected articles of colonial legislation, the decree declared that "the fundo legal [and] mission property restored to the pueblos and vacant lands in their environs shall be reduced to private property to the exclusive benefit of the Indians."[36] Its authors undermined the stated intention of the law, namely, to authorize small private holdings for indigenous households, by recognizing the legitimacy of only titled property. In practice, then, Decree 89 hastened the transfer of communal resources into private hands and deepened the divisions between Indians and *vecinos* in the pueblos, as illustrated by the land claims documented above for Mátape and Batuc. The Sonoran legislature passed six different decrees between 1831 and 1835, based on the law of 1828, that facilitated private claims to so-called vacant lands (*baldíos*, or what colonial authorities had referred to as *tierras realengas*). State officials acted on these provisions, measuring and dividing village lands in Arizpe, Bacoachi, Sinoquipe, Banámichi, Huépac, Aconchi, and Baviácora, in the Sonora river valley, and in Saguaripa, Santo Tomás, Pónida, and Arivechi in the Eudeve pueblos of eastern Sonora.[37] By the close of the decade, the process of privatizing the commons had become irreversible in the region north of the Yaqui valley to the Pimería Alta, dispersing the landed patrimony that had comprised the *común* and the colonial legacy of the mission pueblos.

Land and Livestock: Corporate Patrimony in Chiquitos

The opening decades of the nineteenth century were characterized by the apparent continuity of colonial mission structures in the pueblos of eastern Bolivia. Indeed, the catalog of mission products, centered on wax and cotton cloth, seemed to reinforce the economic circuits established by the Jesuits nearly a century earlier, based on the supervised labor of indigenous men and women. The Bolivian national state and the prefecture of Santa Cruz recorded the income generated annually from the sale of mission produce as part of public revenues that figured in budgetary tables and inventories of individual missions.[38] State authorities left the day-to-day supervision of the mission pueblos to the diocesan priests and lay administrators—now called *ecónomos*—who had replaced the Jesuit missionaries after 1768, but they enforced their nominal jurisdiction over the pueblos and their assets as part of the public domain.

The detailed descriptions of Santa Cruz de la Sierra and the Chiquitanía compiled by Alcides d'Orbigny during his extensive tour through the province in 1830–31 provide a window for viewing the pueblos during this transitional period.[39] D'Orbigny made his way eastward from Valle Grande toward Santa Cruz during the month of November, when he encountered continuous rain. He remained in the provincial capital from January to June 1831, waiting for flooding to subside in order to cross the Río Grande and pass to San Xavier, the first of the Chiquitos missions. Well advised to follow the customs of gift exchange that characterized relations between the Chiquitanos and Cruceña society, d'Orbigny loaded his baggage with a stock of gifts for the Indians before departing Santa Cruz. D'Orbigny described the geographical location, spatial layout, principal productive activity, and the resident Indian population of each mission. Their assets included cattle *estancias*, as well as the towns with their complement of workshops, patios for refining wax, a chapel, a *colegio* for the resident priests and Indian officers, and the urban grid of indigenous housing. Concepción, for example, retained its preeminence from colonial times in population and in the variety of its products, including cotton, indigo, wax, and vanilla. San Ignacio, farther east, boasted nearly 3,300 souls, although its economy rested solely on gathering honey and wax; San José, the largest of the southern missions, supported 1,800 resident Indians at the time of d'Orbigny's visit and produced hammocks, cotton textiles, tamarinds, and fine-quality wax.

The two decades following d'Orbigny's visit of the Chiquitanía brought critical changes in the relations between the missions and Cruceña society, which affected the physical landscape and expanded the role of the state in the frontier economy of the province. Chiquitos suffered over four years of drought in the mid-1840s, an ecological crisis that had repercussions for the artisanal and agricultural output of the pueblos and their subsistence. The governor and prefect attempted to ameliorate the hunger experienced throughout the province by shifting resources to the neediest villages and importing livestock from the northern province of Moxos. Furthermore, the governor traded cattle from the missions to the Brazilian settlement of Casalbanco for maize to distribute to the Indians of Chiquitos under his charge.[40] During this same period, in the winter of 1845, the scientific expedition led by the French count Francis de Castelnau passed through the Chiquitanía and Santa Cruz, en route from Brazil to Peru. Castelnau and his colleagues took special note of geological formations, streams and rivers, and vegetation, as well as of the composition and work routines of the pueblos they visited. Evidence of the early stages of drought east of the Río Guapay appeared in Castelnau's description of wide riverbeds and arroyos that were nearly dry, although the pueblos of San Ignacio and San Miguel benefited from small lakes or lagoons that filled during the rainy season.[41]

Castelnau's itinerary extended from the Brazilian border westward to Santa Cruz, passing through the pueblos of Santa Ana, San Ignacio, San Miguel, Concepción, and San Xavier. The distances between the pueblos were broken by numerous cattle ranches, such as San Gregorio, where Castelnau's party found shelter and forage for their animals. Castelnau negotiated directly with the Indians, often trading necklaces, earrings, and beads to the women in return for meals, maize to feed their horses, and their husbands' labor.[42] All but two of these *estancias*—Parubio and Santa Rosa in the environs of Santa Ana, which were properties of the state—belonged to the Indians in common who cared for the livestock and reported to the administrators of their respective missions. These common herds, organized by sections (*secciones*), provided a form of currency similar to the thread and cloth produced in the missions that circulated throughout the province. Meat from butchered cattle served as payment to Indian men and women for their labor to spin and weave, gather and clean wax, clear roads, carry the mail, or serve as guides for foreign travelers. Prefects and the governor authorized moving livestock on the hoof to trade for grain, as we have seen above, to support frontier defense, and to facilitate road building.

Prolonged drought in the mid-nineteenth century, leading to hunger, loss of livestock, and population movements, weakened the corporate structures and communal life of the pueblos at the same time that the productive resources of the missions—cultivable land, pastures, and cattle—became the object of dispute between ecclesiastical and civil authorities. Scattered references to *denuncias* of *baldíos* in Chiquitos were overshadowed by officially sanctioned projects to lease common lands in the valleys lying south and west of Santa Cruz de la Sierra. Concurrent with public interest in the productive value of land and its potential for colonization, one documented case of measurement and adjudication of "the place called Espejo," in 1847, signaled growing competition among Cruceño families to acquire property.[43]

Demetrio Manuel Soruco, representing himself and his six brothers, filed a claim to "the uninhabited and almost inaccessible forested lands called 'de Espejo,' where I have it on good reference that there are places apt for cattle breeding and agriculture." Soruco requested an extension of two leagues by one league, presenting as his guarantor Manuel José Justiniano Mercado, another long-time Santa Cruz *vecino*. The Sorucos' *denuncia* laid claim to woody hills broken by the Espejo stream that fed into the Piraí river and to flat lowlands with natural pasture that could support livestock. Previously, two aspiring competitors, Manuel Antonio Salvatierra and Ignacio Antonio Cuéllar, had registered a petition for part of de Espejo, approximately one-quarter league. The prefect of Santa Cruz authorized the measurement of the Sorucos' claim to an extension of two leagues north-south by one league east-west and the placement of *mojoneras* to indicate the limits of their property, giving them preference over Salvatierra and Cuéllar in view of their potential "in society, with capital, with knowledge to undertake agricultural enterprises and, finally, with the means and elements to improve the land and people in it with other equally useful establishments." The land adjudicated to the Soruco brothers in de Espejo lay adjacent to the pueblo of Porongo, in the province of Cercado, immediately northwest of Santa Cruz. The commissioners in charge of measuring the land summoned the priest of Porongo, José Ramón Suárez, to clarify the limits of his parish, but Suárez declared that he did not know the precise boundaries of the lands of Porongo and was unable to find documents to corroborate the founding of the pueblo and the extension of its lands. Thus the boundary was set according to the Soruco brothers' specifications.[44]

In 1852, the prefecture and commandancy general of Santa Cruz—with jurisdiction over the provinces of Chiquitos, Cercado, and Cordillera, as well as Santa Cruz de la Sierra—initiated an investigation into the occupation and use of mission lands. Prefect Francisco Ybañez's directive to the Bolivian Ministry of the Interior, copied to the bishop of Santa Cruz, made it clear that his purpose was to convert the open lands customarily held by the mission pueblos into rental properties. He proposed establishing a schedule of rents from which the state could derive income and dedicate a part of the anticipated revenues to the upkeep of the cathedral of Santa Cruz and other parish churches. Ybañez argued not only that these fertile lands were underused but also that the priests assigned to the spiritual ministry of the missions abused their posts by planting the land for their own benefit and exercising their influence to allow only those *vecinos* of their liking to pasture their herds or cultivate plots of land on mission holdings.[45]

Not surprisingly, the vicars assigned to these scattered parishes mounted a spirited defense of their administration of mission lands and herds. Their responses to Prefect Ybañez's challenge, even if self-serving, reveal the informality of customary landholding arrangements and the patchiness of grazing and cultivation in the eastern Cordillera and the Chiquitanía. In the provinces of Cordillera and Cercado, where the Indians no longer directly supported their priests, the pueblos retained minimal communal holdings. Father José Ramón Suárez of Porongo reported that the Indians of this pueblo had occupied what was left of mission lands "since time immemorial" with subsistence cultivation and livestock. While he defended the principle that the Indian peoples should rightfully be considered proprietors of these village lands, Suárez pointed out that *vecinos* of Santa Cruz possessed "spacious fields and large extensions of uncultivated bosque" where their cattle herds grazed, as well as large clearings planted with sugarcane and other crops. In addition to these substantial holdings, squatters had appeared in the mission villages, but they did not endure as smallholders, in part because of hostilities between these poor *vecinos* and the Indians. Suárez described them as temporary settlers who brought their few dairy cattle, horses, and service animals to the province seasonally, without a firm resolve to stay and prosper. For this reason, and because of the abundance of unoccupied land in the Department of Santa Cruz, he predicted that the prefect's project to establish land rentals would not prove successful in furthering settlement or raising revenues.[46]

The vicar of northern Cordillera, in charge of three mission curates, reported that his pueblos retained only the "league and a half" that had been assigned to them under the auspices of the colonial Recopilación de Indias. These restricted lands sufficed for grazing the herds that belonged to the missions, the Indians' livestock and shifting cultivation, and the plots set aside for the support of the priest. Pasture lands that had once belonged to the pueblos of Abajo and Piray had already been awarded to private *vecinos* because these missions no longer had herds of their own.[47] Juan Miguel Montero, vicar of the southern Cordillera, had jurisdiction over eight curates, of which seven were missions. Montero distinguished among four types of rural property: lands adjudicated to private owners; lands belonging to the pueblos; lands occupied informally (*indistintamente*) by Indians and *vecinos*; and lands vacant and unproductive (*absolutamente baldíos y yermos*). The latter three categories, Montero explained, belonged "to the nation." Indians and non-Indian settlers cultivated small plots and grazed their cattle in the lands of the pueblos; in these same extensions the priests claimed their right, as tutors of the Indians, to use the land for their own support.[48]

Montero recognized that disputes had arisen over damages caused to the Indians' crops by cattle belonging to the *vecinos*, but he assured the bishop that they were resolved in ways that reconciled opposing interests. Lands that were possessed *indistintamente* by local Indians and non-Indians encouraged by the Bolivian state to settle in the eastern frontier were held *en mancomún*, a kind of communal regime that recognized de facto use and occupation more than formal property demarcations. Montero associated the "baldíos y yermos" lands with those areas occupied by nomadic Indians—probably Chiriguanos and Guaycurús—whom he called "barbarous . . . and enemies of order and Christianity." He firmly recommended against renting out any of these lands and reminded the bishop of the Indians' prior rights to the resources necessary for their own subsistence. To that end, he underscored the need to document the pueblos' "ancient possessions," due to the loss of colonial papers during the political revolts of the independence era.[49]

At mid-century, then, the frontier between the Andean cordillera and Santa Cruz held a mixed Indian and non-Indian population with a landholding regime that combined individually and communally held lands. Vestiges of the colonial mission order could be seen in the ecclesiastical administration by the vicars and in their recognition of the antiquity of the pueblos under their charge. The viability of establishing land rentals in the Cordillera and Cercado provinces was in doubt because of the ecological conditions that limited commercial production and the ethnic and social mosaic of their inhabitants. Permanent water sources were necessary for cultivation even as private and communal interests vied for the productive resources that could sustain farming and herding.

Francisco Ibañez's proposal to convert Chiquitos pueblo lands into rental properties struck at the foundation of the indigenous economy that relied on communal labor. In accord with the province's traditional dependence on Indian labor, all public works and services, including the upkeep of the churches, the opening and maintaining of roads, mail delivery, and work for the state three days each week fell to the native peoples of the mission towns. Father Gregorio Basan, vicar of the northern Chiquitanía (San Ignacio, San Miguel, San Rafael, and Santa Ana), warned against implementing the prefect's initiative, arguing on the basis of ecological conditions, migratory habits, and historical precedents in the province. In contrast to the Bolivian highlands, agriculture in the Chiquitano lowlands did not benefit from irrigation, nor was cultivation enhanced by the plow. Natives relied on slash-and-burn techniques to clear small plots that yielded maize, bananas, and yuca for two years. Should the region fall victim to drought, as had occurred in Basan's memory for four consecutive years, then all crops would wither.[50] The pueblos had already suffered the effects of the Indians' out-migration in bands that passed through the provincial capital or moved across the border into Brazil. This movement would increase markedly, Basan predicted, should the Indians learn that their lands were put up for rent.

The southern Chiquitos province, including the pueblos of San José, Santiago, San Juan, and Santo Corazón, comprised pasture lands with watering holes (*aguadas permanentes*) sufficient to water and graze livestock. These were occupied by mission herds, divided into sections that corresponded to the Indians' commonly held patrimony and to privately owned cattle. At a greater distance from the pueblos, and extending all the way to the Brazilian border, were national lands of potential use for stock raising, planting, and mining. Salt deposits located at a fair proximity to San José de Chiquitos, for example, had been exploited at different times since the turn of the nineteenth century. Successive prefects of Santa Cruz and governors of Chiquitos had promoted salt extraction, using Indian labor, as a means of raising public revenues.[51]

In 1833, the national government had introduced the notion of private ownership of livestock linked to the status of citizenship for the Indians of Chiquitos by distributing three head of cattle to each household with the promise of giving out the remaining herds to native families should this initiative prosper. Governor Marcelino de la Peña reported to the prefect of Santa Cruz that the allotment of livestock to "all the Indians" was completed during the first three months of 1834.[52] The Chiquitanos, however, "never took that livestock as their property"; rather, the cattle came under non-Indian management, and the herds grew or diminished according to the individuals in charge of the livestock sections assigned to each *parcialidad*. Vicar Basan noted that from 1840 to 1852, the "principal stock" had been drastically reduced due to the effects of drought and poor management of the herds. Should rentals of land and cattle be auctioned, he questioned rhetorically, what would there be left to lease? Furthermore, without surplus production of crops or livestock, how could there be income sufficient to generate annual rental payments?[53]

Despite these warnings, by early fall 1852, the prefect's order to lease the cattle of San José by sections to the highest bidder was put into effect. The auctioneer received a total of 2,112 head from four sections, or *estancias*, that had supported the herds of San José. That same year, nearly 8,000 head of cattle from the mission pueblos of San Xavier, Concepción, San Miguel, San Ignacio, San Rafael, and San José were auctioned to six bidders for a period of six years. The rent was calculated at twenty-five pesos for each lot of 100 head received and capitalized, subtracting losses (*rebajas*) and calves less than a year old. Theoretically, the pueblos figured as the creditors of the leased herds, and the lessees were obligated to distribute to the Indians the head of cattle customarily given out for slaughtering on feast days at different times of the year. The Indians themselves would supply the labor, for which they were to be paid in tools, and the cattle so leased would continue to graze on the *estancias* belonging to the pueblos. At the end of the rental period of six years, the herds should be returned to the Indians, their rightful owners.[54]

Gregorio Basan's assessment of the Indians' aversion to the privatization of communal resources through the rental of mission lands and herds implies a competing principle supported by additional evidence: namely, that the Chiquitanos associated the corporate assets of the pueblos with the physical maintenance of their churches and the religious life of their communities. Documented controversies over ownership and management of mission livestock centered on San José and the southern pueblos of Chiquitos, where the Indians' voices emerged briefly in the reports that circulated from the pueblo to the bishop and prefect of Santa Cruz, and from Santa Cruz to national authorities in Sucre. The issue was openly debated in 1858, when it was necessary either to renew the leases or to return the cattle to the Chiquitanos. Lessees and provincial administrators of the Chiquitanía admitted that the proceeds of the livestock rentals that should have been used to purchase tools for the Indians had been diverted to the coffers of the Department of Santa Cruz; these lapsed payments had led to the Chiquitanos' verbal protests and threatened uprisings.[55]

The following year Domingo Riberto, a native official (*natural y cacique*) of the pueblo of San José, denounced the abuses of the system of livestock leases that had impoverished his people. Directing his protest to the national government, Riberto laid out the history of the political emancipation of the Chiquitanos as citizens in 1833, and their communal custodianship of mission herds in sections, referring to the earlier *parcialidades* from that date until their leasing by public auction in 1852. Calling the rental of livestock a "scandalous robbery," he gave vent to his anger over the government's failed promises to supply tools for the Indians and sufficient head of cattle for the Chiquitanos' subsistence and the maintenance of their hospital. Riberto demonstrated that he understood well the intricacies of classifying portions of the herd as *terneros* (veal calves) under one year of age and thus "dead" or exempt from rental payment. He accused the lessees of cheating the Indians by claiming as *terneros* head of cattle that had already been branded and, therefore, should have been counted for the rental fees.

"Greed and gold" had overcome the rightful claims of the Chiquitano peoples. Domingo Riberto warned that should the system of rentals continue, "it might be impossible to control the desperate anger of the Indians, who are tired of these injustices."[56] Ignoring Riberto's protests, the prefecture of Santa Cruz auctioned the herds of the same six pueblos for rental a second time in 1859. The Indians of San Juan did indeed rebel against the lessee of their cattle in 1860, "gravely wounding him" and forcing a cessation of the rental of that mission's livestock. Their "tenacious and constant opposition" to the leasing of the pueblo's herds derived from the miniscule benefits they received, having been promised metal tools as their part of the payment of the semiannual rents. The small size of the herds and the enormous distances of the *estancias* from markets in San Juan precluded the lease's yielding monetary or material rents.[57]

For over a quarter century provincial authorities had assumed the prerogatives of the state to make use of the pueblos' livestock—*el ganado de las secciones*—for purposes alien to the well-being of the communities themselves. The system of leases put in practice, beginning in 1852, depleted mission herds and impoverished the communal patrimony of the Chiquitos pueblos. State-sponsored auctioning of livestock rentals denied the Indians' accumulated knowledge of herd management, first acquired under Jesuit tutelage, and transferred their collective assets into private hands. The loss of livestock curtailed long-standing methods of remuneration and the reciprocal exchange of meat for measured labor output that had supported the moral economy of mission life.

Conclusions: Territory and Community in Sonora and Chiquitos

From Ocuca to Santo Corazón, and from Batuc to San José, the changes in land tenure documented for Sonora and Chiquitos during more than a century show distinct patterns of conflict and negotiation over territory. Contrasting geographic features and different cultural patterns created distinct spatial concepts and historical practices concerning proprietary boundaries and land-use rights in each of these frontier provinces. The history of land concentration and fragmentation among peasant smallholders, exemplified by the presidio and villa of Pitic in eighteenth-century Sonora, cannot be documented for Chiquitos, where corporately held pueblo lands and private *estancias* overlapped in unevenly defined territorial spaces until after the mid-nineteenth century. It is equally striking to notice the disputes over the enclosure of the commons in Sonora and the procedures established there for claiming title to measured pieces of land, juxtaposed with the debates over the leasing of common lands and cattle herds instigated by the Bolivian state in Chiquitos.

We return to the concept of social ecology to underscore the different meanings of *territory* in relation to the natural environment and to explain the significance of supposed land scarcity in Sonora and land abundance in Chiquitos. In both Sonora and Chiquitos the basis of the colonial and early republican economies was agropastoral; in the former, drought threatened the livelihood of Indians and Hispanics alike from one year to the next; in the latter, seasonally occurring floods interrupted communications and trade from the Chiquitano villages to the administrative and commercial hub of Santa Cruz and the roads leading to the cordillera. The shift from communally organized, irrigated horticulture to private land tenure documented for Sonora from the late eighteenth to the nineteenth century marked a transition from indigenous and European garden crops to extensive herding, followed by fencing and the enclosure of water sources. Livestock breeding and grazing played an important role for strengthening the private sector in the Chiquitanía as well, notwithstanding the endurance of the corporate economy of the Indian pueblos—sustained by gathering and weaving—until the mid-nineteenth century. Furthermore, the substratum of swidden cultivation in small clearings cut out of the *bosque* continued to provide subsistence for the Chiquitanos, although food production became increasingly marginal to the private plantations of sugarcane that gradually came to dominate the landscapes surrounding Santa Cruz. The provinces of Cercado and Cordillera immediately west of Santa Cruz resembled Sonora, to some degree, in the competition for arable land between private landowners and renters and the remnants of Indian missions. In the Chiquitanía proper, however, the land itself was not directly in dispute until the leasing of mission land and cattle in the 1850s.

The social and demographic components of changing territorial boundaries and claims to land and its resources play an important role in the different stories of land tenure told here for Sonora and Chiquitos. Clearly, the pace and density of Spanish settlement in each of these provinces is closely linked to the historical process of changing territorial boundaries. The emergence of a mixed population of *vecinos* and Indians who had moved into the Spanish economic sector in Sonora, living in close proximity to Indians who sustained the communal life of the pueblos, heightened demand for private landholdings and hastened the process of enclosure in accord with Bourbon policies aimed to convert Indian commoners into individual laborers and petty producers. In contrast, the relatively sparse population of Hispanic settlers in Chiquitos reduced pressure on the land. Merchants and landowners based in Santa Cruz de la Sierra, as well as the ecclesiastical structure established in the missions, sought enrichment and the accumulation of assets through livestock and the commercialization of the Indians' labor. A land market developed in Sonora during the transition from colonial to national rule through juridical changes in the institutions of land tenure propelled by the social, economic, and political pressures of population growth and the concentration of governing power in the hands of a commercial elite. Cruceña elites successfully gained control of the commercial circuits based in the missions before seeking to appropriate their resources of land, water, and livestock, but they could not silence the Indians' protests and reiterated claims to their territorial patrimony.

The politics of land tenure remain inseparable from the material environment. Just as institutional innovations that define the legitimacy of use rights and proprietorship over bounded territories effect changes in the physical environment, so do modifications in the land and its resources, stemming from human agency and climatic fluctuations, alter the meaning of the legal structures governing land tenure. Iberian definitions of measurement and enclosure—such as leagues, *sitios*, *suertes*, and *estancias*—created spatial categories of dominion and usufruct over designated properties. Yet these very categories and the territorial divisions they implied were contested repeatedly through social practice, such that land tenure became a foundational component of community, as well as of property. The cultural landscapes described here for Sonora and Chiquitos emerged historically from this dialectic between land and society, community and territory, in multiple and intersecting processes of change.[58] The following chapter brings us a closer view of the societies that developed in each of these frontier provinces through their comparative histories of mixture and segmentation along the lines of ethnicity and gender.

CHAPTER 4

Ethnic Mosaics and Gendered Identities

> Thinking historically is a process of locating oneself in space and time. And a location . . . is an itinerary rather than a bounded site—a series of encounters and translations.—JAMES CLIFFORD, *Routes: Travel and Translation in the Late Twentieth Century*, 1997

The previous chapter began with an affirmation of the formative bonds between human communities and the land, focused on the historical construction of territories in Sonora and Chiquitos. This chapter turns to the communities themselves, and to the changing meanings of ethnic identities over time and space. It develops historical approaches to the concepts of ethnicity and gender in the public sphere of colonial relations and the intimate sphere of domestic communities. Ethnicity—understood as a sign of belonging or a common point of allegiance—and gender—signifying the social construction of prescribed roles for men and women in relation to sexuality, family formation, and work—constitute the basic working concepts of this chapter. The closely related themes of hybridity and *mestizaje* help us to understand the cultural fabric of both these frontier societies. *Mestizaje* refers to racial mixture arising from biological procreation among persons of different origins or from migration and the adoption of visible signs—such as clothing, language, and behavioral patterns—that combine different traditions or form new ones. *Hybridity* expresses the historical production of new cultural formations set in motion by colonial power relations through multiple transactions of selection, adaptation, and endurance.[1]

The gendered experience of colonialism emerges differently in Sonora and Chiquitos. As we saw in chapter 2, gender defined the division of labor in the Chiquitano missions, while in Sonora, gender proved less pronounced in the organization of mission economy but held significance for family formation and *mestizaje*. Gender figures prominently, too, in the histories of seizure, ransom, and sale of human captives that spanned the preconquest and colonial periods of both regions. Gendered differences intersected with ethnicity and distinctions of rank and prestige among peoples for whom social inequality was not based on the accumulation of property.[2] This chapter addresses these themes comparatively in both experience and representation and relates gender to a broad discussion of identity. It examines allegiances that crossed ethnic boundaries, creating moving frontiers of internal cohesion and external exclusion among communities. Tribal names and native languages, like *Chiquitos* or *Opata*, are themselves amalgams of diverse ethnic lineages and signs of the cultural commingling that occurred in both provinces through the mobility of Indians, Europeans, and Africans in and out of missions, villas, presidios, and mining centers.

Our discussion examines ethnic mosaics in the world that Indians and Spaniards created through the Jesuit mission enterprise and in the landscapes they molded from the eighteenth century to the end of colonial rule. It focuses first on the pueblos, and then widens its lens to view the colonial society that enveloped Indian communities in webs of interdependency, by comparing different rhythms of ethnic mixture in Sonora and Chiquitos. Parallel processes of diversification and hybridization took place in both regions, but fragmentation of ethnic identities remained a prominent feature of Chiquitos, while Sonoran ethnographic maps showed a pattern of consolidation. The emphasis on movement, displacement, and reinvention questions the ethnographic labels culled from documentary sources, repeated from one text to another, and used as guideposts for field observations in ways that reassess the complex signs of cultural identities.[3]

"We, people here . . . ," rendered in thousands of indigenous languages, is nearly a universal translation for the self-identifications of different local ethnicities.[4] The O'odham of Sonora and the M'oñeyca of Chiquitos are living ethnic designations that connote both the fulfilled condition of humanity and linkages to place.[5] It is important to note, however, that the association of particular peoples with specific territories is not set or binding over time. People establish their relationship to the land historically, through the landscapes they build and their use and replenishment of resources. The spaces they occupy and the boundaries that distinguish one from the other are contested, fluid, and changing, despite the illusion of fixity that cartography brings to the mapping of ethnic territories. Contrasting conditions of subsistence and material culture frequently serve as markers of ethnic differentiation between nomads and sedentary village dwellers, irrigation and swidden farmers, or herders and hunter-gatherers. Nevertheless, distinct ecosystems do not presuppose equivalent ethnic identities, for different groups of people share similar subsistence patterns and create linkages of interdependency with their environments.

Geography and locality, then, do not determine ethnicity, nor do ethnic identities conform neatly or wholly to specific languages. Formal linguistic affinities, as well as living languages in both written and oral forms, are, without a doubt, closely related to culture and to particular communities. It is well to remember, however, that the linguistic-ethnic linkage was established, in part, by imperial practices that defined ethnicity in terms of the dialects that missionaries and secular officials set out to codify in written grammars and vocabularies in an effort to make sense of the tribal affiliations they sought to control.[6] Similar names were frequently applied interchangeably to linguistic and ethnic groups, such as Guaraní in the greater Río de la Plata, or Pima-Tepehuán in northwestern Mexico, thus reinforcing their apparently synonymous character. The Guaraní languages and territorial polities extended widely throughout the Paraná, Uruguay, and Paraguay rivers, and the term became a generic identification that referred to many different peoples. Guaraní-related dialects formed part of the cultural matrix of the Chiquitos missions, and present-day Guaraní communities maintain an important presence in the Izozog drainages of the Parapetí river, on the northwestern margins of the Gran Chaco.[7] "Guaranized" languages expanded through the mission system, but not all Guaraní speakers shared the same ethnic identity.

The Pima-Tepehuán ethnic and linguistic chain comprises a somewhat analogous case for northern Mexico. Pima and Tepehuán speakers belong to the same family of languages, and colonial documents of varied origins seemed to recognize a widely extended territorial band of different Piman (O'odham) and Tepehuán (Ódami) peoples along both the western and eastern flanks of the Sierra Madre Occidental. Yet each of these designations subdivided into numerous ethnic groups, and their various boundaries were not permanent or clearly drawn. Moreover, *Pima* and *Tepehuán* are externally derived ethnic designations; the former referred to Pima-speakers' characteristic use of the negative (*pima*) to express key concepts in their language, and the latter was a composed term in Nahuatl (*tepetl* plus *hua*) that may be translated as "mountain people."[8]

Anthropological classification systems built on a presupposed symmetry between language, and ethnicity, in turn, reflected the colonial censuses that listed different native populations in an effort to establish fiscal control over them and document the success of the missionary orders' evangelizing efforts.[9]

Cultural geographers like Carl O. Sauer and Ralph L. Beals followed the scholarly traditions set by nineteenth-century naturalists, such as Adolph Bandelier and Carl S. Lumholtz, who relied on early conquest narratives to guide their explorations through the Sierra Madre Occidental and the foothills and deserts of northern Mexico.[10] Sauer's mapping of ethnic groups for northwestern Mexico on the eve of conquest, together with his estimates of population and probable location, drew on a detailed reading of Jesuit *annuas* (annual reports), as well as his own field observations during the early twentieth century. The successive publication of ethnographic maps based on Sauer's research have lent permanence and stability to categories culled from the documents which mask a more dynamic picture of changing physical and social spaces occupied by diverse peoples claiming particular identities in different historical moments.

In a similar way, tabulated lists of languages and tribal groups based principally on Jesuit reports and linguistic manuals described a highly fragmented ethnic mosaic in eastern Bolivia. Lorenzo Hervás y Panduro compiled a six-volume catalog of languages, published at the turn of the nineteenth century, in which he undertook to classify diverse "nations" according to their dialects.[11] French linguist Lucien Adam produced numerous grammatical studies of Caribbean and South American languages, and, in 1880, he published the *Arte y vocabulario de la lengua chiquita* in collaboration with Victor Henry, bringing together in one volume four Jesuit manuscripts.[12] The classification systems used in these catalogs influenced Alcides d'Orbigny's ethnographic notes on his travels through the region, as well as Erland Nordenskiöld's anthropological studies of the riverine lowlands and the eastern Andean foothills. Alfred Métraux incorporated the languages and peoples enumerated by both Hervás and Adams and Henry in the *Handbook of South American Indians*, and these reappear in recent histories of Chiquitos.[13]

This literature has served to systematize and make legible the apparent cultural profusion of native peoples in the imperial frontiers of northern Mexico and inland South America. The works cited here, among others, have contributed beyond a doubt to the knowledge of languages and ways of life. Nevertheless, they have reinforced the acceptance of ethnic categories as facts and tended to favor the descriptive placement of ethnographic names over historical processes of change and confluence.[14] Without denying the value of this tradition as a necessary reference point for further study, the examination of ethnicity undertaken here for Sonora and Chiquitos emphasizes change over continuity and the mutually formative influences of ethnic allegiance, gender, and social status in the creation and recreation of identity.

The reference to itineraries and translations in the opening epigraph of this chapter underscores the idea that ethnic identities are processes of becoming, and that the journey may prove more important than the destination. Ethnic affiliations are closely linked to family lineages, but their meanings unfold in historical sequences that do not have a singular origin; nor do they constitute permanent formations. The fission or consolidation of different groups due to migration, intermarriage, and conflict alters ethnic and gendered identities. Diverse social allegiances emerge through parallel processes of consensus and contention, creating ethnic mosaics through cultural contact, hybridity, and fragmentation.

Emerging Identities in Colonial Borderlands

Tribal names like *Opata* and *Chiquitos* represent colonial constructs, hardened into racial categories invoked to enforce social control. Yet colonized peoples reworked these same categories into cultural identities that connoted lines of commonality and affiliation, as well as boundaries of difference and exclusion—expressed through territorial claims, linguistic patterns, and societal norms.[15] Native communities on the frontiers of empire became entwined with the development of colonial societies, themselves partial transplants of Iberian cultural and political norms placed in ecological and cultural settings that contrasted strikingly with their metropolitan centers, whether Rome, Seville, Madrid, or the viceregal capitals of Mexico City and Lima. Ethnic differences in these colonial societies became magnified as they served increasingly to maintain social distance among so-called Indians, Spaniards, Blacks, and mixed-race groups collectively termed *castas*.[16]

The mission communities and colonial settlements of Sonora and Chiquitos developed historically as partial societies.[17] Transitory contingents of governors, militias, and missionaries who represented colonial authority were often divided among themselves and lacked the full panoply of royal and ecclesiastical officialdom that graced their viceregal capitals and European metropoles. Colonial settlers—miners, merchants, livestock breeders, commercial agriculturalists—helped to propagate frontier societies characterized by physical and social mobility, even as they clung to racial labels to set themselves apart from

Carte du Mexique, 1771. "Map of Mexico or New Spain, including New Mexico, California, and part of the adjacent country." Courtesy of Map and Geography Information Center, Centennial Library, University of New Mexico. The map includes Louisiana, at this time under the imperial aegis of Spain, and broad ethnic categories of indigenous peoples.

the mixed peoples of color who composed the laboring class of mining towns, ranching estates, and mission communities. Their pretensions to superior status rested only partially on their putatively Spanish bloodlines, and more assertively on their claims to property and their ability to command the labor of others. Thus these frontier societies simplified the complex categories of *casta* stereotypes that developed in the urban settings of New Spain and the Andean highlands.[18]

Native peoples, who comprised their own partial societies and moved in and out of these colonial spheres, understood ethnicity in a variety of ways. Linguistic differences worked their way into the ethnic labels employed by colo-

Mexico or New Spain, in which the motions of Cortés may be traced, 1795. Robertfonis, *History of America*. Courtesy of Map and Geography Information Center, Centennial Library, University of New Mexico.

nial officials to identify segmentary groups, referring often to band or lineage affiliation. Alternatively, they could signify social distinctions between chiefs and commoners and gendered dialects that distinguished forms of proper address for men and women.[19] As suggested above for the terms *Guaraní* and *Pima-Tepehuán*, the labels and their meanings, some originating in precontact times and others emerging after conquest, changed significantly under colonialism. Fusion of separate languages, whose speakers constituted fragmented minorities due to the impact of epidemics and migrations, produced several dominant languages elevated to the status of a lingua franca in the missions, with written vocabularies and doctrinal catechisms. This process of amalgamation took place in Sonora, where different dialects named in the early

Jesuit accounts—such as Nebome, Heve, Tegüima, Eudeve, Akimel, Sobaípuri, and Tohono O'odham—coalesced into Opata, Pima, and Pápago by the mid-eighteenth century.[20]

In eastern Bolivia, the common designation of Chiquitos circulated in the province by the late sixteenth century. *Chiquitos* stood for the liturgical and colonial language of commerce and governance, but numerous ethnic categories persisted in the written records of missionaries and provincial governors. The Chiquitos, Ayoreos, and Guarayos—recognized by the Spaniards as discrete peoples—further subdivided into different dialects and groups, Manazica, Manapeca, Paiconeca, Paunaca, Mococa, Morotoca, Zamuco, Covareca, Piñoca, and Guarañoca, among many others. One suspects that what the missionaries heard and transcribed reflected categories that changed according to the status of the speaker and the circumstances of encounter, because these names are repeated but often applied inconsistently throughout the documentary records.[21] Native peoples attached importance to these markers of identity, but they may have used them differently as indicators of deference or reciprocity, alliance or hostility, affiliation or exclusion, according to their histories of interethnic warfare and colonial contacts.

Indigenous peoples of both Sonora and Chiquitos developed new ethnic and social identities to deal with the colonial experience. Indians adopted the very language of Spanish colonizers to distinguish between Christians who were baptized and familiar with Catholic liturgical practices and *gentiles*, those who lived beyond the pale of the organized mission or visited the pueblos only from time to time. This distinction of religious identity, sealed by the Catholic sacrament of baptism, proved central to the dialogue that Indians were obliged to engage in with colonial authorities, and it was their best defense against enslavement or arbitrary service under the *encomienda*. It became internalized in the unequal treatment Indians gave one another within mission villages and among mission settlements in different provinces. It corresponded, in part, to the degrees of difference between nomadic hunter-gatherers and sedentary agriculturalists that further distinguished mission residents from wandering bands who passed through the missions but avoided the discipline of full participation in the religious and political life of the pueblos.[22] By the late eighteenth century, these ethnic and cultural identities intersected with gendered roles and incipient class distinctions meaningful for the material and religious cultures of mission life.

Colonial Institutions and Ethnic identities

Colonial institutions weighed heavily on these processes of social differentiation from early European-Indian encounters to the consolidation of mission pueblos and Hispanic settlements. Two institutions that shaped ethnic classifications and channeled political and cultural relations among different native groups and between them and Spanish authorities and colonists were the *encomienda* and the mission.[23] *Encomienda*, meaning literally "entrustment," belied the benign connotation of the term under the conditions of conquest. Developed as an institution of both dominion and guardianship during the long reconquest of Muslim territories by Christian military forces in the Iberian peninsula, *encomienda* was transposed to the Antilles and later to the Spanish American mainland, where it was adapted to the political cultures of different native peoples and to the conquistadores' demands for material rewards and titles of nobility. *Encomienda*, in principle, regulated political and economic relations among the crown, Spanish expeditionaries, and subjugated Indian communities. During the early phases of conquest in each of the colonial theaters—Hispaniola, New Spain, and Peru, for example—*encomienda* represented a form of indirect government through which the monarchy authorized conquistadores to appropriate for themselves and redistribute to their followers grants of Indians from whom they had the power to exact both labor and tribute. As the crown attempted to establish an imperial bureaucracy in the core viceroyalties, it took on a power struggle over the entitlements claimed by the first generation of *adelantados* and *encomenderos* to the services of "their" native charges. The institution generated tensions over private and public jurisdiction and became the object of juridical and theological debates that, in turn, set the foundations for colonial legislation and policies throughout Spanish America.[24]

In central New Spain and the Andes, the claims of *encomenderos* yielded gradually to governmental regulation for the collection of tribute and the assignment of paid Indian laborers under the terms of *repartimiento*, with priorities for mining and public works. *Encomienda* endured, however, as a form of obligatory service in the frontier regions of northern Mexico and greater Río de la Plata. The mining and ranching economy of Nueva Vizcaya sustained a lively demand for laborers through *encomienda* alongside the work allotments of *repartimiento*, and in the northeastern provinces of Nuevo León, Coahuila, and Nuevo Santander, the nomadic lifeways of native peoples and the dependence of colonial settlers on stock raising and the products of hunting and gathering prolonged *encomienda* as a way of generating wealth and the principal nexus between Spaniards and Indians.[25]

Documentary evidence for rival claims to *encomiendas* during the sixteenth and seventeenth centuries yields a proliferation of names for different "nations" and *rancherías* that were displaced and resettled on the estates of the *encomenderos* they served. It is plausible that the fragmentation of larger entities—such as Conchos, Julimes, and Sumas—into smaller named ethnic groups occurred precisely through the allotment of particular caciques and their followers to different *encomenderos*. The longevity of the institution and the multiplicity of indigenous nations coincide especially in those regions where Spanish civil colonization preceded the advance of Jesuit and Franciscan missions.[26] Two important questions remain for our discussion of colonialism and ethnic identity: What were the bases of ethnic lineage or affiliation? And, what were the bonds of coercion or loyalty that brought Indians to work for their assigned *encomenderos*? The operation of *encomienda* in Paraguay and Chiquitos, summarized below, sheds light on the history of the institution and its importance for the ethnohistory of north-central and northeastern Mexico and of the interior lowlands of South America.

In northwestern New Spain, north of the villa of San Miguel de Culiacán, Spanish *adelantados* were unable to impose the institution of *encomienda*, due in large measure to the Indians' resistance and to the colonists' greater interest in the more lucrative prospects of the mining centers that expanded through Durango, Chihuahua, and the western provinces of Sonora, Ostimuri, and Sinaloa.[27] For the purposes of the present discussion, then, *encomienda* did not have a lasting impact on ethnic identities or on Spanish-Indian relations in Sonora. Two forms of bonded labor, however, gave rise to new social categories there. Labor allotments sent periodically to the mines and Spanish estates under the terms of *repartimiento* affected Pima and Opata villagers, especially during the formative period of mission *reducciones* in the seventeenth century. Native governors were responsible for recruiting workers for the mining *repartimiento*, and ranch owners negotiated with missionaries for temporary hired laborers in the harvest season.[28] Outside the formal terms of *encomienda*, captives brought into the province through the northwestern ecological and cultural frontier of the Pimería Alta constituted a servant class, named as such and gendered female. The word *nijoras*, identified as a tribe in early colonial maps and documents, referred to captives separated from their communities of origin. Women and children, taken for ransom in wars between O'odham- and Hokan-speaking groups, were traded among Indian pueblos and Spanish settlements. The practice later extended to the capture of Apache prisoners in the increasingly bitter wars between Spanish presidial forces, with Opata and Pima auxiliaries, and the Athapaskan nomads of the sierra.[29] The *nijoras*' presence in Sonora was documented consistently in mission records, and, as we shall see below, they appeared in late colonial censuses.

The longevity of a service *encomienda*, by way of contrast, had a strong formative presence in Tucumán, Paraguay, the Gran Chaco, and in Chiquitos. It proved central to patterns of interethnic warfare and trade, as well as to the evolution of social and gender relations between Spaniards and Indians throughout the interior provinces of Spanish South America. We may recall that Alvar Núñez Cabeza de Vaca, *adelantado* for the Río de la Plata in 1541–45, sought out alliances with different bands of Guaraní in order to sustain his explorations north of Asunción de Paraguay. The Guaraní, for their part, allied with the Spaniards against their Guaycurúan adversaries of the Gran Chaco. Cabeza de Vaca lost the governorship at the hands of his subordinate officers when these alliances failed him over conflicting claims to indigenous labor.[30] His successor, Domingo Martínez de Irala, formalized the administration of *encomienda* a decade later by awarding the Indians' personal service to colonists loyal to his governorship. As the century progressed, *encomienda* in Tucumán and Paraguay shifted from an alliance marked by gift exchange to bonded servitude through increasing levels of violence. Native rebellions or refusals to work were forcefully repressed, and Indians placed in perpetual service as *encomendados originarios* were relocated to Spanish settlements and estates. Women played a key role in the service *encomienda* as conjugal partners given to Spaniards, following Guaraní customary alliances, and as captives taken in raiding and warfare.[31]

The institution of *encomienda* had a long but ambivalent history in Chiquitos. Formal *encomienda* grants date principally from 1585 to 1630, but the Spanish and Creole colonists of Santa Cruz relied on the forced labor of native servants and rural workers for their economic survival and social status until the mid-eighteenth century. Although Indians in the *encomienda de pueblo* theoretically remained in their own communities and owed stipulated services

Tabula geographica Peruae, Braziliae, et Amazonum regionis (1730).
Courtesy of Map and Geography Information Center, Centennial Library, University of New Mexico.

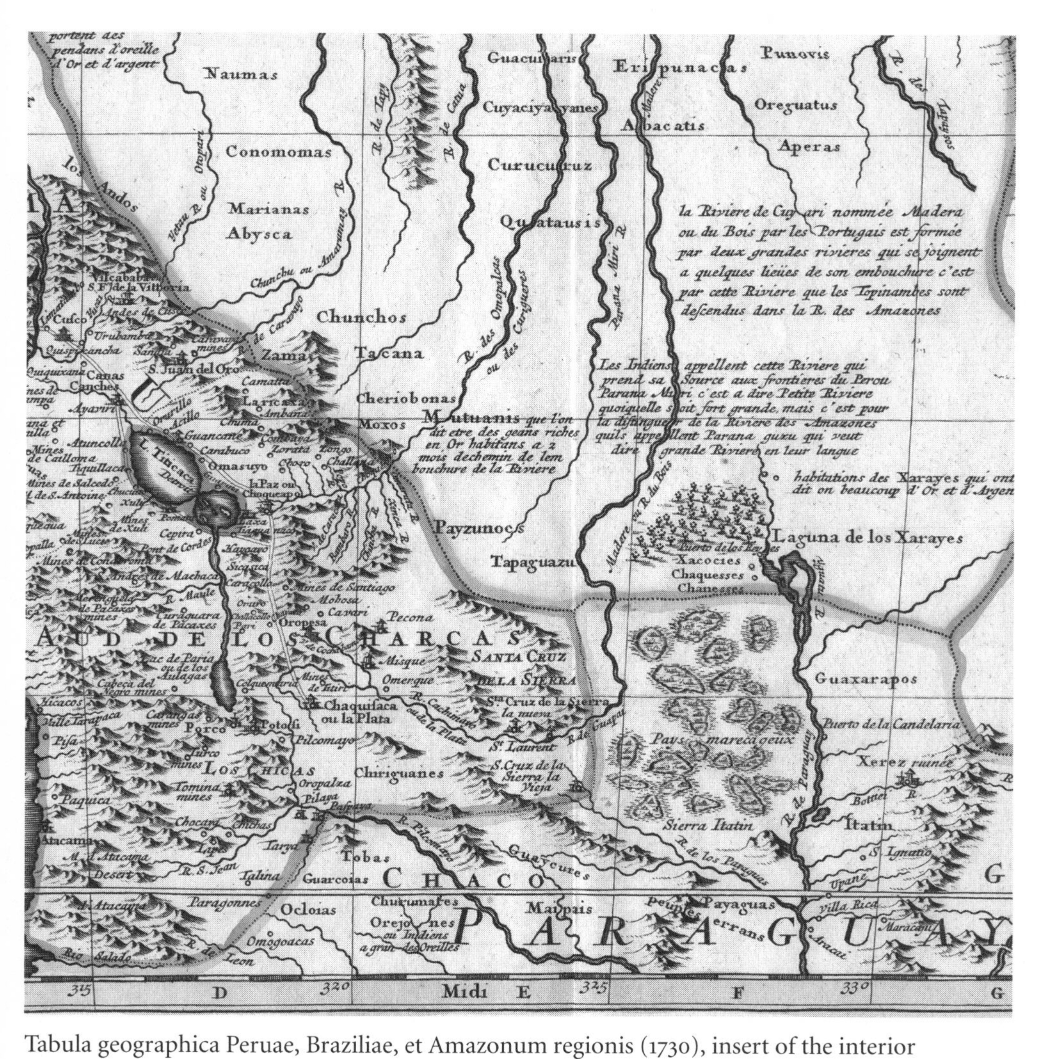

Tabula geographica Peruae, Braziliae, et Amazonum regionis (1730), insert of the interior borderlands, including the province of Chiquitos, identified as the *pays mareca geux*. Note the detailed representations of different environments, such as swamps and low forests, as well as the text that places different ethnic groups along particular rivers. This 1730 map reproduced place names along the Paraguay River and the hypothetical Laguna de los Xarayes according to the sixteenth-century chronicles of exploration.

for a fixed period of time to their *encomendero* as a form of tribute, governors of Santa Cruz made captives taken prisoner in wars and punitive raids on native villages permanently bonded laborers who were placed on the estates (*chácaras*) and in the households of Cruceño colonists. These *indios de servicio* became further confused with the *piezas de rescate* captured directly from their communities or seized from their Chiriguano captors who had first enslaved them.[32] The phrase *pieza de rescate* ("ransomed pieces") denotes the dehumanizing treatment and terminology that marked these bonds of servitude.

Warfare supplied the frontier society of Santa Cruz with servile laborers. Cruceño militias, with the active support of provincial governors and the colonial council (*cabildo*), took war to native villages through aggressive expeditions, *malocas*, not unlike the Portuguese *bandeiras* or *caça a o indio*—with the purpose of enslaving Indian captives. These predatory raids marked Spanish-Indian relations from approximately 1590 to 1720, with periods of greater and lesser intensity despite repeated prohibitions issued by the viceroy and the *audiencia* of Charcas, and confounded the early evangelizing efforts of the Jesuits to congregate Indians in mission pueblos. Local authorities attempted to justify their raiding expeditions by denouncing native cannibalistic practices (a highly dubious accusation meant to legitimate their enslavement), by claiming to purchase or ransom indigenous prisoners who otherwise would have been killed, and by asserting the need to defend a dangerous border against the Portuguese *bandeirantes*.[33]

Documentary records for both *maloca* and *encomienda* provide some clues to the origins and destinies of different native peoples during more than a century of nearly constant warfare in the lowlands bounded by Mato Grosso, Paraguay, and Santa Cruz. It appears that Cruceño raiding expeditions had the effect of dispersing and fragmenting the native communities of Chiquitos and other indigenous nations, a process further complicated by the movement of Spanish frontier settlements from Santa Cruz la Vieja westward to San Francisco de Alfaro and San Lorenzo de la Barranca, approaching the Andean foothills. Repeated stories of violent encounters among native and imperial warring parties suggest not so much that distinct Indian nations were predisposed as enemies of one another, but, rather, that war had become the means for gaining—and selling—human captives. Ethnic identities for whom the Spaniards devised classificatory tribal names entered the documentary record through warfare and the trade in enslaved captives. As in Paraguay,

TABLE 3 Male Indian Tributaries in Santa Cruz

1575	16,000
1585	11,000
1595	5,000–9,000
1620	1,200–3,000

Source: José María García Recio, *Análisis de una sociedad de frontera: Santa Cruz de la Sierra en los siglos XVI y XVII* (Sevilla: Diputación Provincial de Sevilla, 1988), 157.

women, adolescents, and children composed the bulk of the *piezas de rescate* who then labored in Cruceño homes, cane fields, and on cattle ranches.

Archival references to the *encomienda* are not sufficiently systematic to quantify demographic losses or estimate rates of native population decline. They do, however, provide the outline of demographic, ethnic, and gendered patterns of change in the Indians held in *encomienda* and captured as *piezas* in the environs of Santa Cruz. The early phase of *encomienda*, from approximately 1585 to 1630, described a period of severe population decline, despite imprecise numerical estimates of native households and communities. Some *encomiendas* granted during this period were designated *por noticia*, that is, on the reputed existence of Indian pueblos theoretically awarded to serve particular Spanish colonists but not physically subjugated. Thus it is probable that the figure of forty thousand *indios encomendados* associated with the earliest grants issued by Ñuflo de Chávez did not represent actual persons held under Spanish dominion. Four censuses of male Indian tributaries in Santa Cruz dating from 1575 to 1620 provide the above estimated counts.

The dramatic decline in registered *indios de encomienda* suggested by these numbers may be explained in several ways. First, the high mortality rates stemming from the contagion of Old World epidemics undoubtedly precipitated repeated demographic crises.[34] Second, the population losses stemming from biological causes were compounded by the violent displacement of Indians captured in the *malocas* and by the drop in natives formally assigned to *encomienda* grants in contrast to the *indios de servicio* and *piezas* distributed to Cruceño households, and possibly not included in the census. *Encomendero* grantees in Santa Cruz and San Lorenzo frequently sold their charges "to Peru," to serve in the mines or to the agricultural estates in Tarija, Cochabamba, Cliza, and Mizque. Third, Indian captives may have fled their bondage, either returning to the forest or migrating westward to the Andes, where conditions for their subsistence were possibly more favorable than in Santa Cruz.[35] Formal *encomienda* grants declined in the lowlands after the mid-seventeenth century, although raiding expeditions for new captives continued or even increased. By the turn of the eighteenth century, Jesuit *reducciones* in Moxos and Chiquitos successfully exempted from *encomienda* service "[those Indians] the missionaries have ransomed from the surrounding nations because they have reduced them to our holy faith, leaving them in complete liberty."[36]

Partial censuses and documented disputes over the possession of *encomiendas* and *indios de servicio* provide names of ethnic designations among different native peoples of the Santa Cruz lowlands, but without linguistic or geographical references. Indigenous nations included general categories, such as Moxos, Chiquitos, Chanés, Yuracarés, and Chiriguanos, as well as what appear to be specific ethnic groups: Quibichicocíes, Tamacocíes, Jores, Parecíes, Xamares, Xarayes, and Casachionós. In the early eighteenth century, Jesuits protested the capture of two thousand Itonamas in a *maloca* directed by the governor of Santa Cruz, securing a royal band prohibiting future raids and directing the *audiencia* of Charcas to investigate this incident. Their efforts succeeded in freeing the Itonamas, who formed the core population for the mission of Santa María Magdalena, in the province of Moxos.[37] A list of Indians held in *encomienda* in the province of Santa Cruz dating from 1717 indicated ethnic origins for Chiquitos (111), Umuanas (40), Yuracarés (17), Sirionós (6), and Tubayonós, Baures, Mobimas, Mojos, Tapacuras, Totorocos, and Canacurés, who numbered 31 altogether. The Chiquitos represented slightly over half of these *indios encomendados*, who totaled 205, far less than the combined *encomienda* and tributary Indians reported a century earlier for Santa Cruz and San Lorenzo (see table 3).[38]

What were the motives of indigenous peoples who became entangled in the *encomienda*? Lowland tribal peoples acted, at one level, according to the practices of warring skirmishes centered on the taking and ransom of captives that had defined political alliances and hostilities before Spanish conquest. In the decades of prolonged European encounters and territorial rivalries following the Domingo Martínez de Irala and Ñuflo de Cháves expeditions of the mid-sixteenth century, traditional boundaries and conditions of servitude changed irrevocably as native peoples found themselves caught in the pincers between Cruceño *malocas* and Portuguese *bandeirantes*. Old enemies like the Chiquitos and Chiriguanos, for example, sought cross-ethnic alliances *and* the status of tributaries in their negotiations and armed confrontations with Spanish forces. At a second level, indigenous combatants explored alternative sources of power, whether in Christianity or in the materials and technologies of the Europeans. Metallurgy, livestock, and firearms revolutionized methods of swidden horticulture, domestic crafts, and warfare in ways that heightened rivalries among different indigenous groups.[39]

Native peoples resisted enslavement at the hands of Spaniards and the Portuguese through flight, armed combat, and legal protest. Three cases dating from the eighteenth century show the ways in which Indians and mixed-race peoples used legal institutions and manipulated different categories of ethnicity and social status to gain degrees of freedom in the colonial order. While these constitute singular cases not statistically representative, they speak to the contested values that complicated ethnicity in colonial Chiquitos. Leonor de Reina petitioned the *audiencia* of Charcas to be liberated from the bonds of *encomienda* in 1705, on the basis of her condition as a mestiza. Leonor was the natural daughter of Antonio Reina, "a free soldier of the King," who had died in a military skirmish with the *mamelucos*, and of María Mero, an Indian woman who lived in the household of Doña Thomasa de Sosa, in the city of San Lorenzo, as part of her *encomienda*. Leonor carried her father's surname, although her parents had not married, and demonstrated that he had publicly recognized her as his child "from the earliest days of my birth." Leonor had lived with her mother until her fourteenth year, when the governor of Santa Cruz, Don Pedro de Gálvez, freed her from the Sosas' service. Two years later, for reasons unknown to her, Gálvez reversed his earlier ruling and forced Leonor to return to the Sosa household and submit to their *encomienda*. Leonor then turned to the *audiencia* "to seek my rights"; the judge who reviewed her case found in her favor, freeing her from the *encomienda*, because of her status as a mestiza.[40]

Indians also protested *encomienda* service on the basis of their freedom to choose conjugal partners stipulated in the codification of colonial law that guided court decisions, as shown in the following case.[41] In 1751, ten Itonama men and women were assigned to the *encomienda* of Ignacia Ortiz, by governor Francisco de Arangoytia, who cited "the merits and services of her father . . . in military functions and punitive campaigns against the barbarians who infested and oppressed the province." Thomasa, one of the Itonama women who had been allotted to the Ortiz *encomienda* along with her children, sought marriage with Bacilio, an Indian who came from "a lost or diminished *encomienda*." Over the objections of Doña Ignacia, Thomasa was granted her freedom so that she might marry freely and live with her husband.[42]

The prolonged legal struggle between two Chiquitos brothers, Eugenio and Jacinto Masavi, and Cruceño Lorenzo de la Roca illustrates in dramatic terms what was at stake in the contested claims to freedom and privilege that surrounded the institution of *encomienda* in Santa Cruz in the mid-eighteenth century. Both parties claimed the principle of paternal lineage and the relationship of vassalage that formally bound them to the crown: Roca, in his pretension to exercise his "feudal right" to keep an allotment of Indians in his household, and the Masavi brothers, in their determination to gain freedom for themselves and the rest of their family. The Masavis claimed their status as "soldiers of his Majesty, may God keep him, and now tributaries . . . like all those of our nation."[43] Don Lorenzo demanded recognition of his hereditary right to the "second life" of the *encomienda* granted to his father, Joseph de la Roca, in 1703 in recognition of his military services.

The Masavi family history, covering three generations, illustrated well-established patterns of migration between their villages and the Spanish settlements of Santa Cruz. Eugenio and Jacinto, together with the witnesses whose testimony they solicited to support their case, identified them only as "of the Chiquitos nation," without specifying a particular ethnicity or place of origin. The Masavis' grandfather, Gregorio, had left his pueblo and arrived in San Lorenzo during the 1690s in search of a master whom he could serve. He first settled in the household of one Alonso Pardo, but later moved to the house of Joseph de la Roca where he married Lucía Choe, a Chiquitos woman who lived and worked in the Roca *encomienda*. The Masavi-Choe couple, with their children and grandchildren, had remained in the Roca household until the death of Don Joseph in 1742. Their situation worsened when Don Lorenzo assumed control of his father's *encomienda*. Eugenio and Jacinto argued that Lorenzo's harsh treatment, in violation of the terms and conditions of an *encomienda* grant, forced them to seek their freedom. Advised, perhaps, by the protector of Indians, a royal official appointed for the Province of Santa Cruz, they accused Lorenzo de la Roca of physical punishment (beatings) and the sexual abuse of several women (of different generations) in their family, as well as of neglecting his duties to supply them with food, clothing, and religious instruction.

The centerpiece of the Masavis' legal petition concerned the illegality of Don

Lorenzo's claim to the *encomienda*. They insisted that their grandfather had migrated to San Lorenzo of his own free will—he was not a *pieza de rescate*—and that they, as his descendants, should enjoy the same liberty. Eugenio and Jacinto argued further that they could not be held in bondage, for, as befitted the Chiquitos nation, they were free from the conditions of servitude reserved for war captives and ransomed prisoners: "For the Chiquitos were not to be brought in at the point of war . . . since the time that raiding expeditions ran [through the province] they themselves delivered the piezas that the residents of this city needed. It is laughable to look at the nominal list of the encomienda granted to Don Joseph de la Roca, deceased not so long ago, when we consider how much time has passed since those wars took place."[44] The identity that the Masavi brothers established for their family lineage in order to defend their liberty was inseparable from the history of violence that had enslaved so many lowland Indian peoples. The *cabildo* of San Lorenzo and the governor of Santa Cruz denied the Indians' petition and imprisoned Jacinto to intimidate him in his appeals to the *audiencia*. Furthermore, the *cabildo* punished the Masavi family by forcing them to work in the restoration of the convent of the Mercedarians in San Lorenzo. Jacinto managed to escape his confinement and traveled to La Plata in order to present his case in person before the *audiencia*. Acting on its authority as an appellate court, the *audiencia* issued three provisions, from March 1751 to August 1752, ordering the *cabildo* to free the Masavi family from Roca's household.

The arguments debated between the *cabildo* and the *audiencia* reveal a power struggle between local authorities and the officials who set out to implement royal policies. While the *cabildo* of San Lorenzo allied with Lorenzo de la Roca in a defense of "feudatory rights" over the labor of native servants, the *audiencia* of Charcas disavowed any such hereditary rights to *indios encomendados* in favor of contractual agreements between Spanish (and Creole) *vecinos* and free Indian men and women. The freedom that the Masavi brothers sought so passionately, in this context, meant the freedom to choose their patron and to move from one master to another, in recognition of their status as tribute-paying vassals of the king. The favorable decision they received from the *audiencia* did not, however, release them from the obligation to work; nor did the Masavi family opt to return to the forest or to seek refuge in the mission pueblos.

This case illustrates the ambiguity of the rules that governed the *encomienda* and the elasticity of ethnic identities in Santa Cruz. It suggests that a significant population of Indians lived outside the missions and sought a livelihood in the Spanish sphere of colonial society and economy. Furthermore, it points to a network of communication through which the Masavi brothers gained knowledge of the law and access to sponsors or protectors who helped them along the way—including Jacinto's trek of over 200 kilometers, rising over 2,500 meters, from Santa Cruz to La Plata. Through these networks and the Cruceños' demand for labor, the world of Leonor de Reina, Thomasa, and the Masavi family intersected with the Jesuit *reducciones* that became emblematic of Chiquitos culture and community.[45]

Ethnic Identities in the Missions

During roughly a century following initial European contact, the multiple impacts of disease, dislocation, and forced labor shaped parallel processes of ethnic fragmentation and amalgamation. Those processes led to complex ethnic mosaics through the formative experiences of the missions for material culture, work, and religious ritual.

Chiquitos

If the descendants of Gregorio Masavi had migrated to one of the mission pueblos to escape the abusive conditions of Roca's *encomienda*, how would they have appeared in the mission rosters? Eugenio and Jacinto claimed affiliation with the Chiquitos nation in the document trail they created, noting that their ancestors had participated in the forays that brought forest peoples into Spanish service in Santa Cruz and had populated the mission compounds. The process of *reducción* gathered different native groups together in the same villages and created the texts that rendered ethnic categories and languages legible to the missionaries and to the colonial bureaucracies.

Periodic censuses taken of the Chiquitos pueblos under Jesuit and clerical administration identify numerous linguistic and kin affiliations distinguished spatially in separate residential sections and designated, in Spanish, as *parcialidades*. This term appears in different areas of the Spanish American empire to refer to subunits with distinct residential and ethnic identities that either formed part of larger communities or were brought together by colonial policy in consolidated *reducciones*. In southeastern Mesoamerica, for example, *parcialidades* signified social divisions associated with specific sections of a town or village, expressed as *calpulli* in Nahuatl or as *calpul* or *cah* in Maya. In the Andes, the term referred variously to the *ayllus*, lineages that claimed descent from a common ancestor and shared territorial use-rights, or to their subdivisions, known in Quechua as *hanansaya* and *hurinsaya*.[46] The proliferation of ethnic identities through the *parcialidades* may well represent a cultural response of indigenous peoples to the fiscal and social pressures of the colonial regime. It is important to distinguish between the category as used by Spanish officials to designate a tribute-paying unit or a residential area and the changing indigenous meanings of *parcialidad* to establish reciprocal relations of support and political alliances.

The tribal names recorded in Spanish documentation reflected the kin-based social and political organization of the Chiquitano peoples. Missionaries and provincial governors recognized different *parcialidades* in their reports, and these names appear in testimonial documents referring to specific grievances or uprisings that occurred in the pueblos. It is probable that some of the named *parcialidades* emerged during the colonial order and that social markers of difference, both in and out of the missions, did not correspond neatly to ethnic designations. Cultural identities and political affiliations changed over time through concurrent patterns of congregation and migration that brought different linguistic and kin groups into the pueblos.[47]

Mission records provide comparative, if partial, profiles of the numbers and distribution of native groups who lived in the missions, as tables 4 through 7 below show. Their lists of *parcialidades* overlap across the missions, adjusted to the catechismal categories of *confessant* and *communicant* that mattered for the priests' administration of the religious life of the *reducciones*. Gendered and generational distinctions among children, adolescents, married couples, and unmarried (single and widowed) persons were duly noted, as these carried weight for the economic and social controls that priests and secular officials attempted to impose on the mission communities.

The apparent stability in the total number of persons recorded as living in the pueblos over time masks the repeated occurrence of epidemics and the spatial mobility so deeply embedded in lowland cultures. Formal Jesuit administration of the missions, with regularly kept documentation, began over a century and a half after the violent disruptions in native settlements occasioned by *encomienda* and slaving raids. Additional sources that recorded diseases, plagues, and climatic events like frosts, fires, droughts, and floods provide telling stories of natural and social crises that threatened the formal organization of permanent *reducciones*. Jesuit records preserved for 1739–65, for example, reported severe epidemics in San Xavier, causing elevated deaths in 1739 for all age groups, and in San Miguel the following year. An outbreak of dysentery in San José in 1743–44 struck adults and children, male and female, but showed the greatest losses in the adolescent population. Twenty years later, San José again registered unusually high deaths among children.[48]

The quantitative data compiled in the censuses should be read in the light of qualitative evidence that signals discontinuities in the specific ethnic groups that peopled the missions and in the meanings of the labels used to identify them. An especially thorough enumeration of resident mission population—distinguished by *parcialidad*, gender, and age categories—was carried out in 1767–68, during the transition from Jesuit to diocesan administration of the Chiquitos missions. Table 4 summarizes the data, using categories that remained fairly consistent for both the Jesuit and post-Jesuit years. The category of *Casados y casadas* (married men and women) corresponds roughly to the number of households, although the census does not provide a description of actual household composition. *Viudos* and *viudas* are widowed men and women; the latter usually outnumber the former. *Solteros* and *solteras* represent unmarried adolescents or young adults; *párvulos* and *párvulas* are children, and the terms usually signified that they were not yet of an age to partake of the sacraments, or that they were beginning instruction in the catechism.

Each mission had more than one *parcialidad*, and Concepción housed as many as eleven different ethnic bands. The whole province came to be known as Chiquitos, yet we may ask—who were the "Chiquitos" people? The term has given rise to a number of speculations concerning its origins and meaning. If, indeed, it derives from the Spanish word for "small" in the diminutive (*chico, chiquito*), the first Jesuits who entered the province in the closing years of the sixteenth century explained that the designation Chiquitos referred to the doorways of the houses built so low that anyone wishing to enter them had to crawl through their entryway. Fathers Diego Martínez and Diego de Samaniego concurred in explaining this practice as the Indians' protection against the cold (for they wore very scant clothing), from their enemies, and from mosquitoes and other insects.

An anonymous eighteenth-century study of the Chiquitos language offered an alternative explanation, suggesting that the term came from indigenous vocabulary: *chiqui-s*, meaning "egg" or "testicles." The author proceeded to

TABLE 4 Chiquitos Population Data, 1767–68

Pueblos	*Year*	*Parcia-lidad*	*Families*	*Widows*	*Widowers*	*Single Boys*	*Single Girls*	*Souls*
San Xavier	1767		720	31	51	890	789	3,201
Concepción	1767		713	20	41	998	793	3,278
San Miguel	1767		245	8	20	419	436	1,473
San Ignacio	1767		731	4	34	797	837	2,734
San Rafael	1767		562	20	26	798	778	2,746
Santa Ana	1767		367	8	34	481	530	1,787
San Joseph	1767		618	3	46	780	650	2,715
San Juan	1767		425	10	19	559	515	1,953
Santiago	1767		410	4	58	363	369	1,614
Santo Corazón	1767		532	9	32	560	622	2,287
Total			5,323	117	361	6,645	6,319	23,788
San Xavier	1768	7	770	25	42	244	168	2,022
Concepcion	1768	11	745	15	61	741	620	2,913
San Miguel	1768	4	322	18	24	387	380	1,373
San Ignacio	1768	6	588	7	39	489	422	2,183
San Rafael	1768	3	438	62	35	548	525	2,046
Santa Ana	1768	4	388	4	34	481	476	1,771
San Juan	1768	6	430	5	31	439	464	1,770
Santiago	1768	5	420	6	13	332	403	1,578
Santo Corazón	1768	3	546	0	44	561	582	2,287
Total		52	5,255	144	352	4,693	4,390	19,981

Source: Archivo y Biblioteca Nacionales de Bolivia, Mojos y Chiquitos, vol. 24, folios 1–2.

Note:
The numbers appear incorrect for "souls"; however, this category does not reflect the total of the numbers in the preceding columns. It represents the persons for whom the missionaries were responsible fiscally and spiritually. "Families" in this table stands for the terms *casados* and *casadas* (married men and women) used in the original census. The numbers of *parcialidades* were not given in the census for 1767.

explain that the Indians thus named (perhaps by others) strove to change their designation to M'oñeyca, meaning "the men" or "the people."[49] Early Spanish references to the tribal wars between Chiriguanos and Chiquitos recorded that while the Chiriguanos feared the Chiquitanos' lethal use of poisonous herbs, they called them Tapuymiri, meaning "slaves of small things," later rendered in Spanish as Chiquitos.[50] It is unlikely that a common name existed for the Chiquitos in one of the native languages, for the people who came to be known by that term did not constitute one nation; rather, they represented an

TABLE 5 Selected Chiquitano Ethnic Names and Meanings

Puizoca/Pisoca	canoes
Quimomeca	bow and arrow
Yirituca	hilltop
Booca	*chacras* (farmers)
Piñoca	weavers
Taus	clay (potters)
Piococa	fish

Source: Roberto Tomichá, *La primera evangelización en las reducciones de Chiquitos, Bolivia (1691–1767)* (Cochabamba: Editorial Verbo Divino, 2002), 248–52.

ethnically diverse population. Contemporary Chiquitanos have sought to adapt a word from their language to express their identity; one selection is Besiro, meaning "correct speech" or "the right way."[51]

Lists of *parcialidades* compiled by different Jesuit missionaries in the course of their efforts to create a unified written language for the Chiquitos province conformed to a pattern in which as many as seventy-seven ethnic groups were linked to four Chiquitos dialects and six non-Chiquitos languages: Arawak, Chapacura, Guaraní, Otuquis, Tunacha, and Zamuco. Roberto Tomichá's admirable synthesis of colonial sources and modern ethnographic studies for the region is reproduced in the tables below.[52] It is doubtful, however, that all of these ethnic designations represented discrete groups of people who existed in the missions at any one time. Rather, as different bands were brought to the pueblos—either by force or of their own accord—their identities were transcribed (and perhaps modified) to distinguish them from communities already settled in the pueblos. All of these groups shared similar material cultures with livelihoods based on hunting, fishing, foraging, and farming. They differed in the variety of languages and dialects they spoke, in specific ritualistic practices, and in their levels of political consolidation. In some cases their distinct ethnic labels referred to family lineages or to specific caciques who led them at the time of their first contact with the missions; often the names by which specific *parcialidades* became known were applied to them by other groups or by the Spaniards—as occurred for the Chiquitos designation itself. Some of the names listed for the Chiquitos missions have occupational or geographic connotations (table 5)—yet another indication that the *parcialidades* took on their identities through the process of *reducción*.

Reading these tables comparatively, we begin to understand the ethnogra-

TABLE 6 *Parcialidades* Related to the Chiquitos Languages

Dialect	*Ethnic Group*	*Pueblos, 1745*	*Pueblos, 1767*
Tao	Aruporeca		Concepción
Tao	Booca, Boococa	Concepción	Concepción
Tao	Boros, Parayaca	San José, San Juan	San Juan, Santo Corazón
Tao	Pequica	San Miguel	San Miguel
Tao	Piococa		San Ignacio, Santa Ana
Tao	Purasis	San Xavier, Concepción	San Xavier, Concepción
Tao	Quiviquica	San Xavier	San Xavier
Tao	Tabica	San Rafael	San Miguel, Santa Ana
Tao	Tanipica	San Miguel	San Ignacio
Tao	Taus, Caotos	San Rafael, Santiago, Santo Corazón	
Tao	Tubasica	Concepción	Concepción
Tao	Xamarús	San Miguel, San José	San José, San Ignacio
Manasí	Cusica	Concepción	Concepción, San Ignacio
Manasí	Manasica	Concepción	Concepción
Manasí	Quimomeca	Concepción	Concepción
Manasí	Tapacuras		Concepción
Manasí	Yirituca	Concepción	Concepción
Manasí	Yurucareca	Concepción	Concepción
Piñoco	Guapaca	San Xavier	San Xavier
Piñoco	Piñoca	San Xavier, San José	San Xavier, San José
Piñoco	Piococa		San Xavier
Piñoco	Poojijoca		San Xavier
Piñoco	Quimeca		San Xavier
Penoquí	Piñoto, Penoquís	San José	San José

Source: Roberto Tomichá, *La primera evangelización en la reducciones de Chiquitos, Bolivia (1691–1767)* (Cochabamba: Editorial Verbo Divino, 2002), 170.

phies that missionaries compiled for the Chiquitos province. Fathers Lucas Caballero and Juan Patricio Fernández began recording different dialects and ethnic groups in the course of their *entradas* into the forests to bring native *rancherías* to mission life. Some of these names endured, while others were added during the expansion of the *reducciones* eastward. Their significance surfaced in the *parcialidades* that became the basis for internal governance in the pueblos, as we shall see in greater detail in chapter 5.

The spatial distribution of *parcialidades* and the languages they spoke suggest the outline of ethnic territories, but without clear or fixed boundaries. Within the Chiquitos family of languages, Tao-speakers appeared in every mission, while Manasí-speakers were confined almost exclusively to Concepción, and

TABLE 7 *Parcialidades* Related to Non-Chiquitos Languages

Language	*Ethnic Group*	*Pueblos, 1745*	*Pueblos, 1767*
Arawak	Batasica	San Rafael	San Rafael
Arawak	Baurés	San Xavier	San Xavier
Arawak	Cupíes, Tupís	San Rafael	
Arawak	Paiconeca	Concepción	Concepción, San Xavier
Arawak	Parabaca, Tarabaca	San Miguel	San Miguel
Arawak	Parisis		San Ignacio
Arawak	Paunaca	Concepción	Concepción
Arawak	Quidaboneca	San Rafael	San Rafael
Arawak	Veripones, Vejiponeca	San Rafael	San Rafael
Arawak	Xarayes, Saravecas	San Rafael	Santa Ana
Chapacura	Napeca	Concepción	Concepción
Chapacura	Quitemas	Concepción	Concepción
Chapacura	Tapacuras	Concepción	Concepción
Guaraní	Guarayos, Itatines	San Xavier, San Miguel	San Xavier, San Miguel
Guaraní	Guadores, Guarades		San Ignacio Concepción, Santo Corazón
Otuqui	Carabeca	San Miguel	Santo Corazón
Otuqui	Currucaneca	San Rafael	San Rafael
Otuqui	Curuminas	San Rafael	Santa Ana
Otuqui	Ecobarés, Ecorabeca	San Rafael	Santa Ana
Otuqui	Otuquis	San Miguel	Santo Corazón
Otuqui	Tapis	San José, San Ignacio Zamucos	San José
Otuqui	Tapurica		San Ignacio
Tunacha	Tunachos		Santiago
Zamuca	Cururarés	San Juan, San Ignacio de Zamucos	San Juan, Santo Corazón
Zamuca	Imonos		Santiago
Zamuca	Morotocos	San Juan	San Juan
Zamuca	Ororobedas	San Juan	San Juan
Zamuca	Panonos	San Juan	San Juan
Zamuca	Tieques		San Juan
Zamuca	Tomoenos	San Juan	San Juan
Zamuca	Ugaraños	San Ignacio de Zamucos	San José, Santiago
Zamuca	Zamucos, Ayoreode	San Ignacio de Zamucos	San Juan, Santo Corazón
Zamuca	Zatienos	San Ignacio de Zamucos	
Zamuca	Zeriventes		San Juan

Source: Roberto Tomichá, *La primera evangelización en las reducciones de Chiquitos, Bolivia (1691–1767)* (Cochabamba: Editorial Verbo Divino, 2002), 276–77.

the Piñoco dialects to San Xavier. Among the non-Chiquitos languages, ethnic groups linked to Arawak were found in all the western, central, and northern missions, from San Xavier to Santa Ana. Chapacura-speakers were concentrated in Concepción; Otuqui-speaking *parcialidades* tended to coalesce in the central pueblos of San Miguel, San Rafael, and Santa Ana, as well as in the southern pueblos of San José and Santiago. Zamuca-related ethnic groups—including the present-day Ayoreóde—who clustered in the southern and eastern missions of San Juan, Santiago, and Santo Corazón, counted among the most nomadic of the native peoples who experienced mission life. Jesuits founded the pueblo of San Ignacio de Zamucos in 1724, deep in the arid grasslands of the Gran Chaco, after numerous attempts to draw Zamuco-speaking tribal groups into the missions of San José and San Juan, where they were especially vulnerable to epidemic contagion. Twenty years later, in 1745, San Ignacio de Zamucos was disbanded due, in part, to ethnic conflicts between the Zamucos and Ugaraños. The mission was reestablished as San Ignacio de Chiquitos in 1746, some eight leagues north of San Miguel, with diverse speakers of Tao, Arawak, Guaraní, Otuqui, and Zamuca dialects.[53]

Sonora

The history of ethnic identities in Sonora followed comparable processes of differentiation and consolidation, recognized by Indians and Spaniards alike through shifting patterns of alliance and conflict. Population studies for northwestern Mexico have emphasized demographic decline and ethnic amalgamation over the colonial period. A history of epidemic and endemic disease, as dramatic as it is undeniable, challenged the survival of indigenous households and communities in this frontier region as in all areas of Indian-European encounters. Repeated episodes of smallpox, measles, typhus, typhoid, malaria, yellow fever, dysentery, and influenza were often compounded by hunger and dislocation following droughts, floods, and crop failures. A number of these subsistence crises corresponded to pandemics that swept through all of New Spain, as occurred during the *matlazáhuatl* scourge of 1737 and the smallpox epidemic of 1781.[54]

Colony-wide counts of families and persons registered during the seventeenth and eighteenth centuries show a steady decline in the estimates of the Indians considered to be under the missionaries' spiritual and temporal care. Missionaries accounted for the sacraments administered, which served as signs of their neophytes' advancement in Christian doctrine and practice. Records of baptism, confirmation, confession, communion, marriage, and burial open a window onto life-course events like birth, puberty, marriage, and death, but they do not provide an exact correlation to demographic profiles. Colonial mission censuses periodically took stock of the population resident in the pueblos, but missionaries could offer only rough estimates of the bands and *rancherías* who visited the missions, remaining beyond their control. Sonoran missions, smaller in size but more numerous than their Chiquitano counterparts, did not list separate *parcialidades*. Each mission district, however, had a head village with two, three, or occasionally more *visitas* that often comprised more than one ethnic group; for example, Opata and Eudeve, Opata and Jova, or the different subdivisions of the *Pima* nation. Missionaries recognized ethnic differences, linked in their minds to native languages, and tended to coalesce the populations they served into larger, more generic categories as time progressed. Their representation in the documents may reflect the political and cultural absorption of smaller groups by larger ones, or simply the blurring of distinctions among related dialects, but it does not necessarily point to their biological extinction.

Village-dwelling farmers found in the missions the material and cultural resources to rebuild their communities, despite demographic losses, but the consolidated towns that constituted the cornerstone of the mission program of *reducción* existed within a social and ecological environment of smaller, mobile *rancherías*. The ethnic configuration and size of colonial Sonoran communities varied due to migrations, subsistence crises, and conflict among different native groups and between Indians and colonists. The persistence of native piedmont villagers was overshadowed by the more aggressive population growth of Spaniards and mixed-race *vecinos*, beginning in the last third of the eighteenth century. Mining *reales* and presidios interspersed with the missions produced a dynamic of population mixture and actually blurred the distinction between Indians and *vecinos*.[55]

The endurance of Sonoran peoples, albeit with changing ethnic identities and village locations, contrasted markedly with the disappearance of nomadic tribes reported for northeastern New Spain, in the provinces of Coahuila and Nuevo León, for parts of Nueva Vizcaya, and for the arid peninsula of Baja California.[56] Within the province, the Seri and Apache frontier wars established the conditions for an increasingly sharp and conflictive distinction between villagers and nomads. Ethnic differences evolved in the contested relations among native groups, and between them and the colonial regime, for control over resources and political ascendancy.

A number of internal frontiers emerged in Sonora, where the convergence of ecological and cultural differences translated into ethnic distinctions between nomadic hunter-gatherers and sedentary farmers. The western boundary of the Sonoran piedmont, along the San Miguel river and the arroyo of Zanjón constituted one of these ethnic frontiers. Here bands of Tohono O'odham, Eudeve, and Seris lived in the missions of Pópulo, Los Ángeles, Opodepe, and Nacameri, from the late seventeenth to the mid-eighteenth century. When the Indians' lands in Pópulo and its *visita* of Los Ángeles were violently dispossessed by order of royal visitor José Rodríguez Gallardo in 1748, in order to establish the military presidio of San Miguel de Horcasitas, Pimas and Seris rebelled, leading to long-term warfare and raiding that extended westward to the desert coasts of the Gulf of California.[57] A second internal frontier developed from the troubled intermingling of Opata villagers and Jova nomadic foragers in the highland pueblos of Bacadéguachi, Nácori, Arivechi, and Saguaripa, and in numerous *rancherías* dispersed in the valleys and intermittent mountain ranges, further complicated by the Apache wars of eastern Sonora.[58]

The role of nomads in the missions assumed vital importance in the Pimería Alta, where diverse bands of Pimas, distinguished linguistically and by their different degrees of sedentary or nomadic lifeways, moved in and out of the pueblos and had frequent contact with non–Uto-Aztecan peoples, such as the Yumas and Cocomaricopas. If the Hymeris or Akimel riverine villagers of the Magdalena, Santa Cruz, and Asunción valleys constituted at one time the heart of the northern Pima missions, Sobaípuri peoples migrated from the San Pedro valley into the pueblos of Bac, Suamca, and Guebavi, and the Tohono O'odham periodically settled in the missions of Tubutama, Atil, Pitiquito, and Caborca in the western Pimería Alta. These migrants filled Franciscan mission registers, and their labor sustained late eighteenth-century church construction. This was the main conduit for the redistribution of *nijoras* in the missions, where they were absorbed into the resident population, and in Spanish settlements, where they formed the bulk of domestic servants.[59]

The sedentary-nomadic divide made for a contact zone as much as for a boundary, creating moving ethnographic maps of convergences and separations. The dynamic quality of these cultural frontiers gave rise to new identities and the adaptation and reformulation of cultural practices. Ethnic distinctions named in the colonial records evolved from episodes of conflict and changing affiliations among living peoples. Ethnic identities merged as different groups of people entered the missions; at the same time, the Indians' exodus from the missions, whether seasonally or permanently, occasioned the splintering of communities and the coining of new tribal names. Pimas who fled the missions of Ónavas, Movas, Nuri, and Suaqui in central Sonora and Ostimuri became known as Sibubapas (or Suvbàpas), and suspected as potential rebels. Since the Sibubapas often turned to raiding, local presidio commanders watched their movements closely in the environs of the mining towns of San Antonio de la Huerta, Río Chico, Trinidad, and Baroyeca.[60]

Similar movements of Indians and *vecinos* between the missions and Spanish settlements created a second important zone of contact that brought native Sonorans and *vecinos* into close proximity with one another, but in conditions of inequality. Through intermarriage, ritual sponsorship (*compadrazgo*), and informal conjugal unions, many Hispanic and mestizo settlers inserted themselves into indigenous kinship networks, living in the pueblos and squatting on mission lands. Native residents and officers of the missions accepted these Hispanicized migrants, as they did the seasonal visits of nomadic tribesmen, in the belief that they would contribute to the religious and economic life of the pueblo. They were disappointed in their cultural expectations when non-Indian spouses of mission Indians gained entitlement to the usufruct of arable land but claimed the status of *gente de razón* to demand exemption from the obligations of communal labor that fell to the indigenous inhabitants of the pueblos. In a telling episode, the governor of Sonora, Juan Claudio de Pineda, in 1767 admonished the Pima governor of Ures not to assign the mestiza wives of mission Indians to community tasks. Pineda addressed the Pima officer as *hijo* (son):

> Hijo, governor of the pueblo of Ures: Luis, Cayetano, and Nicolás, hijos of this pueblo, have informed me that they are married to women de razón. Because of this, I order you to assign them work loads equal to those of the other mission Indians, without obliging them to work harder or to demand labor of their wives, for they are exempt from community service. . . . You must permit them to sow their land and farm where it is best for them to support their wives. I trust you to carry out my orders and keep the peace in your village, so that I may receive no further complaints.[61]

Pineda's directive reinforced the superior status of Spaniards and *gente de razón* over the Indians. We may speculate that Luis, Cayetano, Nicolás, and others like them who had married non-Indian women, as well as mestizo men who had taken Indian wives, sought the support of the Sonoran governor to attain a privileged position in Ures. Indians who migrated to mining encampments and Spanish estates or served as auxiliary troops in the frontier presidios began to seek the status of *vecinos*. In 1784, the bishop of Sonora, Antonio de los Reyes, lamented the unplanted communal fields and rising numbers of "Indians who live in the class of vecinos" in the missions.[62] By the close of the eighteenth century, the Indian-Hispanic divide had become a zone of cultural hybridity and a site of conflict over land, local governance, and the meaning of community.

Mestizaje and Ambivalent Ethnicities

The preceding sections of this chapter focused on the cultural and ethnic diversity of different tribal peoples within the colonial category of *Indian*, pointing to the emergence of new identities through the institutions of *encomienda* and mission. This final section examines the social divisions and networks among the *vecinos*, who coveted their status as subjects of the Crown and citizens of a particular town or province. The complex ethnic categories they espoused reveal the tensions inherent in their mixed origins and in the emerging class structures of the late colonial period.

Sonora

The parish censuses of 1800–1803 for Santa Ana de Tepache and San Miguel de Oposura in northeastern Sonora illustrate these translations of ethnic status. Tepache developed at the convergence of the mission and mining frontiers between the western branch of the Bavispe river and the Oposura valley. Jesuits registered baptisms at San Joaquín y Santa Ana de Tepache as early as 1636, a *visita* of the predominantly Eudeve mission of Batuc. By the mid-eighteenth century, Tepache was a center of silver and lead mines, Arroyo, Nacatóvori, Lampazo, Santo Domingo, La Coronilla, Las Guijas, and San José del Alamo among them. The non-Indian population of Tepache was sufficiently large in 1784 to figure as the seat of one of the deputies of the *alcaldía mayor* (district magistrate) of Sonora.[63] At the turn of the nineteenth century, the demographic profile for Tepache reflected its mixed origins in a heterogeneous population of *indios de pueblo* (Indians of the pueblo), mestizos, including *indios laboríos* (Indians who worked for contract, separated from the communal regime of the missions), and Spaniards.[64]

Table 8 summarizes the census of 1803, showing proportional figures for the three ethnic categories indicated, arranged by household (married couples), unmarried sons and daughters, widows and widowers. Over half the total population of 1,636, distributed across 335 households, appeared under the broad category of *mestizos, castas e indios laboríos*; over one-third were Spaniards, and one-eighth were *indios del pueblo*. The ratio of children to household is predictably higher for Spaniards than for *castas* or Indians, although the range of difference (2.18 to 3.0) does not suggest dramatic contrasts in the rate of survival. It may, however, point to higher incidences of child mortality among Indians or the separation of adolescents from their parental homes and their entrance into the labor force as day laborers or servants. The male-to-female ratios are lowest among Spanish and Indian widowers and widows, but higher among the children of mestizo and Indian households—perhaps another indication that unmarried girls worked outside their homes as domestic servants.

Turning to the census itself, we find a number of clues to this microcosm of frontier society in the parish seat of Tepache. The priest had arranged the seventy-eight households of the Pueblo of Santa Ana by the ethnicity of the head of household; he did not acknowledge mixed conjugal unions across the ethnic divide. All heads of household and spouses were named, but all other dependents were merely listed as children (or, in one case, a sibling) of the household head, with their ages. He gave Spaniards the honorific titles of *don* and *doña* and placed the most prominent families at the head of the census. Gregorio Ortiz Cortés, married to María del Carmen Padilla, had served a decade earlier as *teniente de alcalde mayor*, and one of his kinsmen, Hugo Ortiz Cortés, held office as *subdelegado* (local Spanish administrator) to the intendant.[65] Members of the landowning Ortiz Cortés extended family appeared in four of the twenty-one Spanish households of Santa Ana. The mestizo category that embraced twenty-one households in the pueblo included "other *castas*" and Indians who, as *laboríos*, counted among the *vecinos*. Three women of the same generation, Ana María, Serafina, and Antonia, with the indigenous surname of Soqui, appeared in three mestizo households, as did the Spanish

TABLE 8 Household Census for Santa Ana de Tepache, 1803

	Households	Persons	Mean Household Size
Spaniards	21	101	4.8
Mestizos	21	89	4.2
Indians	36	102	2.8
Totals	78	292	3.7

Source: Archivo Diocesano de Hermosillo, vol. 3, exp. 9.

Note:
The average of the sum of the three MHS calculated for Spaniards, Mestizos, and Indians is 3.9. The greater number of Indian households slightly reduces the overall mean (292/78).

surname Yguera; a few family names, like Acuña and Ozejo, crossed the divide between Spaniards and mestizos. Thirty-six Indian households were headed by individuals who had indigenous surnames—Tacachi, Tachu, Chiborro, Gopochi, Soypa, Vicocame—or, as occurred mostly among the women, only their Christian names. One Indian head of household, Xavier Maiordomo, may have had a caretaker or supervisory role in the parish church (*mayordomo* may signify someone in charge of other workers or an official of a religious *cofradía*, a lay Catholic brotherhood), or he may simply have inherited the name. None of the men listed in the census, however, held title to a *cabildo* office.

The household *padrón* (official house-to-house census) for Santa Ana de Tepache, when compared with the census numbers reported for the extended parish, highlights a process of ethnic change that mirrored social inequalities among landowners and laborers, peasant smallholders and servants. Ethnic categories themselves proved unstable as people moved physically from one community to another and created new social networks. The parish seat represented one-quarter of the households reported for the entire curate and less than one-fifth (17 percent) of the enumerated population. Indians comprised the single largest category in the pueblo, but Spaniards and mestizos, taken together, outnumbered them: forty-two to thirty-six households, and 190 to 102 persons. Mixed races comprised the fastest growing ethnic category in the valley of Tepache, their numbers increasing through the Indians' passing into the mestizo class and through internal demographic growth. Indians retained a cultural presence in the pueblo of Santa Ana, but they represented smaller families with less resources than those of the *gente de razón*. The ratio of Indian households to mestizos and Spaniards is 0.85, but for persons it is 0.53.

The 1803 census for the neighboring parish of San Miguel de Oposura comprised eight different communities, as shown in table 9. Oposura and Cumpas represented the two core pueblos of what had been the mission of Oposura, founded in 1644. Three privately owned haciendas—Pivipa, Jamaica, and Buenavista—were followed by three mining and ranching settlements: Gécori, Teonadepa, and Toiserobavi. Pivipa, Jamaica, and Gécori had become mixed communities of over 150 inhabitants each. The census of 1761 ordered at the time of the episcopal visitation of the entire diocese of Nueva Vizcaya listed the mission of Oposura with the head village and two *visitas* of Cumpas and Térapa—the latter did not figure in the 1803 enumeration—and seven mixed communities, including Santa Ana de Tepache, that came under its religious jurisdiction. At that time the Opata, understood as Indians of the pueblo, numbered 378, while the combined population of Spaniards, *castas*, and *indios laboríos* included 1,466, of whom some 200 lived in the two mission villages.[66] Half a century later, in 1803, the parish embraced a total population of 1,468 (excluding Tepache), of whom 210 (14 percent) were counted as Indians. The category of *mestizos* included a broad range of mixed-races *and* Indians who had left the communal regime of the pueblos and, in the language of the time, lived as *laboríos* or *vecinos*.

At the turn of the nineteenth century, the pueblo of Oposura had become a center of Spanish settlement, while in Cumpas *mestizos* outnumbered Indians two to one and Spaniards by four to one. The population of these two mission villages comprised 264 Spaniards, an almost equal number (258) of mestizos, and 210 Indians. Conversely, the category of *Indians of the pueblo* did not appear in the six haciendas and *rancherías* because it signified a civic as well as an ethnic status linked to the religious and economic life of the missions. Two of the smaller communities, Buenavista (69 persons) and Teonadepa (76 persons) showed only *indios laboríos y otras castas*—the working population of a hacienda or ranching settlement. The demographic profiles for each of the communities, expressed as average household sizes and percentages of each ethnicity in the total population, reinforce predictable patterns of larger households for Spaniards and mestizos than for Indians and, thus, of more vigorous population growth for the Spanish and mixed categories than for the Indians. A number of significant observations for the cultural and social meanings of ethnicity emerge from a close reading of the censuses across the eight communities of Oposura parish.

The demographic survival of Indians and mixed races, as inferred from average household sizes of 3.3 or higher, and the growing numbers of *indios*

TABLE 9 Demographic Profiles of the Parish of San Miguel de Oposura, 1803

Community	*Ethnicity*	*Household*	*Population*[a]	*Average Household Size*	*Percentage by Ethnicity*
Oposura		90	411	4.6	100
	Spaniard	43	218	5.0	53
	Mestizo[b]	16	75	4.7	18
	Indian	31	118	3.8	29
Cumpas		79	321	4	100
	Spaniard	10	46	4.6	14
	Mestizo	40	183	4.6	59
	Indian	29	92	3.2	29
Hacienda Pivipa		44	208	4.7	100
	Spaniard	12	55	4.6	24
	Mestizo	32	153	4.7	76
Jamaica		32	173	5.4	100
	Spaniard	11	70	6.3	40
	Mestizo	21	103	4.9	60
Buenavista					
	Mestizo	21	69	3.3	100
Ranchería Gécori		35	165	4.7	100
	Spaniard	15	75	5.0	45
	Mestizo	20	90	4.5	55
Toiserobavi		10	45	4.5	100
	Spaniard	4	24	6.0	53
	Mestizo	6	21	3.5	47
Teonadepa					
	Mestizo	14	76	5.4	100
Total Figures for the Parish		325	1468	4.5	100
	Spaniard	95	488	5.1	33
	Mestizo	170	770	4.5	53
	Indian	60	210	3.5	14

Source: Curato de Oposura, Archivo Diocesano de Hermosillo, vol. 3, exp. 9, folio 1803-1.

Notes:

[a]Population breakdown and average household size by ethnicity exclude single-person entries.

[b]The category *mestizo* includes *indios laborios y otras castas*. Indians are designated as *indios del pueblo*.

laboríos point to the reproduction of the labor force. If indeed the two mission pueblos represented half the total population (729 of 1,468) and all of the Indians, the mestizos of the haciendas and *rancherías* (512) nearly doubled those living in the pueblos (258). The haciendas seem to have drawn a population composed of couples at the beginning of their marriages with young children or no offspring. In Jamaica, for example, under the category of *mestizos*, Julián Bázquez and Josepha Fimbres, aged twenty-five and twenty-eight, respectively, had a toddler son two years of age; Victo Suviate (aged twenty-four) and his wife María Ignacia Ramírez (aged twenty-five) were listed with a baby daughter. Juan Angel Cota and Joseph Ygnacio Cota, two household heads in their mid-twenties, were possibly brothers; Juan Angel and his wife Ignacia Flores (eighteen years of age) had no children, while Joseph and his wife María Llanes (twenty-five years of age) had a baby son. Similar examples of young families occur in the pueblos of Oposura and Cumpas, to be sure, but it is tempting to speculate that couples in the initial stages of household formation sought residence and employment on the haciendas. This stage of the life cycle may have been a point of transition for some people from *indio del pueblo* to *indio laborío*.

Men and women over fifty years of age were well represented in the parish, either widowed or as heads of reconstituted families that came together through second (or third) marriages. Widows and widowers lived with adult children, not listed with spouses or children of their own, or with younger cohorts of children and adolescents who may have been grandchildren or nieces and nephews. In the smallest community of Toiserobavi, Andrea Martínez formed a two-person household with her son, and Juana Miri lived with three adult daughters. Two exceptional Spanish widows from the wealthy Escalante family appeared in the larger settlement of Gécori: Doña Anna María, aged eighty, listed separately, was followed by Doña Feliciana Escalante, aged forty-four, who headed a household of four sons and four daughters ranging in age from six to twenty-four years of age. Anna María, perhaps the elderly grandmother, probably enjoyed the support of the younger matron and her family. A less secure prospect may have awaited Anna Acuña, a mestiza widow of thirty years of age, who lived with two daughters, ages fourteen and seven, or Loreta Montaño of Oposura, a widow sixty years of age who lived with her granddaughter, only six years old.

It is rare to find persons listed singly, but in this same pueblo, seven Spaniards and two mestizos (five women and four men) appeared as unmarried adults; the women were further identified as orphans (*soltera huérfana*). Six of the nine (two men and four women) belonged to the large and powerful Moreno family, who controlled much of the arable land in the valley. More commonly, however, elderly men appeared as heads of household with wives fifteen or even twenty years their junior, followed by dependents of varying ages who may have been the offspring of different unions. Miguel Durazo, age fifty, and Josepha Montaño, age thirty-five, a Spanish couple of Oposura (but not distinguished with the honorific title of *don*), headed a household of ten children, whose ages ranged from twenty-four to two years of age.

In our attempts to visualize living communities through the census, it is important to remember that the census taker counted conjugal units sanctioned through the Catholic ritual of marriage, not through their physical homes.[67] We may, however, infer probable patterns of coresidence by observing the sequence of family surnames and the proximity of two generations of adults who may have formed working units. Victo Suviate, one of the young heads of mestizo households mentioned above in the hacienda of Jamaica, may have been the son of Francisco Suviate and Luz Villa, ages sixty and fifty-eight, who had three adult children and one small daughter living in their household. Dominga Suviate, age twenty-six, lived in Gécori with her husband Juan Joseph Bázquez and four children. The Grijalva family spread across five households and at least three generations of mestizos in the pueblo of Cumpas. Widowed Matrona Grijalva, age ninety, was listed alone, while widower Josef Francisco Grijalva had two young adult sons and a daughter aged seven. Octogenarian Clemente Grijalva and his wife Ignacia Gurruel, age fifty, lived with five children, nine to sixteen years of age; Nicolás Grijalva (aged sixty-eight) and Candelaria Cruz (aged twenty-two) had two very young daughters (ages four and two). Balentina Grijalva (fifty years of age) and her husband Juan Antonio Barga, ten years her senior, lived without children. Antonio Grijalva (aged forty) and Dolores Salazar (aged thirty-eight) formed a marital union with four daughters, ranging in age from two to sixteen. Joseph Luis Grijalva (aged thirty-two) and his wife María Miranda (aged thirty) had three sons between five and ten years of age.

The census takers did not acknowledge multigenerational extended families; rather, they depicted large households as nuclear families composed of parents and children. A comparison of the age ranges of children and women in their reproductive years leads us to surmise that some of the offspring are grandchildren, perhaps the issue of unmarried adult sons and daughters, or other young dependents of the household head.[68] Josef Chacari, age fifty, and his wife Valvina Aldano, age thirty, headed the list of Indian households in Cumpas. Theirs may well have represented a composite household, with one son of twenty and three younger children under five years of age. Large households of over eight members, however, appeared exclusively among Spaniards and mestizos. Joseph Antonio Quijada (aged forty-seven) and Leocadia Domínguez (aged forty) were listed twice, both in the Spanish and mestizo populations of Cumpas. Their eight children ranged from infancy to twenty years of age. Joseph Moreno (forty-seven years of age) and Francisca Manzo (forty years of age) headed the leading Spanish household of Toiserobavi with two sets of children: three daughters and a son between the ages of fourteen and twenty-two, and two daughters and a son, from one to four years of age. A similar case stands out in the census for Pivipa: Joseph Bázquez (aged forty-eight) and Josepha Montaño (aged forty) presented seven children, four sons and two daughters, ranging in ages from fourteen to twenty-three years, and a much younger daughter of five years of age.

Ethnic identities revealed significant degrees of ambivalence and overlap, as indicated by the double listing of the Quijada-Domínguez household in Cumpas. Census enumerators in both curates of Oposura and Tepache settled for broad categories, without specifying types of *castas* or Indian groups. They consistently portrayed married couples with the same identity, perhaps in conformity with the Royal Marriage Pragmatic of 1776 that imposed standards of uniformity and parental consent for marriages among couples under the age of twenty-five. Promulgated in New Spain in 1778, the pragmatic was applicable principally to non-Indian families that aspired to high status and expressed the crown's concern with maintaining a racialized social order in support of the imperial regime. Sonoran documentation for this period does not highlight the pragmatic as a major source of contention, but it is probable that ecclesiastical census takers judiciously made their parishioners fit the current royal standards.[69]

This general portrait of ethnicity, constructed on widely inclusive labels, was occasionally punctuated by specific, individual identities. In Pivipa, Mateo, age twenty-five but no last name, was married to Juliana Opata (age twenty-two) with one son three years of age. This is the only explicit mention of the Opata, who formed the core of the founding population of the mission pueblos of Oposura and Cumpas. Three Yaqui households appeared in the census: Marcos Yaqui (aged forty) and Rita Cruz (aged thirty-eight) in Teonadepa had three children seven years and younger; in Jamaica, Mathías (forty-five years of age) and Dominga Yaqui (forty-three years of age) headed a household with five children, followed by a much younger, childless couple, Francisco Yaqui and Ygnacia Martínez. One person was identified as Apache: in Oposura, Barbara Apache (aged thirty-six) and Joseph Ríos (aged forty) formed a household with five children between the ages of four and sixteen years. Two *nijoras* appeared in both the Indian and mestizo categories. Josef Bamori, age twenty, and Rita Nisori, age fifteen, formed a young childless couple among the indigenous households of Cumpas. Francisco Xavier Nifora (aged thirty) and Lucía Soqui (aged thirty-six) figured among the *indios laboríos* of the hacienda of Buenavista, with three dependents between the ages of seven and twenty. The references to *nijoras* implied a veiled recognition of servile labor, and one intriguing notation for the hacienda of Jamaica pointed to the social condition of enslavement. Simón Juraqui, age forty-five, had married Gertrudis Moreno Esclava, ten years his junior, and together they headed a household of six children, ranging from adolescence to infancy. Gertrudis may have been the slave (either purchased or a ransomed captive) of one of the Moreno households, thus carrying their family name.

Social and ethnic differences embedded in the 1803 census of Oposura parish closely mirrored status qualifications of property and lineage as these were consolidated in late colonial Sonora. The Therán, Moreno, Peralta, Montaño, Grijalva, and Escalante surnames represented veritable clans of landed Spanish families who intermarried with one another and formed kinship networks that extended into the mestizo households of former mission villages and haciendas. Juan Joseph Therán, at age thirty-nine and single, was the relatively young owner of the hacienda of San Pedro de Pivipa. Nicolás Peralta, age thirty, served as Therán's *mayordomo*; he and his wife Luisa Moreno (aged twenty-six) had one infant daughter. Thomas Moreno, age forty-eight, who owned the hacienda of San Josef de Buenavista, figured as *teniente de justicia mayor*, a judicial and administrative post. He and his wife María Francisca Bustamente y Therán, age thirty-six, headed the list of forty-three Spanish households that resided in the pueblo of Oposura. The census did not name the owner of the hacienda of Jamaica, but Ygnacio Moreno (age thirty-three) and his wife Juana Zerrano (age thirty) with their five children headed the Spanish households listed there. Independent documentation permits us to sketch the history of private ownership of Jamaica, beginning in the second decade of the eighteenth century, when presidial captain Gregorio Alvarez

Tuñón y Quiroz established the hacienda and *real* of Jamaica. By mid-century, Salvador Julián Moreno and his wife, Beatriz Vázquez, owned Jamaica, as well as properties in Tonibabi and Toiserobavi. The tenant families who settled on Jamaica at that time formed the core community that grew to 173 persons by 1803. Blas Peralta, a resident of Jamaica, secured title to two portions of the hacienda in 1789, which he had received through inheritance and purchase.[70] Peralta family members appeared in numerous households throughout the eight communities included in the 1803 census of Oposura parish, and Blas Peralta, age sixty, with his wife Rosa Rivera (age forty-five), headed a household of ten members in Jamaica. The Peralta-Rivera union figured among the Spanish families of the hacienda, but without the honorific title of *don*.

The interface between provincial elites and mixed populations of mestizos and Indians illustrates poignantly the cultural construction of community. Santa Ana de Tepache and San Miguel de Oposura combined different kinds of settlements—pueblos, haciendas, mining *reales*, and *rancherías*—in which ethnic identities intertwined with gendered and generational differences of status. The parish censuses of 1803 provide only partial clues to the cultural texture of these communities. We cannot specify the differences between *indios del pueblo* and *indios laboríos* in terms of dress, language, or other outward signs of ethnicity. It is probable, furthermore, that Spanish and mestizo *vecinos*, as well as Indians, took part in the religious festivals that marked the liturgical calendar, linking it to the natural seasons of want and plenty, planting and harvesting, hunting, and the onset of summer rains in Sonora.

The meaningful distinction between *indios del pueblo* and *indios laboríos* emerges rather from the political and material dimensions of culture, viewed historically through their participation in the internal governance of their communities and the different ways they gained access to land. Three native officers of the village council were identified in the census, all in the pueblo of Oposura: Juan Bautista, governor, Hernando Santa Cruz, *alcalde*, and Juan Alonzo, *mador* (missionary's direct assistant). Neither Juan Bautista, age forty, nor his wife María, age fifty, had surnames; they lived with one son, age fifteen. Hernando Santa Cruz (age thirty-five) and his wife María del Rosario Sianagui (age not listed) had a young family of two children, and Juan Alonzo (age twenty-five) lived with his wife Barthola (age twenty-two) and an infant son. These individuals represented a young-to-middling range of adult leaders in the community, and their residence in the head village of Oposura placed them

TABLE 10 Parish Census of the Pueblo of Ónavas, 1801

	Spaniards	Castas	*Indians*	*Totals*
Boys	2	28	292	322
Girls	5	24	230	259
Bachelors and widowers	0	7	109	116
Single women and widows	7	16	96	119
Married men	3	12	269	284
Married women	3	17	264	284
Total population	20	104	1260	1384
Male-female ratio	0.3	0.82	1.14	1.09
Percentage by ethnicity	1.5	7.5	91	100

Source: Archivo Diocesano de Hermosillo (ADH), vol. 3, exp. 7.

in a parallel, but subordinate position to the Spanish *teniente de justicia mayor* Thomás Moreno.

Not all secularized mission parishes followed the same path to *mestizaje* that typified Santa Ana de Tepache and San Miguel de Oposura. The census of 1801 for San Ignacio de Ónavas reveals a highly indigenous demographic profile. Ónavas had long figured as a central pueblo of the Pimería Baja, although by the late eighteenth century it was surrounded by mining *reales*, chief among them Río Chico, Soyopa, Trinidad, and San Antonio de la Huerta. Indians constituted over nine-tenths of the total population, followed by *castas*; only three Spanish households lived in the pueblo of Ónavas, as shown in table 10. Juan Antonio Alegre, the reporting priest, acknowledged the itinerant habits of his parishioners in his explanation to the bishop: "The enclosed census of the parish under my care in fulfillment of your order cannot be as exact as one would desire, for the wandering habits of the native peoples in recent years allows them to leave their villages with liberty, making it difficult to know their whereabouts at any time."[71]

The movement of persons from one community to another and the social networks that crossed different ethnic categories were closely linked to environmental changes in the built environment of late colonial Sonora. The ostensible purpose of the diocesan census ordered by bishop Francisco Rouset de Jesús in 1796 was to establish the basis for collecting the tithe set at a minimum of one-half *fanega* of maize per household.[72] It marked an important turning point in land tenure and changing agrarian landscapes in Sonora, for the tithe was meant to substitute for the communal labor that had supported the missions and their resident priests under the regime first established by the Jesuits. As we have seen in chapter 3, the 1794 directive issued by Commandant-General Pedro de Nava to divide mission lands and distribute arable plots among the resident Indian and *vecino* families made increasing amounts of land available for purchase and entitlement through public auction. The measurement of people in the censuses was inseparably linked to the measurement of land in ways that augmented royal revenues and transferred productive resources from communal usufruct to private ownership. *Indios laboríos* grouped together with mestizos and *castas* composed a growing class of tenants and smallholders who eschewed the protections and obligations of the mission regime to seek their livelihood in the marketplace as laborers and petty producers. Social interdependency and inequality between *vecinos* and *indios del pueblo*—opened through intermarriage, physical movement, or formal petition—created the conditions for transculturation in a rapidly changing political scenario of contested claims to wealth and status.[73]

Chiquitos

Mestizaje across the Hispanic-Indian divide played only a minimal role in the colonial frontier of Chiquitos. Indeed, new ethnic identities were understood by contemporaries and analyzed later by historians and anthropologists in terms of the native *parcialidades* that defined political representation and cultural spaces in the mission pueblos. Spanish and mixed-race colonists comprised a small minority in the Chiquitano pueblos, centered around the governance of the missions. Following the expulsion of the Jesuits, they included priests, lay administrators, the provincial governor and his small staff, as well as soldiers stationed at the frontier garrisons. Spaniards, Portuguese, and Blacks (*negros*) passed through the pueblos, but without residing permanently in the province. It is possible that the Chiquitos missions at the close of the eighteenth century resembled the demographic profile shown for Ónavas, in Sonora (table 9), but the categories of *español* and *casta* or *mestizo* did not appear in the periodic counts of mission populations. The well-published observations of French traveler and naturalist Alcides d'Orbigny in 1831 noted lists of indigenous *parcialidades*, but made no mention of non-Indians living in the Chiquitos pueblos. Francis de Castelnau, who traveled through the Chiquitanía in the mid 1840s, named a number of cattle ranches through which he passed between the mission pueblos, and traded directly with the Indians who managed them, taking little note of Creole residents there. Cultural and racial *mestizaje* produced a more complex social fabric in the province during the formative decades of the Bolivian republic through continued practices of commerce and labor recruitment tied to the missions, as well as through new patterns of settlement and entrepreneurial exploration. These processes and their implications for the natural and cultural environments of Chiquitos will be developed more fully in chapter 7 on postcolonial landscapes and the transition from colony to republic.

Conclusions

This chapter has addressed a complex subject viewed through the bifocal lens of historical comparison. Our analysis of different kinds of evidence that bear on ethnic and gendered identities in both Sonora and Chiquitos has shown that the elusive concept of cultural identity remains inseparable from the social networks that describe living communities and the landscapes they create through conflicting claims to resources, labor, and community. It suggests that ethnicity is not so much a fixed category as a historical process of changing identities and translations. Ethnic itineraries, to paraphrase James Clifford, evolve in the course of intercultural contacts, precipitated by colonial conquests in both Sonora and Chiquitos. Moreover, the social and cultural meanings of ethnicity point to processes of amalgamation and hybridization in Sonora, and of fragmentation and differentiation in Chiquitos.

Why is this so, and why is it important? First, the natural environments of each region explain, in part, the different processes and rhythms of ethnic formation. Tropical swidden agriculture and the need to maintain shifting territorial boundaries in the greater Amazonian region reinforced the local identities of multiple communities. A somewhat different picture emerges from the nomadic-sedentary divide of Sonora, which established a cultural boundary that coincided, in part, with the ecological transition from the coastal desert to the riverine valleys and ranges of the foothills and cordillera. Second, the institutions of colonialism in both areas profoundly shaped ethnic identities and their gendered content, through *encomienda* and the missions, by forceably separating and regrouping different bands and domestic communities. Finally, the proximity of colonial settlements to native villages and the webs of relationships—both exploitative and interdependent—that enveloped both of these frontier provinces created different patterns of amalgamation and fragmentation on the borders of empire. These ethnic mosaics and gendered relationships have telling implications for the configuration of secular and spiritual power within native communities and across the colonial divide. Ethnicity and gender intersect with political culture, as we shall see in our examination of local governance and armed conflict in the following chapter.

What do the histories of intersecting identities in northwestern Mexico and eastern Bolivia teach us about the concept of ethnicity? Classic theories of ethnic nationalisms have long held that identities are defined at the boundaries of contested territories and political spheres. Furthermore, historical ethnographies of diverse colonial encounters have endeavored to show the colonized Other as a foil, a refracted image of the colonizers' aspirations to power and domination.[74] Yet these boundaries are not precise or fixed; rather, they emerge as the historical artifacts of conflicts, fissures, alliances, and transactions among diverse sets of actors and, as such, are subject to change. Ethnic mapping, as I noted in the introduction to this chapter, does not necessarily coincide with language or habitat, and the boundaries it describes are not simple linear divisions, but zones of confrontation, exchange, and commingling. Even as we refer to ecotones as zones of ecological transitions, we can imagine ethnic boundaries as transitional zones of cultural hybridity. Figuratively speaking, these are the edges of social forests, where the invention and readaptation of political allegiances take form as ethnic identities.[75]

The colonial divide is a central feature of our comparative stories of ethnic mosaics in Sonora and Chiquitos. We have seen through multiple examples, however, that in both regions social distinctions were not restricted to the boundary between Spaniards and Indians. Europeans in both regions attempted to group many diverse peoples into a single colonial figure: *los naturales* (the natives). Despite the frequent use of the term in colonial documents, officials and settlers discovered that they could not reduce colonized peoples to one homogeneous category. They were forced to deal with multiple and shifting ethnic polities in negotiations for trade and labor, and in warfare. The frontier societies of Sonora and Chiquitos traced enduring boundaries of ethnic identity, yet these very boundaries became zones of cultural change and permeation.

Gendered inequalities shaped colonial understandings of ethnicity and constituted unmistakable markers of difference. *Encomienda* in both regions demanded labor tribute from men and women; however, the social categories created around the practices of captive labor gendered the condition of servitude as female. The accusations of the Masavi brothers of Chiquitos against their erstwhile *encomendero* Lorenzo Roca document the humiliations and sexual abuse that servants and laborers endured at the hands of the colonial masters. In the missions, where missionaries attempted to discipline the sexual practices and conjugal coupling of Christian Indians, gender intersected with the institutions of power that intervened in the intimate relations of colonized native peoples and in the moral definitions of proper behavior. Turning to Sonora, Governor Pineda's admonition to the Pima governor of Ures denied this officer's authority even to assign labor tasks in his pueblo to non-Indian women married to Pima men and living in the village. Pineda's rebuke privileged the racial status of *gente de razón* over the Indian *común* and established gradations of inferiority that subjugated women to men and distinguished among different ethnic categories of colonizers and colonized.

The censuses analyzed for the parishes of Tepache and Oposura point to the racial and gendered ambiguity of terms like *indios laboríos* and *castas* in the contrasting social circumstances of villages and haciendas. The picture of ethnically mixed households that emerges from the censuses shows that indigenous wives followed their husbands into mestizo communities, where they became part of a rural laboring class. The dilution of cultural autonomy in late colonial Sonora, closely related to the shift in control over productive resources that favored the Hispanicized elite of landholders and merchants, had direct implications for provincial governance. Distinct indigenous *parcialidades* sustained a complex mosaic of identities in Chiquitos, yet gendered patterns of labor and the sexual subordination of women constituted central features of the institutional structures and social practices that supported these missions. The following two chapters develop the profound cultural significance of these themes for the secular and spiritual realms of power in the environmental and social settings of Sonora and Chiquitos.

CHAPTER 5

Power Negotiated, Power Defied: Political Culture, Governance, and Mobilization

The Jesuit Joseph Sánchez Labrador made an arduous journey, in the rainy season of January 1767, across the cultural and ecological borderlands that separated the Guaycurúan and Chiquitano mission provinces of the upper Paraguayan watershed. Four years earlier, a failed attempt to reduce a new Guaycurúan community to mission life had resulted in the death of missionary Antonio Guasch at the hands of his erstwhile converts, leading to violent skirmishes between the Chiquitos and Guaycurúan bands. In a memorable episode, the Chiquitos had ambushed a village of three hundred Mbayás (a Guaycurúan ethnic group), killing many of them and taking as many as one hundred prisoners to their village of Santo Corazón. In a climate of distrust with reprisals among different groups of Guaycurús and between them and the Chiquitos, Sánchez Labrador worked hard to persuade one of the Guaycurúan groups (the Eriguayeguis) to go with him to the land of Chiquitos, to allay their fears. In the face of rumors that the Chiquitos had killed all their prisoners, Father José Chueca of Corazón sent a letter to Sánchez Labrador by way of a Guaná captive, who gave the letter to an Mbayá cacique, affirming that the Guaycurús were alive and content in the Chiquitos pueblo. Relying entirely on the Guaycurús' knowledge of the terrain to guide him through the swamps and running streams of the tropical wetlands of the *pantanal*, Sánchez Labrador worked his way through a network of languages and the conventions of diplomacy among regional caciques to arrive, hungry and wet, in the sandy arroyos and forests surrounding the pueblo of Corazón. Followed by the wary Eriguayeguis, the Chiquitanos' celebrations moved the missionary to tears.[1]

Father Sánchez Labrador's account of *reducción*, warfare, and the intricate channels of negotiation in the Guaycurúan-Chiquitano frontier illustrates the tenuous quality of Spanish dominion. This chapter examines the linkages between power and ethnic identity through the indigenous figure of the cacique and the colonial native council, or *cabildo*. It argues that the political culture of Chiquitanos and Sonorans was shaped by the competing spheres of authority that arose within their communities and between them and the

Hispanicized sectors of both frontier societies. As indigenous peoples learned to use colonial institutions, they adapted preconquest modes of leadership to the offices that Spanish military, civic, and ecclesiastical authorities established in the missions and, in the process, molded these offices to their own purposes. Building on the discussion of ethnic identity and gender developed in chapter 4, this chapter explores the political culture of native communities by examining the separate roles of men and women in the pueblos, where the formal institutions of authority were gendered masculine, and the differences in status ascribed to ethnicity.

My interest in the political ambiguities centered in the *cabildo* arose from documents, which frequently name native officers and their titles, and from my field work in eastern Bolivia. The numerous and widely dispersed *comunidades* (rural hamlets) in which most of the Chiquitanos live at present are linked among themselves and to the mestizo towns that control the architectural spaces created by the colonial missions through political officers whose titles are derived from the *cabildo*. Among the men and women of the Chiquitos communities, the *cabildo* officers have the power to convoke assemblies and assign work tasks, in accord with long-standing practices of reciprocal labor (*minca*);[2] they struggle to gain political recognition in the municipal and prefectural institutions that dispense public monies, award commercial licenses, and authorize the leasing and sale of rustic property. The *cabildo* constitutes the center of Chiquitano political and religious cultures, sustained and reenacted through ceremonies linked to Catholic liturgical rituals.

Community councils became enmeshed in the uneven power relations established by colonialism, acted out on a daily basis through discipline and surveillance, negotiation and compromise, flight and defiance.[3] Colonized indigenous societies, whose preconquest political cultures were not bound by centralized state authorities, pushed against limits that Spanish authorities set to their village autonomy. Moreover, colonial officials in regions like Sonora and Chiquitos, distant from the centers of viceregal power, found it necessary to negotiate the boundaries of imperial dominion within the *reducciones* and in the conditions of frontier warfare.

As the principal institution of internal governance and social allegiance in the missions, the *cabildo* cannot be separated from the imprint of ethnic identities on the colonized indigenous communities. In Chiquitos, the mission councils were shaped largely by the different *parcialidades* that defined the spatial division of the pueblos and their ethnic representation. In Sonora, where different groups settled in the same villages, there is little mention of their separate representation on the *cabildos*. Furthermore, here the dual pressures of Spanish settlement and raiding by nomadic bands of Apaches and Seris shaped the social bonds of community and influenced the internal governance of the missions. Indian men were enlisted in the garrisons stationed in the province, and tensions arose between the council officers who held authority in the missions and the captains appointed to lead native troops, whose allegiances tied them to Spanish military hierarchies.

In this chapter, *ethnic polity* refers to indigenous communities that confronted colonial authorities and took action through petitions, negotiations, protests, and open defiance. Ethnic polities were institutionalized in the two frontiers that form the corpus of this research through colonial mission communities and, within these, the offices of internal governance that merged Iberian precepts of public order with indigenous practices of leadership, reciprocity, and redress. The cultural practices of ethnic polities were crafted historically as Indian peoples shaped and tested the Spanish legal formula of *república de indios* (native polity) established in the frontier missions. Throughout the Spanish American colonies, the theoretical distinction between the "republics" of Indians and those of the Spaniards sought to establish two separate categories of colonial subjects, each tied to the monarchy with distinct bonds of fealty. Spanish towns and native pueblos alike were instructed to elect councils for their internal governance, and the *cabildo* remained an important fixture of political identity for both Hispanic and colonized native communities.[4] The physical, cultural, and economic separations between these two republics blurred in practice—through migration, labor demands that brought Indians into the Hispanic sphere of colonial society, and the growth of ethnically mixed populations.

The political cultures of both the colonial powers and the colonized developed historically through the tensions created between the Hispanic imperial project and the ethnic polities it confronted in the evolving cultural landscapes of Sonora and Chiquitos. It was the objective of Spanish imperialism to establish the dual institutions of monarchy and the church, thereby legitimating the use of power to reinforce economic modes of production and social conventions of deference and hierarchy. Images of the Catholic Trinity and saintly intermediaries, together with personal linkages of vassalage to the king, informed the routines established by the institutions of governance with meaning and reinforced the claims to authority asserted by missionaries and lay administrators. Nevertheless, the values and symbols that underlay the imperial order were not easily imposed; rather, they faced contestation, challenge, and realignment. Iberian imperialism in the seminomadic frontiers of North and South America returned to its roots in medieval Spain, in the legal frameworks for mixed Christian and Muslim communities created in the interstices of warfare during the long centuries of Reconquest. Rhetorical pretensions to absolutism by the Habsburg and Bourbon monarchies and their viceregal officers and councils in the colonies gave way to principles of legal pluralism in the adjudication of disputes over jurisdiction, labor discipline, and property.[5] Native *cabildo* officers constructed a political discourse of mediation that employed the very language of loyalty to the Catholic deity and the Hispanic crown, but they endowed it with alternative meanings.[6]

Five basic categories of thought, action, and social position took shape in these frontier provinces through colonial domination, native responses, and negotiation.[7] These principles became part of indigenous political culture even as they supported the colonial regime.

Catholicism: The acceptance of Catholic sacraments, rituals, and religious discipline became a central part of the identity of colonized Indians and definitive for the relationship between the crown and its colonial subjects.

*"Barbarian" versus settled native life (*reducción*)*: Closely related to Catholic status, living in a polity accorded legitimacy to native peoples. The reconstitution of indigenous communities in the mission pueblos constituted a key element of the imperial project. "Reduced" Indians were distinguished from nomads and freed from the threat of enslavement as *piezas de rescate*.

Hierarchies of ethnicity and gender: Colonial rule throughout the Iberian colonies established ranked social orders based on racial categories and ethnic origins that intersected with gendered roles for men and women. Social conditions of inequality permeated frontier regions like Sonora and Chiquitos, despite the leavening effects of mobility and migration, but prescribed behaviors of deference and servitude were contested and negotiated through both written and verbal exchanges across the colonial divide.[8]

Cacicazgo, *authority personified*: Preconquest forms of political leadership among nonstate peoples—including the many different tribes of Sonora and Chiquitos—conditioned the mode of conquest and laid the foundations for the village polities established under the colonial regime. Indigenous rulers were dubbed caciques by the Spaniards, who spread the term from the Caribbean throughout the American colonies. The term persisted in the political codes of the Spanish American empire (and to the present day), but its meanings and functions changed over time and in different places.

Warfare: Violent encounters of many kinds, including slaving raids, skirmishes, native attacks on colonial outposts, rebellion, and punitive expeditions, punctuated the imperial borderlands. Yet the ideologies and technologies of warfare changed as they became reoriented from the early conquest encounters to the defense of territorial boundaries. Spanish dominion depended on native allies, as we shall see below.

These principles are applicable to other peripheral regions of Hispanic America—the deserts of northeastern Mexico or the grasslands and forests of Chile and Argentina, for example—but differ from a similar list that might be drawn up for the core areas of Mesoamerica and the Andes in three important respects. First, tribute payment in money, goods, or labor provided the nexus between Indian villages and the colonial regime in the core regions. Tribute brought with it significant material obligations burdening both commoners and caciques, but, at the same time, tributary status became a central tenet of the Indians' defense of village lands. As I have noted in previous chapters, formal payment in goods or money did not figure prominently in either Sonora or Chiquitos, where tribute was substituted by labor commitments in the missions. Second, warfare against nomadic peoples, which conditioned life on the frontiers for Indians and Spaniards alike, intruded to a lesser degree in the daily life and public administration of the central colonies. Third, the formal defense of noble lineages tenaciously upheld among the Mayan, Mixtec, and Nahua peoples of New Spain, or again by the kurakas (lords) of the central Andes, figured less forcefully in the figure of the cacique in Sonora and Chiquitos.

What constituted the limits of domination in these two provinces, and what did the pathways of negotiation and resistance look like? The colonial regime endured with remarkable stability, notwithstanding repeated episodes of local rebellion, because indigenous peoples became stakeholders in its judicial and cultural institutions and, no less significant, forced compromises in the political and territorial ambitions of the imperial project. My reading of the actions and words of caciques, warriors, and native officers on the village councils suggests a nuanced appropriation of dominant values that modified the five principles outlined above in different settings of political engagement and ritual enactment. Colonizers and colonized understood power in different realms of experience and fields of action. Power penetrated the political and religious spheres of native peoples (and of enslaved, indentured, and freed persons of African descent) in ways that blurred the boundaries and hierarchies of colonial society. Colonized Indians opened transitional zones between the institutions dominated by their overlords and the spaces, both real and imagined, that lay beyond the imperial domain in the worlds of magic and witchcraft.

Materially and figuratively, then, the different groups of historical actors who constituted colonial society created landscapes of power. In Sonora and Chiquitos, the representations of power and their material ramifications experienced in specific locations and points in time were closely linked to nature and to the built environments that shaped these two frontier provinces. The five principles of Catholicism, *reducción*, hierarchy, governance, and warfare became enforced in the dominant features of colonial landscapes that defined the physical spaces in which colonized peoples lived and moved. Mission compounds, as described in chapter 2, created architectural forms and established daily routines of ritual observance and physical labor through which the tenets of Catholicism became the central feature of identity and sociability for the native peoples living in them. Within the mission pueblos, the churches and convents (*colegios*) established privileged sites for liturgical worship, the priests' residence, and the assembled *cabildo*; schoolrooms and workshops, as well as the plazas and processional routes, opened public spaces that connected the church and convent with the village. The *reducción*, or settled polity, extended farther to the irrigated fields, gardens, and livestock corrals that supported mission economy. In Sonora, colonial landscapes produced by military presidios, grain and livestock haciendas, and mining centers with excavated tunnels, forests cut for timber and fuel, and patios built for processing the ore both complemented and rivaled the missions. In Chiquitos, as private cattle *estancias* and sugar plantations extended eastward from the environs of Santa Cruz de la Sierra, territorial and labor demands forced readjustments in the communal spaces and routines established in the missions.

Notwithstanding the enduring imprint of these colonial landscapes, they became reconfigured through the technologies, cultures, and physical movements of the Indians and mixed-race peoples of Sonora and Chiquitos. The built environments surrounding the missions, mines, haciendas, and presidios—and the roads and streams that connected these different colonial spaces—were not merely reproductions of Iberian norms. Rather, their meanings evolved through conflictive processes of conquest and encounter, resistance and submission, contestation and negotiation. The physical distribution of mission compounds, for example, reflected Iberian preferences for urban grid patterns of spatial design, but they gave way to the exigencies of local environments in terms of water sources, flooding, building materials, arable land, and fear of disease. Labor routines in the missions and Spanish enterprises were interrupted by the seasonal rhythms of native agrarian cycles, hunting and gathering, and ritual pilgrimages to the desert and forest. Finally, as we shall see below, indigenous practices of oratory, communal consensus, and warfare shaped the political life of the institutions originally intended to define the colonial order and enforce the obedience of the colonized.

Native Councils and Mission Communities

What, then, was the indigenous council (*cabildo*), and how did it function in the colonial mission? It was the missionaries' intention to reproduce Hispanic institutions in the frontier communities as an integral part of their evangelizing and civilizing mission. Indigenous council officers—bearing titles of *alcaldes*, *fiscals* (enforcement officers), *topiles* (officials similar to constables), and *gobernadores* molded on Hispanic norms of municipal government and carrying canes of office as insignia of their authority—enforced law and order in the mission pueblos. Missionaries governed through the councils, in a form of indirect rule, and their presence was indispensable to implement religious observance and work discipline; that is, for Christian indoctrination and the production of surpluses destined for circulation among the missions and for sale in colonial markets.

The councils established in the Sonoran missions adapted long-standing traditions of local governance from Iberia and Mesoamerica. Native villagers of central and southeastern Mexico recognized different ranks of nobility and offices with both judicial and administrative functions from preconquest times. Colonial *reducciones* grouped several villages under the political jurisdiction of head towns, where indigenous *cabildos* were installed to collect tribute, complete labor drafts, and enforce Catholic observances. In some cases, as is documented for Cuernavaca, Nahuatl terms persisted well into the colonial period to describe council offices.[9] Comparison of indigenous *cabildos* in central Mexico with similar institutions established in the frontier missions show that different offices were ranked in a definite hierarchy, that most of the offices were concerned with tribute collection, and that often the same office combined secular and religious responsibilities. Indians living in the northern missions remained exempt from tribute payment; however, Indian males were subject to periodic labor drafts under the terms of *repartimiento*, and their recruitment fell to native governors and *alcaldes*.[10]

The governor made for the most important officer of the indigenous *cabildo* in Sonora, with judicial and administrative responsibilities. Over time, native governors, viewed as the guardians of the *común*, the communal land and productive assets of their mission villages, came to represent their pueblos in dealing with colonial authorities. *Alcaldes*, in Iberia, were judges of the first instance. In the colonial missions, their position in the councils was second to the governors and fulfilled both judicial and administrative roles. Governors and *alcaldes* gave testimony, defended the boundaries of village lands during property surveys, and signed written petitions and legal complaints. Within the pueblos these officers resolved internal disputes and meted out punishments to convicted offenders. *Alguaciles* and *topiles* were constables, with different but overlapping duties related to law and order. An *alguacil*, in Spain, was a lower-ranked magistrate than an *alcalde*, or again, an administrative officer who oversaw community resources. For example, an *alguacil del campo* safeguarded communal fields from damage. The *topile*, named in both ecclesiastical and civil documentation for colonial Sonora, is derived from the Nahuatl term *topilli* (pl. *topileque*), "one who bears a staff of office."[11] Both of these officers carried out the orders of the *alcaldes* and governors, enforced attendance at public meetings, and distributed tasks for fieldwork and maintenance of the churches and other mission buildings.

Three additional officers called *fiscales* performed duties directly related to liturgical rites and Christian doctrine and had the authority to oversee compliance with ecclesiastical norms and obligations. The first of these, called *mador*, worked as the missionary's direct assistant who served as ecclesiastical notary to keep the records of baptisms, marriages, and burials. Second in line, the *temastianes* were in charge of teaching the catechism to children and adults up to the age of marriage, serving also as sacristans for the celebration of the Mass. Finally, choirmasters counted among the few literate adults in the missions, and they thus frequently served as council notaries. Father Juan Nentvig observed that the Opatas of his missions learned masses and chanted liturgies by heart, "for regularly the faculty of reading and writing does not go beyond the chapel master who, in his political functions, is the cabildo scribe."[12] Books of *actas de cabildo*—the minutes or official decisions of the councils—have not been found for the Sonoran missions, but they have provided rich material for studies of colonial Nahuatl texts in central Mexico.[13]

The Jesuit Ignaz Pfefferkorn, who lived among Pima and Opata-Eudeve peoples, reported on the selection and duties of native councilmen called *justicias*, or magistrates.

> Certain Indian magistrates were put at the head of each village. It was their duty to assist the missionary in fulfilling his office, to share with him the supervision and care of the Indians, and by vigilance, by the good reputation he enjoyed, and by good example to keep the other Indians in good order. In view of these responsibilities, Indians who were best fitted for the position and who seemed also to be true and pious Christians were appointed as magistrates.
>
> In each village there were also one or two *mayoris*, or as the Spaniards say, *madores*, who supervised the grown children and also cared for the sick. This office was filled by Indians who from their reputations could be expected to be faithful, careful, and diligent.[14]

Father Juan Nentvig was even more explicit in his portrayal of the missionaries' supervisory role in the formation of the "senate or cabildo of these Indian republics." According to the royal provision issued by the *audiencia* of Guadalajara in 1716, and reiterated thirty years later by the viceroy of New Spain, missionaries were to direct the Indians' election of their village magistrates (governors and *alcaldes*), while these, in turn, appointed the disciplinary officers (*alguaciles* and *topiles*) who enforced the Indians' attendance at Mass, Christian doctrine, and work stints in the missions.[15]

The history of internal governance in the missions of Chiquitos follows a similar pattern, but with important differences from the Sonoran model. The figure of the cacique was ubiquitous among the Chiquitano and Chaco peoples, but he represented different degrees and modes of authority. Father Fernández reported that among the Manazicas, a nation of twenty-two *rancherías* whose combined territory formed a pyramid through the forests and savannas north of the mission San Xavier, political leadership comprised a hierarchy of *capitanes* (captains) under the command of a "principal cacique." Fernández described the relationship between the cacique and the common people as one of vassalage; the former authorized hunting and fishing expeditions, receiving a portion of the prey and of the cultivated harvests. Women rendered obedience to the cacique's first wife, and public ceremonies and feasts occasioned by visits among the *rancherías* marked the ascendancy of Manazica notables: "The cacique takes the first place, the second goes to the priests, the third to doctors, the fourth to the captains, and then the rest of the nobility is seated." The position of cacique passed from father to son, after a trial period during which the heir governed the youths of his *ranchería*.[16]

The caciques' authority among the eastern and southern Chiquitano peoples was less structured and more contingent on the course of events than

Fernández described for the Manazica nation. Father Knogler, who served briefly in San Xavier and principally in Santa Ana, interpreted the cacique's political role in relative terms. "The Chiquitos, more than other nations, respect persons of advanced age and position. Although there are no social classes among them, each band [*nación*] has a cacique, an accomplished person with particular prestige. Their language expresses this, since caciques are called 'fully men [men properly speaking]': ma onycica atonie."[17] Caciques were equally important among the Guaycurúan bands of the Chaco, but the endurance of their authority depended on the consensus of their followers and on their skill to make war and peace and to deal with missionaries and Spanish colonists.[18]

Caciques provided the foundation for governance within the missions, from the initial phase of *entradas* to the maintenance of law and order in the settled *reducciones*. Knogler further reported that missionaries enhanced the prestige of the caciques by giving them special ceremonial clothing, an elevated seat in the church, and a cane of office that they carried in all public processions. The essential role of the caciques in accomplishing the missionaries' goals through indirect rule appears especially in their power of convocation to Mass, religious instruction, and daily work tasks in the mission. To the extent that caciques personified the ethnic polity of their separate bands, their elevated position in the missions defined the structure of the colonial Indian *cabildo*.

Large hierarchical native councils were a central feature of the mature phase of the ten reductions of Chiquitos, although the Jesuits never ceased using the strategy of forest *entradas* to populate and maintain these missions. The formal structures of native councils were not instituted at one time, but developed gradually as the missions reached a stable population of different resident bands. Missionaries relied on the *cabildos* to enforce discipline and to support the increasingly complex and voluminous production of surplus goods for trade with the Andean colonial cities and mining centers. The caciques, for their part, reasserted their authority over their separate bands through the *cabildo* offices and the visible signs of their investiture. Collectively, they were known as *jueces* (magistrates; similar to the *justicias* referred to in Sonoran documentation), but their separate duties were ranked and designated by different titles. As in the neighboring province of Moxos, culturally and ecologically similar to Chiquitos, the caciques "merged with the cabildo," that is to say, traditional authority figures had preferential access to political offices in the missions.[19]

Corregidor, a term transferred from royal oversight of municipal administration in Spain to colonial governance in the Americas, referred to the highest male officer of the mission, with both administrative and judicial authority. The *corregidor* was assisted by a *teniente*, followed by the *alferez real*, a standard-bearer, and by two *alcaldes*, a *comandante* (commander), a *justicia mayor*, and a *sargento mayor* (sargeant). These officers were assisted, in turn, by the *alguaciles*, *fiscales*, and *regidores* (council members at large), titles that appeared in the Indian pueblos of New Spain and in the Sonoran missions as well. The choirmaster and sacristan, directly related to training Chiquitano musicians, liturgical celebration, and the maintenance of the sacred vessels of the church, enjoyed elevated positions. *Cruceros*, so designated by the crosses they wore, carried out further vigilance of the *parcialidades* resident in the missions, relaying council orders and reporting back to the *cabildo* and priests cases of illness, births, deaths, and other events that required attention.[20] Direct vigilance of productive labor fell to different captains in charge of carpenters, iron- and silversmiths, weavers, wax refiners, muleteers, leather workers, and cowboys; the *mayordomo de colegio* kept the warehouse, provisioned the missionaries' refectory, and oversaw the distribution of meat rations and other goods to the Indians.[21] Each of these officials was distinguished by a silver-tipped cane of office, a baton, a wooden cross, or the keys to the warehouse and the chapel, symbolizing their authority and responsibilities.

In both of these frontier provinces, the obligations incumbent on all mission Indians to attend catechism and liturgical ceremonies, as well as to perform communal labor, were enforced by moral suasion and by the threat of physical punishment, usually whipping or confinement to the stocks. Punishments were carried out by indigenous officers under orders from the missionary or from the governors and magistrates. Abuses of physical punishment, often the source of bitter complaints, are recorded for both Sonora and Chiquitos, although they seem to have been more tenacious in the Chiquitanía, especially under the clerical regime following the expulsion of the Jesuits. Fear and resentment of physical punishment punctuated Indians' written protests, providing a recurring theme in their acts of resistance and open rebellion.[22]

Certainly, from the missionaries' point of view, the *cabildo* served as a vehicle for social control. The collaboration between missionaries and Indian officers, at times carefully orchestrated, was not merely theatrical staging. Governors and *alcaldes* were empowered by the missionaries' dependence on them to assign daily tasks in mission fields and workshops and to assure attendance at Mass and catechism. More directly, and just as visibly, indigenous officers who held keys to the pueblos' granaries and warehouses played a central role in the semiannual distribution of food, clothing, and tools among families resident in the missions. Furthermore, the hierarchy of offices created through the *cabildo* established a ranked order of privileges that defined concrete benefits in the form of additional rations of food and gifts, honored places to sit and stand in religious services, and access to the priests' quarters in the convents or *colegios* that comprised the architectural center of the missions. The elite status of privileged officeholders was underscored by the observation, in both provinces, that literacy was reserved only for a few male council members, most often choirmasters and catechists.[23]

The solemnity of annual elections reinforced the *cabildo's* authority and the elite status of its members.[24] Mission commoners observed the day-to-day presence of *cabildo* officers in the missions not so much as a sitting council, but as they moved through the pueblos—exhorting, visiting, distributing food and other gifts, and enforcing the rules for worship and work. The missionaries' candid statement that they appointed indigenous magistrates, governors, and *corregidores* belies any appearance of village-based democratic elections. Nevertheless, the priests' words do not convey the different meanings their mission neophytes ascribed to public office or the ways in which they internalized the hierarchies established in the pueblos. We are left to ask how the Indians insinuated themselves into the selection process: how did they convince the missionaries of the relative virtues of certain individuals to hold office? The authority vested in the *cabildo* overlapped with native criteria of association and leadership based on kinship networks, ethnic alliances, and linguistic affiliations in ways more complex than the missionaries acknowledged.

The association between native councils and distinct ethnic groups living in the missions is more marked for Chiquitos than it is for Sonora. Their *cabildos* were structured to accommodate representation for each of the *parcialidades* by the caciques, who acted as intermediaries between their kinfolk and the ecclesiastical and civil authorities of the missions. Nineteenth-century documents and ethnographic testimonies show persuasively that the importance of the *parcialidades* for the internal organization of Chiquitos villages outlived the colonial order. The governor of the province of Chiquitos, Marcelino de la Peña, began his comprehensive report of 1832 with the following observation: "The governing regime in this province and the way of life of its inhabitants is that the pueblos are formed by parcialidades, each parcialidad has a principal judge under the name of corregidor, teniente, alferez, and so forth."[25] That Governor de la Peña, who resided in Santa Ana and was no stranger to the Chiquitano pueblos, conflated the *parcialidades* and the *cabildo* offices reflects their importance for the political culture of the province during the early republican era. Similarly, Alcides d'Orbigny, the French naturalist and anthropologist escorted by de la Peña in his travels through Chiquitos in 1830-31, took special care to list and describe the *parcialidades* he found in each mission, as well as the languages spoken among them. D'Orbigny's catalog is a singular impression formed by an outsider to the region, yet one recorded with the detailed precision of a keen observer trained as a scientist. His attempt to fix the ethnic mosaic of Chiquitos into a rational order of nations, tribes, languages, and dialects reflects the overriding interest of his generation of anthropologists in creating taxonomies of flora, fauna, and humans. D'Orbigny's report distorted the historical, changing quality of Chiquitano ethnic identities, but it captured the significance of the *parcialidades* for governing the missions.[26]

The specific names of most of the *parcialidades* listed for the Chiquitos missions dropped out of the written record during the late nineteenth century. By the mid-twentieth century, political power and urban property had passed into the hands of non-Indian Bolivians, and most of the Chiquitanos had dispersed into small hamlets (known today as *comunidades*) scattered in the forest. Nevertheless, contemporary Chiquitanos recall evidence of a different sort that associates residential blocks, or "sections," of their pueblos with family lineages and ethnic identity. Furthermore, native *cabildos* persist to the present day. They provide a unifying structure for the remnant populations of the pueblos and *comunidades* that surround them, maintain their ceremonial life, resolve local disputes, and defend what remains of their land, water, and forest resources.

Musician and long-time *cabildo* member in the pueblo of San Ignacio de Velasco, Ramón Conrado Morón Tomichá, recalled seven sections that extended in four cardinal directions around the church and central plaza of San Ignacio: Piokó, Rúber, Tañimpî, Waraioj, Kusikia, Rámano, Punarr.[27] Morón specified that each section had its cacique and five additional members on the *cabildo*; the full council was comprised of these section representatives plus the *cacique general*, his immediate assistant, four *comisarios*, and three *síndicos*. These officers shared tasks, but their distinct titles indicated their ranked authority. The latter two terms came into the Chiquitano local governance from republican-era municipal councils. The most important functions of the *cabildo*, in his experience, were to direct communal labor, often gathering and cutting savanna grasses and palm leaves for roofing houses, and to organize religious festivals throughout the liturgical year. Cooperative labor to accomplish a particular task, known as *minga*, was closely associated with the governing authority of the *cabildo*.

> Each Saturday the council met in the house of the cacique general to divide up the work. The cacique general passed the orders on to the cacique segundo, and he in turn to the intendants and comisarios. Then, they went to each house to announce what the work was for the following Saturday. Unmarried women were also called to gather grass; the grass was stored in the house of the cacique general until it was used. It was good in those days, when the work was shared in common, with respect.[28]

The passing of orders from the caciques to the intendants and *comisarios*, and then to the people of each section, reflects the principle of hierarchy that upheld colonial hegemony and governed the internal deliberations of the *cabildos*. The work required of men and women living in the sections and in the communities surrounding San Ignacio at times served their internal benefit—for example, to repair a roof or to clean the cemetery or sweep the plaza before a religious procession—but it also provided the labor to clear the dirt paths and roads that connect the small communities with the towns of present-day Chiquitanía. In this way, community members satisfied a municipal tax in kind, the *prestación vial*. Finally, the reference to Saturday as the day designated for *cabildo* meetings and shared community work, reminds us of the weekly labor that Chiquitanos devote to cultivating their own *chacos* and to the peonage of past decades on the private estates that surround their villages.

To summarize, then, with a long-term comparative view of political institutions, the *cabildo* merged with the *parcialidades* that were distinguished spatially in each of the Chiquitos missions by creating a representative structure for their caciques. Male cabildo members exercised disciplinary, ceremonial, and redistributive functions visible to the entire community.[29] The Sonoran mission villages, smaller but more numerous than those of Chiquitos, were not as fragmented internally as the latter, nor did the Spaniards apply the terminology of *parcialidades* to their inhabitants. Nevertheless, Sonoran missions combined families from different *rancherías*; social and ethnic inequalities were expressed in the ascendancy of certain ethnic groups over others and in the primacy of head villages over smaller settlements in each of the mission districts. The Opatas and Eudeves, who held the best agrarian lands of central Sonora, provided the nuclear population of the missions and lorded over more nomadic groups that entered and left the pueblos, notably the Jobas of the sierra and the Tohono O'odham of the desert plains.[30] These core populations coalesced into ethnic polities through their control of mission *cabildos*, their identification with the villages, and their representation of these communities to Spanish officialdom and to colonial society.

Aspiring native leaders in Sonora had two principal avenues to elite status: *cabildo* offices in the missions and military rank in the auxiliary troops that assumed a significant role in frontier defense. The Spanish presidial system, which expanded during the eighteenth century to contain the northern Mexican provinces in the face of widespread raiding by Athapaskan peoples (Apaches) from the Sierra Madre and by Hokan-speaking nomads (Seris, or Cunca'ac) from the Sonoran Desert coast, depended on companies of indigenous soldiers who were salaried and organized under the command of their own captains. Furthermore, the General Commandancy of the Internal Provinces, established in 1779, created a new hierarchy to which indigenous captains appealed for prestige and gifts to redistribute to their warriors. Opata soldiers, in particular, were recruited for numerous punitive expeditions against Apache bands, traveling considerable distances from their home villages to Chihuahua and New Mexico. Three companies of Opata and Pima soldiers manned presidial garrisons at Bavispe, Bacoachi, and San Ignacio.[31] As we shall see below, these two means of social and political ascendancy created conflict in the pueblos, even as the dual presence of mission and presidio proved central to governing colonial Sonora.

Discipline or Defiance? Conflict, Confrontation, and Negotiation

Cabildo officers and militia captains asserted their autonomy by leaving the confines of the mission to address their petitions and demands to provincial governors and military commanders. The leading figures of governor and *alcalde* (in Sonora) and *corregidor* and cacique (in Chiquitos) came to personify their communities—an identity that, no doubt, sharpened in focus through the colonial experience. The following episodes illustrate the contradictory roles of the *cabildo* and its conflictive relationship with ecclesiastical and imperial authorities. In issues involving territory, the distribution of mission harvests, assets, craft production, and political jurisdiction, the native councils operated at times as vehicles of social control and, at other times, as forums of protest. Different expressions of the *común*, a concept of identity that coalesced into the defense of community, emerged from the testimonials elicited by investigations into local conflicts.

Sonora: Land and the Indigenous Común

In 1716, Indians of three small *rancherías* (Xecatacari, Oviachic, and Buenavista) located within the mission district of Cumuripa in the Yaqui valley appealed to judicial authorities. They sought to reclaim land their forebears had occupied before they were violently dispossessed by a Spanish rancher and militia captain, Antonio de Ancheta.[32] Two years before the Indians drew up their petition Ancheta had died, leaving the land abandoned and, by default, allowing the Pimas to return and plant their crops anew. The *alcalde* Diego Camorlinga composed a history of occupance, dispersion, and resettlement in Buenavista based on the memory of six living witnesses he brought before the Spanish magistrate. Their arguments rested on effective occupation of the land and the Indians' status as Christians desirous of living in a settled pueblo. Their petition was supported by a census showing seventy-six Christian households and eleven households of non-Christian adults (*gentiles*) with baptized children. Heading the list of Christian families were four *cabildo* officers: Diego Camorlinga, *alcalde*, married with five children; Lázaro, *alguacil*, married with one child; Sebastián, *topil*, married, with four children; Baltasar, *topil*, married, with one child. All the witnesses spoke in the Pima tongue, but their identity rested not so much on a particular ethnic origin as on their status as *el común y naturales de Xecatacari*, commoners with a shared stake in a particular territory. The *común* had material, political, and religious significance in the cultural production of community under the conditions created by colonialism. In this case, the *común* was linked to arable land, the founding of a pueblo, and the Indians' ties to Catholicism.

Sixty years later, two related sets of written petitions centered on the mission of Opodepe and its *visita* of Nacameri in the San Miguel river valley invoked the political concept of the *común*. Their content reveals competing interests among the Franciscan missionaries who had replaced the Jesuits as the spiritual guardians and administrators of the pueblos, the governors and *alcaldes mayores* of the province, and presidial commanders.[33] Rival native officers comprised governors and *cabildo* members of individual villages, often pitted against the captains general who led troops of warriors and allied with the Spanish military. In the summer of 1777, Ambrosio and Diego, war captains of Nacameri, appealed to the *alcalde mayor* Joachin Joseph de Rivera to intervene on their behalf with the Franciscan superiors. These two Opata leaders, "representing themselves and all the individuals of that pueblo, and on behalf of the justices and the *común* of Opodepe," complained that their missionary, Antonio Martínez, refused to supply them with provisions from the mission's granaries when they were called to military campaigns. They wanted Martínez removed, accusing him of trafficking with mission produce for his own benefit.

Rivera forwarded the Indians' request to the Franciscan commissary Juan de Prestamero and to his own superior, Juan Bautista de Anza, together with a more far-reaching complaint by the captain general of the Opata nation that missionaries and village justices of several pueblos refused to recognize his authority and supply his troops with food supplies for their military assignments. Juan de Prestamero visited Opodepe and Nacameri, accompanied by three Spanish colonists, to observe conditions in the villages and question the Indian men and women living there concerning the gifts of food and clothing they received from their missionary and whether they were happy with him. On the basis of his own investigation, Prestamero defended Martínez and dismissed the complaints raised by the Opata captains as machinations by private ranchers in Nacameri who coveted mission lands. He retorted that if missionaries failed to supply presidial auxiliaries with food, it was because of shortages occasioned by the declining labor force and shrinking land base of the pueblos.

On Christmas Day 1778, Manuel Grijalba, governor of the pueblo of Opodepe, visited Pedro Tueros, the commanding officer of the presidio of Horcasitas, with a special request. Contrary to the demands voiced by the Opata captains of Nacameri a year earlier, Grijalba exhorted Captain Tueros to intervene before the Franciscan commissary to prevent the removal of missionary Antonio Oliva from Opodepe, insinuating that, if necessary, he would appeal directly to commandant general Theodoro de Croix. While expressing the Indians' affection for Father Oliva, Grijalba centered his complaint on the losses of village harvests and the decline of mission assets due, he said, to the frequent dismissal and replacement of missionaries assigned to the pueblos.

The apparent contradiction between these two documents opens an interesting view to a number of questions and themes concerning the ways in which power was contested at the community level. First, native leaders of Opodepe and Nacameri defended the missions as their communal patrimony; second, the *común* emerges as a central concept that bound together the native elite of political officers and the people they claimed to represent and for whom they spoke.[34] Third, Opata captains and governors showed that they knew how to maneuver the separate military, civilian, and ecclesiastical authorities of the colonial order.

Significantly, however, despite the verbal imagery of consensus expressed in the phrase, "in the name of the *común*," these two episodes illustrate the internal contradictions that divided the mission communities. We may recall from chapter 4 that many of the missions supported mixed populations of Indians and *vecinos*. That both indigenous and Spanish partial societies had become entwined in the Sonoran missions is evident by the burden of Franciscan commissary Prestamero's defense of Martínez and his mode of inquiry. Prestamero conducted his investigation into the conflict in the company of three local colonists who had occupied posts as *alcalde mayor* and *juez comisario* (a local judge with authority delegated from the *alcalde mayor*) in the pueblos of Opodepe and Nacameri. His accusation of political maneuvering by private landholders points to increasing pressures on mission land and water throughout Sonora during the last quarter of the eighteenth century.

War captains and presidial auxiliaries, as well as *cabildo* officers, assumed the role of spokespersons for the *común* when either confronting or petitioning colonial authorities. When we observe the content of the petitions, as well as the strategies developed by the native leaders to establish their stature and rightful place in the day-to-day governance of the missions, we seem to witness a series of conversations about competing entitlements to the resources and products of the Indians' labor and about the spheres of authority that justices and captains sought as their proper domain. In their efforts to master the institutions of colonial rule, local communities divided internally and followed several alternative paths. Beneath their assertions to select a missionary of their choosing, Opata leaders staked a claim to redefine the *común* and direct the polity.

These conversations and the conflicting points of view they expressed extended to the options of war and peace. When we compare different episodes of warfare and negotiated peace accords, we observe that native peoples responded variously and, at times, in contradictory ways to the burdens and opportunities of the Spanish colonial system. Indians labored in Spanish mines and ranches in return for clothing, shelter, and cash payments; they defended their pueblos against the nomads and served as auxiliary troops for the presidios *or* they turned to raiding missions and Spanish settlements, moving in and out of the villages. Spanish commanders found to their dismay that it was not easy to distinguish clearly among different groups of friends and enemies, town-dwelling peasants and mounted warriors. Provincial governors and presidial captains labeled the latter barbarians (*bárbaros*) and reserved their worst anger for the Christianized Indians who deserted the missions and turned to rustling livestock, kidnapping, and lightning attacks on Spanish outposts. Yet they recognized that they were dealing with strategizing enemies who developed a viable political economy of raiding on the fringes of the settled colony. Here, the documents leave behind the formal institutions associated with the *cabildo*, but their accounts of raids and punitive expeditions together with attempted negotiations reveal alternative means of political organization that intersected with the norms and values of the Christian polity so piously defended as the foundational principles of colonial hegemony.

Raiding provided dispersed nomadic groups with a livelihood and an effective military tactic. Bands of raiders attacked ranches, mining camps, and mule trains to inflict isolated casualties and escape to the mountains or desert thickets where mounted soldiers were forced to abandon the chase, horses became a hindrance, and poisoned bows and arrows proved more effective weapons than firearms. Raiding enemies stole livestock that deprived Spanish enterprises and missions of essential track animals and food supplies. Athapaskan nomads of the Sierra Madre developed extensive livestock trade networks that reached beyond Nueva Vizcaya and New Mexico to the Comanche tribes of the Great Plains.[35] They observed presidial troop movements, knew which garrisons were poorly manned, and, on daring occasions, they carried off the entire horse herd of northern presidios—as occurred at Tubac on the eve of the carefully planned expedition of 1775 that originated in San Miguel de Horcasitas and combined the forces of both presidios with a civilian caravan to found San Francisco in Alta California.[36] Missions and Spanish estates especially feared random attacks during the harvest season, when individual workers were scattered in the fields.

Raiding bands carried off human captives, predominantly women, who were held for ransom or absorbed into these itinerant *rancherías* as laborers and conjugal partners. Spanish punitive expeditions used the same strategy, seizing *piezas de rescate* who were then traded back to their kinsmen in return for livestock and prisoners or distributed as servants among Hispanic households. The taking of captives and the negotiations for their exchange enflamed military hostilities, to be sure, but they also kept the lines of communication open between colonial officials and nomadic bands. Raiding *and* trading, viewed at times as the violent and peaceful faces of frontier survivorship, defined the parameters of warfare and contested the terms of colonial domination in Sonora.[37]

The alternatives that colonized peoples faced in the mid-eighteenth century did not offer them a clear choice between falling in with the Spaniards or returning to a preconquest political order. Environmental and cultural conditions had changed so radically that a pre-Hispanic status quo ante simply no longer existed. Colonial mining and ranching enterprises had altered topography, vegetation, and stream flow in all three subregions of Sonora: the cordillera, the piedmont, and the coastal desert. The growth of livestock herds, grain haciendas, and increased population among *vecinos* and Indians reduced the arable fields and water sources available for subsistence horticulture and began the enclosure of savannas and scrub forests that had provided a reserve for hunting and gathering. To the pressures of these colonial landscapes on traditional lifeways were added the social and political constraints of *reducción* on the modes of leadership and community organization associated with the *cacicazgo*. Outside the core populations in the mission pueblos, dispersed and seasonal residence patterns in mobile *rancherías* were recorded in the documents as fragmented ethnic groups. This held especially true for separate bands of Pimas, who became known as Piatos, Sibubapas, or Suvbàpas in central Sonora and the Pimería Alta; their ethnic identities coalesced around the Indians' relationship to the colonial regime.[38]

"With a Barbarous Fury": Raiding and Warfare on Sonoran Frontiers

Three major rebellions during the second quarter of the eighteenth century challenged colonial domination in the northwestern provinces: the Cahita and Pima revolt of 1739–41 in Sinaloa and Ostimuri; the Seri and Pima uprisings that began in central Sonora in 1749 and continued as guerrilla warfare to the end of the century; and the Pima rebellion of 1751, compounded by a smallpox epidemic of that same year.[39] This rebellion, as we saw in chapter 3, destroyed the settlement of San Lorenzo, near the mission of Magdalena, and briefly but spectacularly threatened the Spaniards' hold on the Pimería Alta. The sequel of assaults and armed confrontations, burned ranches and pueblos, and thefts and murders in all three uprisings forced the temporary abandonment of mining *reales* and challenged the viability of the Jesuit mission program. These crises strengthened Spanish resolve to retain its imperial boundaries by militarizing the frontier and expanding the presidial system across northern New Spain. New garrisons were stationed at San Pedro de Pitic, at the confluence of the lower San Miguel and Sonora rivers, and at San Carlos de Buenavista, near the village that the Pimas of Cumuripa had reclaimed earlier in 1742. A decade later, in 1753, presidios were established in San Ignacio de Tubac and Santa Gertrudis del Altar. Opposition continued to be heard in discordant voices of defiance, adaptation, and compromise; dispersed rebel bands developed a variety of tactics that whittled away at imperial demands and preyed on the colonial order without overthrowing it. Of the many skirmishes and confrontations recorded in the archives compiled by the late colonial administration, the episodes that follow illustrate well the political maneuvers practiced by Seris, Pimas, and Apaches in their prolonged struggle to test the limits of domination and to set the terms of resistance.[40]

Spanish military forces mounted two major expeditions intended to force the surrender of armed guerrilla bands of Seris and Pimas (Piatos and Sibubapas) that roamed the low hills and arid valleys between the lower Sonora and Yaqui rivers: the campaign led by governor Diego Ortiz Parrilla, in 1750, that invaded Tiburón Island in search of rebel Seris, composed primarily of Pima auxiliaries; and the ambitious expedition of 1768-71, commanded by colonel Domingo Elizondo and directed against the Indians' stronghold in the Cerro Prieto.[41] This desert fortress formed a natural barrier along the eastern edge of the Seris' coastal habitat, comprised of rocky promontories, canyons, and secluded water holes, its vegetation largely thorny thickets. Neither of these punitive expeditions succeeded in dislodging the rebel encampments entirely, although they destroyed some *rancherías*, took prisoners, and resettled some of the Seri and Pima families in the presidio of Pitic and the missions of the Lower and Upper Pimería. Spanish counteroffensives were successful only when they adopted the Indians' tactics of dividing into small and mobile detachments. Rebel forces—weakened by hunger, thirst, and disease, and divided among themselves after the defection of some of their caciques—surrendered as family units.

Numerous small bands of Seris and Pimas raided pueblos and mining *reales* during the two decades between these military campaigns. Their attacks threatened the missions of Tecoripa, Cumuripa, Soyopa, Suaqui, and San José de Pimas, as well as the mines of San Antonio de la Huerta, Aguaje, Aigamé, and San Xavier. Governor Joseph Tienda de Cuervo reported, for instance, that on 30 April 1762, "enemy Indians," some sixty strong, attacked the pueblo of Soyopa, two leagues distant from the *real* of San Antonio de la Huerta. Since most of the Indians of Soyopa were away from the village (at the beginning of the wheat harvest), the few remaining ones could not defend themselves. The raiders killed eight men and women, took one woman captive, and carried off the mission's livestock. Later, they lay in wait along the road from Mátape to San Antonio, where they murdered two *vecinos* from the *real* and their servant; they then followed a circuitous route to their refuge in the Cerro Prieto. From the presidio of San Miguel de Horcasitas, the governor sent his lieutenant to round up soldiers stationed in various pueblos and concentrate them in San José de Pimas, in order to sweep the lower hills of the Cerro Prieto and deter a second raid. Tienda de Cuervo was convinced, from the reports that he had received, that the attack on Soyopa was carried out by Pimas, not by Seris, and feared that the Pimas would frustrate his attempts to reduce the Seris and separate them from their Pima allies.[42] His fears were realized only a month later, on 7 June, when Seris and Pimas together raided Cumuripa, killing two men, taking four boys captive, and stealing all the livestock belonging to the mission and to the Indians. At the same time, a different band crossed the middle Yaqui river and fell on several ranches in Ostimuri, carrying off their cattle. Presidial troops deployed northward in the pueblos of Alamos and Nácori pursued the raiders to the Cerro Prieto, where they engaged as many as a hundred warriors in combat for three or four hours. The soldiers managed to recover over three hundred head of cattle, horses, and mules, and killed two of the enemies, but the rest of them "took refuge in the thick [thorn] forests that surround the hills."[43]

Captives played an important role in these skirmishes of attack, counterattack, and swift retreat. Nearly all accounts of raids on villages or assaults against travelers included the seizure of persons—most often women and children—who were taken back to the raiders' encampments. The motives for taking captives were mixed: to hold them for ransom, to incorporate them into the captors' *ranchería*, to provoke enemy warriors to fight, or, at times, to even the score in retaliation for the Spaniards' imprisonment of *piezas de rescate*.

All parties in this guerrilla warfare—presidial troops, Indian allies, and rebel bands—used the captives as sources of information. Spanish military commanders relied on captives to help them locate the *rancherías* of itinerant bands; in the event, however, that ransomed or escaped captives returned to their home communities, they may have unwittingly delivered the messages their captors wanted them to convey.

One of the women taken captive in the raid on Cumuripa fled her confinement in the Cerro Prieto, and, after a full day and night, arrived in the pueblo of San José de Pimas on 9 July 1762. There, she informed captain Juan Bautista de Anza that the rebel cacique Siaritaca remained ensconced in the most desolate part of the mountainous refuge "with many Pimas," while the Seris had come out of the hills and moved out to the coast to gather the ripened fruit of the *pitahaya* cactus. She contended that Pimas alone had conducted the raid on Cumuripa and warned Captain Anza that Siaritaca had ordered his followers to leave the Cerro Prieto with what livestock remained and travel northward, in the direction of Belén (on the coastal estuary at the mouth of the Yaqui river), because the soldiers stationed at San José de Pimas endangered their hideout.[44]

Three years later, the commanding officer at the presidio of Buenavista, Juan María de Oliva, reported to governor Juan de Pineda that on 6 May 1765, a band of enemies attacked the Yaqui pueblo of Cócorit, leaving thirteen persons dead among adults and children. The Yaquis followed their assailants and fought with them, but apparently fell back and were returning to their village when Oliva found them. He could not persuade any of them to join his small detachment, but he proceeded in pursuit of the enemy with only seven soldiers and one *vecino*. He found their encampment and managed to emerge unharmed from the skirmish and recover over five hundred head of cattle and four captive women. Oliva attributed his good fortune to the miraculous intervention of his divine majesty and the most Holy Mary.[45]

That same day a Yaqui woman from Huirivis, who had been held captive by a band of Seris and Pimas, escaped and appeared in the village of Potam, where Jesuit missionary Juan Mariano Blanco reported her arrival. She had come from a large *ranchería* in the plains of Santa Rosa, several days journey from Potam, where the Seris and Pimas had planted squashes, maize, and beans. The day before she fled, the Yaqui women saw a good many Seris and Pimas leave, each supplied with two horses, to bring back more stolen livestock. She heard her captors threaten to return to the Yaqui villages and "finish them off." They boasted that "the Yaquis were afraid of them, since they had taken away their women and the Yaqui men had fled, they knew [the Yaquis] would not dare seek them out in Santa Rosa." (This was bravado, indeed, because the Yaquis were themselves feared warriors.) What most troubled Father Blanco was the news that "some Pimas who live with the padres [in missions]" had gone to Santa Rosa, proffering tobacco sheaves and asking the renegade Pimas to allow them to join their *ranchería*, "because here the Spaniards punish us harshly," and they had been well received. We may speculate on the Yaquis' reluctance to pursue the rebels of Santa Rosa: they may have been outnumbered and, at the same time, reluctant to return under the command of Lieutenant Oliva, rather than fight on their own. Alternatively, their posture toward the Pimas of Santa Rosa may have been determined by the family networks of the men and women who had ties to the same villages and, thus, ambivalent allegiances that crossed the ethnic categories and enemy lines with which Spaniards mapped the province.[46]

Indeed, Spanish military commanders feared that their friends among native Sonoran peoples might change their loyalties and ally themselves with the enemies. Governor Pineda warned the viceroy Marqués de Cruillas about the treacherous nature of frontier warfare. Pineda, commenting on Oliva's daring counterattack on the Indians who had raided Cócorit, pointed out that these nomadic bands seemed to time their raids and fight with more resolve than they had eight or ten years previously, if one believed the stories *vecinos* had told him. Pineda used these arguments to bolster his request for more soldiers and horses, as the presidial detachments needed to be constantly on the move and the "lack of rain had dried up the pastures," thus weakening the stamina of the horses and slowing their patrols. He closed with a final, disquieting thought: "Sir, I have observed something that has intrigued me. When our Indian allies fight the enemies, it is fearsome to watch. We should be on our guard lest they join with [the enemies], because in that case they will turn against us with a barbarous fury."[47]

Spanish confrontations with the Apaches in northeastern Sonora, in the mountainous terrain that joined the province with Nueva Vizcaya and New Mexico, followed similar patterns of warfare as in the Cerro Prieto, with the important difference that unlike the Seris and the Pimas, the Apaches had not been reduced to mission life. Apache bands attacked missions and haciendas especially during the wheat harvest, when *vecinos* and Indians left the relative safety of the pueblos, and carried away livestock, cloth, and captives. Spaniards learned to distinguish the foot and horse prints of different Apache bands and, thus, to ferret out their trails. Presidial troops from Terrenate and Fronteras routinely mounted punitive expeditions against Apache *rancherías*, killing native warriors, destroying their supplies, and taking women and children captives. Periodically, both parties extended peace overtures and agreed to exchange prisoners.

In July of 1762, an Apache cacique known as Captain Valdés among his people engaged Gabriel Antonio de Vildósola, captain of the presidio of Fronteras, in peace negotiations. The initiative began when four Apache men and women entered the presidio and presented Vildósola with a letter from Valdés, written in Spanish by one of their captives, Juan Estéban González. Valdés declared that he and his people were ready to live in peace; he "wanted to be good, and, with his people they would cause no further trouble in our country, neither thefts nor murders." Valdés had brought his entire *ranchería*—composed, according to González, of eighty men, seventeen women, and five youths—to the edge of the Sierra de Tesocomachi, facing the eastern side of the presidio, and asked Vildósola for food and the full exchange of their captives. Vildósola sent the ambassadors back to Valdés with loads of *pinole* (toasted corn meal), dried meat, and tobacco, in addition to gifts of maize, wheat, tortillas, and meat they received in the town adjacent to the presidio. He welcomed the Apache leader's proposal for peace and invited him to come to the presidio himself, without fear.

The two captains parleyed for three days through interpreters and the exchange of letters, with additional gifts of food and livestock, to determine where the exchange of prisoners should take place. Although Valdés refused to come to the presidio himself, he accepted Vildósola's invitation to allow his people to trade with the *vecinos* of Fronteras. The Apache women moved freely through the presidio and town all one day, trading their tanned leather hides (*gamuzas*) for woven cloth, food, and metal knives. Finally, the two leaders met on the opposite banks of the river that divided Fronteras from the eastern ranges, under a light rain, each supported by an escort. Valdés addressed Vildósola in slightly broken Castilian, announcing that he was a Christian, that he did not want to harm anyone, but that out of fear he could not cross to the Spaniards' side.

Vildósola crossed the river and, after many entreaties, the two men dismounted, approached each other, and embraced. At Valdés's command, his eighty warriors, armed with lances and bows and arrows, shouted their thanks to Vildósola; the Apache women and children watched from a safe distance on a hilltop. Valdés declared that he and his Spanish counterpart were now friends and compadres, said that he wanted to live in peace (*sosegado*), and presented the five captives he had to exchange. Vildósola, for his part, turned over ten captives; each side admitted that there were more, but postponed their exchange to a future date, when the Apaches could produce the four additional captives that Valdés's ambassadors had first offered to return. One of the four, who was known to Vildósola but who had become attached to his Apache master, at first declined to be ransomed. Captain Valdés negotiated his return to the Spaniards, at Vildósola's insistence to recover this "apostate," in exchange for a horse to compensate his master. He then requested that Vildósola allow the Apaches to trade with the Spanish soldiers, and a lively exchange of leather hides and mules for horses ensued. When the two captains parted at nightfall, Valdés again assured Vildósola that he and his people "wanted to live as Christians, they believed in God and Holy Mary, and they would live in peace for many years." Captain Vildósola replied that if Valdés kept his promises and returned the missing captives, then he would do all in his part to help Valdés and his people if they wanted to live in a pueblo, "or other kind of polity," and again enjoined him to come to the presidio.[48]

Notwithstanding this promising encounter, Spaniards and Apaches reverted to observing one another in customary ways. Vildósola sent interpreters and gifts to the river the following day, but when Valdés and his people did not appear, he sent two detachments of soldiers to reconnoiter the area. They discovered that Valdés's band had followed a route along the road (*camino real*) to Caguillona that brought them westward into Sonora, rather than eastward into "their own territory" to retrieve the captives they had promised to return. While Vildósola admitted that the Apaches may have simply been gathering agave roots, he suspected treachery. Vildósola sent word to the presidio of Terrenate and to the pueblos in the Oposura and Sonora river valleys to alert them of the Apaches' movements, referring to them once again as enemies.

This missed opportunity for peace illustrates the tensions that permeated the nomadic frontiers of Sonora, as well as the attempts to negotiate through trade and rehearsed dialogues. Spaniards and Apaches both made use of interpreters to translate between the Athapaskan and Castilian languages—underscoring the important role of captives in this process—and to find a common political vocabulary. In this case, Valdés and Vildósola referred to the symbols of Catholicism, polity, and military rank. Their material exchanges involved trade in commodities of livestock, food, clothing, tools, and tobacco, and the return of human captives. The proposed alliance between the presidial commander and the Apache war leader was frustrated by mutual fears and habits of mistrust, as evidenced by Valdés's refusal to enter Fronteras and by Vildósola's suspicious interpretation of the Apaches' movements along forest trails.

We find in this story several clues to the Apaches' adaptation of Spanish rituals and symbols of authority. Captain Valdés insisted on recognition of equal status with Vildósola and conserved his own prestige by allowing his people, beginning with the women, to trade with the *vecinos* and soldiers of Fronteras but remaining aloof from the bartering himself. In addition, Valdés attempted to master the Spanish language and made explicit, repeated allusions to the Christian god and saints. A separate report of that same year, 1762, referred to an Apache encampment in the Sierra Blanca, near the Opata missions of Oposura and Cumpas, where a hundred Indians under the leadership of four captains had amassed several hundred head of cattle. According to an escaped captive that the Apaches had taken to the *ranchería*, they boasted of having killed a Spanish captain and exhibited as trophies some fine clothes and a silver-tipped cane, the insignia of *cabildo* officers. Governor Joseph Tienda de Cuervo surmised that the Apaches' victim was an itinerant merchant who had disappeared three months earlier on the road between Nacameri and Ures; as the *comisario* of Concha, in Nueva Vizcaya, he carried a cane of office.[49]

What do these stories of hostilities among Seris, Pimas, Apaches, Yaquis, and Spaniards tell us about disputed territories and emerging identities? They relate landscapes of refuge in the mountainous cordillera and in the coastal desert, where rebels and fugitives could hide in thickets or caves and barrancas and conceal their trails from pursuing soldiers. Yet the Seris' Cerro Prieto or, again, the Apaches' route along the Caguillona range offered refuges in a different sense as well, as places where natural springs and reservoirs provided water, game, and edible plants, like the fruit of the saguaro and *pitahaya* or the root of the agave. They were not *naturally* forbidding; rather, the repeated experiences of attack and retreat, capture, and death made these landscapes seem dangerous and turned them into geographies of fear.

"Work Harder Because the *Tenientes* Are Coming": Rebellions in the Chiquitos Missions

Numerous local tumults produced a documentary trail for eastern Bolivia, especially during the period following the expulsion of the Jesuits. These brief but violent uprisings revealed simmering conflicts among the separate ethnic groups that converged in the Chiquitos pueblos. Even in the absence of province-wide rebellions, in contrast to eighteenth-century Sonora, local confrontations occurred often in Chiquitos, and the *cabildos* assumed a central role in their outcome. The following cases illustrate recurring themes that involved distinct *parcialidades* and often pitted different bands of Chiquitanos against their missionaries or the lay administrators that oversaw the economic life of the pueblos. Native *cabildo* officers addressed their complaints to both ecclesiastical and civil authorities, moving from petition to open defiance.

In October of 1779, the Manapecas, one of the *parcialidades* of the mission of Concepción, squared off against their priest, Manuel Zapates, in the central plaza of the pueblo. Zapates was armed with a rifle, accompanied by six Afro-Brazilians armed with bows and arrows; the Manapecas, in turn, carried bows and arrows, drums, and flutes. Fighting ensued all afternoon and evening until midnight on Saturday 16 October; the priest was injured by a *tacuara* (arrow) shot to his stomach, and at least one of the Indians was injured by a rock fired from Zapates's rifle. The Indians were incensed because the priest had arrested and ordered flogged one Simón, a member of the Manapeca *parcialidad*. Furthermore, they accused Zapates of keeping a mistress in the *colegio*, the building that housed the mission's school, craft shops, and meeting place for the *cabildo*. Finally, they resented his retaining freed blacks from the Portuguese colony of Matto Grosso in the pueblo, who served as his personal escort.[50]

In early November, Gabriel Lazcano, the priest of San Ignacio who served as ecclesiastical judge for the province of Chiquitos, arrived in Concepción at the behest of the bishop of Santa Cruz to investigate the incident. Lazcano appointed an ecclesiastical notary and two interpreters to hear and record testimony from two "black Portuguese" men, one of whom had settled in Concepción, two native members of the *cabildo*, and a third Indian who was not a resident of the pueblo but had arrived in Concepción from the village of Porongo to buy cotton.[51] The two blacks and the Indian from Porongo limited themselves to recounting the events of the October uprising. Diego Maeños, *alcalde ordinario* on the *cabildo*, and Pedro Camocini, *mayordomo* of the *colegio* and conversant in both Chiquitano and Spanish, gave a fuller history of enmity between the Manapecas and their priest. Their remarks centered on the woman who lived illicitly with Zapates and on the incarceration and corporal punishment of Simón. Maeños recalled at least one confrontation between Zapates and the Manapecas prior to the standoff and shooting in the plaza, and Camocini emphasized the internal loyalty of the Manapeca nation.

To place these incidents in context, the 1779 tumult of Manapecas in Concepción occurred during the planting season, when Indians attended their plots scattered outside the mission compound. Maeños had not witnessed the tumult, for instance, because he was in his *chacra* that day. Maeños reported that during the first confrontation, the Manapeca rebels had hidden among the cotton trees that surrounded the mission compound and supplied the raw material for spinning and weaving cloth. Their anger erupted from the priest's mistreatment of one of their number and their conviction that he had violated the *colegio*, a space that held cultural and religious significance, through his illicit liaison with a married native woman. In Camocini's words, her husband had united with the Manapecas "as if they were his conationals in the colegio," in a combined effort to rescue Simón and bring the woman out of the priest's custody "at any cost."[52] The drums and flutes used commonly in fiestas and religious processions became part of their arsenal of weapons, accompanying the Indians' shouting during this standoff in the mission plaza. The presence of Afro-Brazilians in the Chiquitos pueblos, noted by more than one witness, alerts us to the mixed character of this frontier province, a zone of refuge and trade that crossed the Luso-Hispanic boundary.

A decade later, the council judges of the pueblo of San Ignacio led an uprising, aroused by inflammatory cross-accusations between priests and the governor of the province, Antonio Carvajal, who disputed control over the economic life of the missions. The tumult began on the eve of Corpus Christi, 5 June 1790, leaving at least four dead among Spanish soldiers and civilians. Indian rebels were armed with bows and arrows and clubs (*macanas*). News of the uprising spread to the pueblos of San Miguel, San Rafael, and Santa Ana in the west and to San José, the principal mission of southern Chiquitos. Rebellious council judges of San Ignacio refused to obey Carvajal's order to appear before him in Santa Ana, and they confiscated the entire store of mission products and trade goods kept in the pueblo. Rumors of further uprisings circulated in two of the neighboring missions, where it was believed that over two thousand Indians were armed with arrows.[53]

Information about the tumult comes from Manuel Roxas, the priest of San José mission, the testimony given by Gregorio Barbosa, a soldier stationed in San José, and investigations carried out by the provincial governor at the behest of the *audiencia*.[54] The rebellion was fueled by the Indians' fear of a more repressive labor regime and their sense of entitlement to the wealth in kind that was produced by their own labor. Native justices were alarmed by the news that Carvajal had brought lieutenants to oversee work routines in the pueblos. News of the uprising in San Ignacio reached San José at least by 19 June, when Barbosa recalled that while he was "lying in his hammock," he learned from two Indians who had returned to the pueblo from working in the cattle *estancia* that their priest, Manuel Roxas, had threatened the council judges, saying, "work, work harder, because the tenientes are coming, who will whip you to death. The tenientes will not hear confession or give Mass, because only we Fathers celebrate the Mass and we are Christs on earth. What do you want governors or lieutenants for?" Barbosa learned more about the events in San Ignacio from a letter the Indians had brought to Felix Ydalgo, the commander of the militia stationed in San José.

The letter, signed by the *corregidor*, *teniente alférez*, and the other *jueces*, was copied in both Chiquitano and Spanish and forwarded to the *audiencia* in La Plata. Beginning with a religious blessing, "Praised be Jesus Christ, Amen," the *cabildo* officers of San Ignacio expressed their anxiety over the turn of events in their pueblo and their fervent hope that the governor would visit them to put things right. They addressed the governor as a political body, desirous of meeting with him face to face, one authority to another under the aegis of the king. The *jueces* reminded Carvajal of his duty to protect the Indians, "as God has commanded," couched in phrases of deference. In terms of the motives for the uprising, the *cabildo* officers cited the sexual abuses committed by both the priest and the administrator, who had taken indigenous women as their concubines, and the flogging of one José, the ironsmith, without proper consultation with the council. Their narration of events revealed a growing tension during the days dedicated to the feast of Corpus Christi, marked by the appearance in San Ignacio of Miguel Roxas, commander of the garrison stationed in Santa Ana (Manuel Roxas's brother), and of additional black soldiers armed even with a cannon. The soldiers' presence and the priest's violent words and actions disrupted the religious solemnity of the Mass. Carvajal took the officers' version of events seriously enough to issue an order prohibiting the entrance of Indian women into any Spaniard's residence and reiterating the customary authority vested in the *corregidores* of each pueblo and the *jueces* that headed each *parcialidad* over the distribution of trade goods and the collection of their products in cloth and wax.[55]

Manuel Roxas gave a different version of the uprising in an angry letter he directed to Felix Ydalgo. Roxas blamed the conflict on Carvajal and his unwise innovation of bringing lay lieutenants to the province, but ignored the accusations of moral impropriety leveled at his colleague, the priest of San Ignacio identified only as Don Simón. Peppered with curses, Roxas's epistle provides names of the dead and wounded and references to the movement of people and rumors from one pueblo to another, showing his intimate knowledge of the province. Born in San Miguel, the son of an artisan who had worked and lived in the Chiquitanía, Roxas grew up speaking both Chiquitano and Spanish.[56] He had served as interpreter for previous governors and was distinguished among other clerics for his ability to preach to the Indians in their language. From his letter, as well as from Barbosa's testimony, we learn that Carvajal had further angered the Indians of San José by taking away the *alférez*'s cane of office because he suspected that the *alférez* had threatened to kill him and the lieutenants. According to Barbosa, however, the *alférez* had been stripped of his title because he and the Indians who had been sent to open a road had not completed the job—but they had slaughtered, butchered, and eaten eight yearlings. The *alférez* was an important member of the *cabildo* who held the banners during processions, kept public order, and supervised the work of Indian men in the *estancias*. Spanish officials recommended exemplary punishment for the leaders of the rebellion, but expressed fear of "losing these neophytes" who might flee the missions and join the "barbarians" in the forest.

The San Ignacio uprising of 1790 combined different layers of meaning with particular significance for the prerogatives that *cabildo* members considered as their entitlement. It occurred at the beginning of the dry season and on the eve of one of the most solemn feast days in the Catholic calendar, Corpus Christi. The drought was particularly severe that year, and it was reported that mission cattle were dying. The *cabildo* officers complained to Carvajal that the herds, an important part of their communal patrimony, had decreased alarmingly since Simón had taken charge of the mission. Faced with these losses and frightened by rumors (perhaps unfounded) that lieutenants would be placed in the pueblos to make them work under the bite of the lash, the officers of San Ignacio seized the goods from the storehouse and launched an attack on Spaniards whom they considered to be their enemies. The priests' lives were spared, despite the Indians' anger at Simón's conduct. The uprising began in two *parcialidades*, but it involved the entire pueblo, and rumors of rebellion spread to neighboring missions, even if fearful witnesses may have exaggerated the number of armed rebels. Lay civilians and soldiers who lived in the pueblos or passed through on extended visits comprised a small but visible minority that often assumed positions of authority and thus posed a threat to the political integrity of the native community.

Conclusions

The political cultures of Sonora and Chiquitos, when compared, suggest a number of ways that colonial polities restructured native communities at the same time that they limited and shaped the imperial project. As the rebellions of Concepción and San Ignacio illustrate, the polities of Chiquitos were more closely tied to the formal structures of the mission *cabildos* than in Sonora. In northwestern Mexico, *cabildo* officers undeniably played an important role in the internal governance of the missions and in cementing the entire process of *reducción* on which the colony depended. Nevertheless, the maneuvers, negotiations, and armed confrontations that characterized relations between Spanish officialdom and nomadic *rancherías* of Seris, Sibubapas, and Apaches tested the limits of colonial power outside the missions and beyond the formal institution of the *cabildo*.

Parcialidades played a more distinctive role in the Chiquitanos' identity and political allegiances than among the Sonoran peoples. Ethnic segmentation, so deeply ingrained in the spatial organization of the Chiquitos pueblos and in the composition of their *cabildos*, may explain the isolated tumults and the absence of province-wide rebellions in eastern Bolivia. Conversely, the Indian communities of Sonora were far more integrated spatially and socially with Hispanic society than in Chiquitos. The proximity and interdependence of the missions and Spanish settlements conditioned the internal governance of the pueblos in Sonora, Ostimuri, and Sinaloa; moreover, frontier defense relied heavily on native militias and presidial auxiliaries. This created a parallel system of ranking for indigenous warriors, especially among the Pimas and Opatas, that both supported and rivaled the *cabildo* officers. Furthermore, it gave them leverage with colonial authorities for defending communal patrimony and negotiating privileges of dress, remuneration, and freedom of movement.

We may summarize the differences between these two colonial frontiers by saying that the *común* symbolized the ethnic polity of Sonora, while the *colegio* represented the political and ceremonial center of the Chiquitos pueblos. Bringing their historical narratives together highlights the complexities and tensions of colonial governance, reflected not only in words and texts but also in the built landscapes of northwestern Mexico and eastern Bolivia. As described above, the architecture and imagery of Catholicism provided the foundational design for the mission compound, designating spaces for worship, discipline, labor, and the privileged status of musicians and councilmen. Colonial agricultural, ranching, and mining enterprises—including the missions themselves—altered the topography, stream flow, and vegetation of Sonora and Chiquitos, albeit in different ways. These changes in the natural environment set territorial limits for the exercise of power and the avenues to resistance.

Even more dramatically, the patterns of guerrilla warfare that intersected the political culture of Sonora tested the boundaries of Spanish domination and forced a redefinition of the terms of negotiation for both colonizers and colonized. Yet no less significant, the *rancherías* of Seris and Apaches that challenged Spanish dictates to live in a polity borrowed the wealth (in commodities and livestock) and the symbols of religious and political status from the colonial order. On the Chiquitos-Chaco borderlands, nomadic bands of Guaycurús appeared frequently as feared raiders, instigating armed conflict with mission dwellers. The 1763 skirmishes between Chiquitos and Guaycurús led to the diplomatic mission of Joseph Sánchez Labrador, through different ethnic and linguistic intermediaries, that was recounted at the opening of this chapter. Violence on this frontier was not reported in the same way as for Sonora, perhaps because the demarcation between nomads and villagers was less clearly drawn and Spanish imperial policies relied less on military garrisons in the greater Paraguayan borderlands than in northern New Spain. Even as the raiding and foraging economies of nomadic bands and their migratory patterns established shifting territorial domains, the roles of villagers and nomads alike proved decisive in shaping the imperial boundaries of both frontiers.

It is clear that northwestern Mexico and eastern Bolivia constituted unfinished imperial projects, open frontiers that remained contested ground.[57] The rich historical and ethnographic literatures on the European-Native American frontiers of recent years have shown not only the massive, and often destructive, impacts of conquest and colonization on the peoples of the Americas but also the complex political realignments of the survivors, their migrations and spatial rearrangements and, in some cases, their territorial aggrandizements. Europeans were not the only empire builders in the Americas after 1492, and it is plausible to argue that the Comanches of the North American Great Plains, the tribal peoples of the South American pampas, and the Iroquois federation of the Great Lakes and eastern woodlands established political and territorial hegemonies that endured for significant periods of time.[58] Sonora and Chiquitos, however, were not theaters for the consolidation of indigenous chiefdoms that controlled extended territories. Although certain ethnic groups, notably the Opatas and Cahitan peoples of Sonora and Ostimuri and the Chiquitos *parcialidades*, recovered demographically and even prospered economically through the mission system, none of them established polities above the village level.

In both provinces, the public spaces for political and military indigenous leadership were gendered masculine. Women figured as the objects of dispute (as we saw in the 1779 tumult of Concepción), and they appeared as both victims and intermediaries in the narratives of frontier warfare and captivity, but they did not hold office in the formal institutions of government. Native perceptions and practices of power transcended the political hierarchies established under the aegis of the colonial regime, however, to comprehend realms of spiritual power that persisted beyond the physical and cultural boundaries of the mission compounds. In these diffuse spaces between culture and nature, life and death, the human and the divine, gendered and ethnic definitions of propriety and subservience did not disappear, but assumed new dimensions. The principles of parallel male and female roles underscored, in different ways, the cosmic orders, symbolic meanings, and ritual practices of Sonoran and Chiquitano peoples. These realms of spiritual knowledge, intertwining shamanic and Catholic religious orders, make the subject of the following chapter.

CHAPTER 6

Priests and Shamans: Spiritual Power, Ritual, and Knowledge

Anauscia Santissimo Sacramento
Naqui anè ycuu Altar
Inta yto Virgen Santa María
Niñemooco oximananene quichezeña
Onumo aybo yi yy nicocinitanna ninahity tacañe
[Blessed be the Holiest Sacrament of the Altar
and the Virgin Holy Mary, who from her
origin is free and pure of all sin]
—Chiquitano prayer

Cuida ma amo niguachi i María Santissima
se amo jidaguade: Sta. María DiosacDe
téma no vepin veu, gua no muc tachi,
gua ne nanadote sé no opaguachi, gua tebu
no jidagua. Ah, Santa María no Dé
[Tell Holy Mary in your language with all your heart:
Holy Mary Mother of God,
pray for me now and in the hour of my death,
and defend me from all my enemies,
and care for and guard my soul.
Ah, Holy Mary my Mother]
—Opata prayer

These verses of adoration and supplication centered on Mary employ the Christian language of prayer, opening an avenue to the divine that was reinforced by music, dance, and the liturgical imagery that adorned the mission churches. Yet Christian practices of prayer and worship constituted only one path of many for those who sought life-giving power in nature and the cosmos. This chapter seeks out spiritual landscapes through the sensibilities of language and gender in the production of culture and in parallel definitions of the sacred in Sonora and Chiquitos. Building on earlier discussions of ethnicity, polity, and identity in chapters 4 and 5, it underscores the overlapping spheres of political and religious authority in colonial indigenous cultures and interprets native adaptations of Catholicism and shamanic spiritual powers—at times in consonance with Christian imagery, at times in opposition to religious discipline—in the missions. Church architecture, processions, and musical celebrations, as well as rituals enacted in forest clearings, caves, and natural springs, created spaces for communion between the human and the divine that, in turn, resonated with the political voices heard on the native councils and in petitions presented before Spanish authorities. These spiritual landscapes were closely related to the physical environments, the corporate indigenous communities, and the translated and transposed religious cultures that developed in both regions through the missions.

The prayerful invocations reproduced above appear to transport phrases lifted from the Catholic rosary and the cult of Mary into two of the indigenous languages that dominated the spoken and written word in the missions of Chiquitos and Sonora. Indeed, Catholic imagery and vocabulary permeate the public expressions of religiosity that comprise foundational pillars for the cultural histories of Sonoran and Chiquitano peoples from the colonial period to recent times. Yet these verses reveal complex fields of symbolic iconography and mediated communication with spiritual powers that transform and enrich both Christian and native religious traditions. They lead us to ask how religious practices impart meaning to the lived experiences and historical memories of Sonoran and Chiquitano peoples. How do rituals create spiritual landscapes? In what ways do sacred places and ceremonial performances condition the negotiation of power across overlapping webs of religious, social, and political relations undergirding colonial hierarchies?

Power is a central concept for the themes developed in this chapter, referring to natural forces, the exercise of agency, and the capacity to mold cultural explanations of nature and society. The use of terms like *negotiation* and *mediation*—which imply transaction, translation, and compromise—underscores both the liberating connotations of power and its fearsome, destructive qualities as these are encountered in the natural world and in the political realms of human conflict. Religion accomplishes the important cultural work of communicating with the life-giving and life-taking powers of spiritual beings through worship, sacrifice, and performance. These acts of intercession are carried out individually, by priests and shamans, and collectively through customary rites and ceremonies.

The varied religious landscapes mapped here imply different approaches to time and space. For this reason our route through different expressions of spiritual power will take us to historical sources, as well as to contemporary ethnographic descriptions of religious practices and beliefs. The narrative builds a spiritual geography around the core themes of priests and shamans; sacred architecture, wilderness, and cosmology; language, musical drama, and translated meanings. Each section indicates different moments of the past and present that represent these enactments of religiosity and the sources on which they are based.

Priests and Shamans

Shamans become the media for spiritual powers that can bring healing, prophetic knowledge, endurance and triumph in the face of hardship, or, conversely, illness, defeat, and death. Widespread practices of shamanism throughout North and South America testify to their antiquity, leading some scholars to suggest that shamanic relics and rituals arrived from the Eurasian continent to the Americas with the earliest human migrations.[1] Shamans operate within ecclesiastical institutions or through other communally recognized religious norms. Although shamanic roles vary among different native peoples, in general, shamans master important skills and esoteric knowledge that empower them to cure illnesses, interpret dreams, prophesy future events, and transform themselves into zoomorphic and spiritual beings. Shamans' role as intermediaries between life and death and their ability to transcend the boundaries of time and space, nature and culture, are expressed in the language of flight, physically leaving enclosed spaces and metaphorically moving beyond the material conditions of human existence. Whether or not shamans exercise political authority, they serve as guardians of cultural traditions and protectors against foreign invasion or alien spirits.[2]

For the native peoples of Sonora and Chiquitos, the frontier between nature and culture was fraught with danger and ambiguity. An important attribute of the shamans was the capacity to transport themselves spiritually into the natural domain of forests and deserts and then return to their communities, to the realm of human culture. The wisdom shamans gained from communing with natural powers, requiring them to cross boundaries of time and space, enabled them to sustain a fragile balance between opposing forces associated, in turn, with the seasons of want and plenty, rain and drought. Shamanic knowledge established cultural norms for regulating hunting, fishing, gathering, sowing, and harvesting. Stories gathered around the restorative roles of shamans who, in turn, created narratives that explained past events emanating from European colonization, Catholic missionization, and the production of wealth for commercial exchange.

Priests and shamans held parallel but distinct roles under the colonial order. The Judeo-Christian meaning of *priesthood* constituted the celebration of sacramental rites that renewed moral covenants between humans and their gods.[3] Roman Catholic missionaries working in Sonora and Chiquitos interpreted what they observed of indigenous religious practices through the veil of their own priestly vows, thus differentiating among priests (*sacerdotes*), shamans (*hechiceros*), and healers (*curanderos*). These terms comprehended different kinds of religious specialists, distinguished by the spiritual powers they claimed to possess. Their appearances in the historical record as mediators, conservators, or innovators of religious traditions were, at times, contradictory, but they often proved pivotal in the tensions that marked the colonial divide. In repeated instances, missionaries referred to shamans in a derogatory manner as "sorcerers," a reference that expressed only weakly the priests' defense against the challenge that shamans mounted to them as healers and guardians of local custom.[4]

Chiquitos

Shamanic powers of interpretation and intervention are central to the present-day cosmology of the nomadic Ayoreóde, based on the principles of balancing opposing forces and maintaining communicative bridges between nature and culture. Their mythologies are peopled with animals possessing moral and spiritual qualities that account for seasonal changes, natural disasters, and human illnesses. These mythical creatures explain historical differences among peoples of indigenous, European, and African descent whose presence is acknowledged in the temporal and spatial dimensions of the Ayoreo world. Significant personages in their cosmos include the sun (*guedé*), who redistributes power among the strong and the weak; the jaguar (*putugútoi*), who symbolizes the power of nature; a small bird (*asojna*) once endowed with shamanic powers, who marks the changing of the seasons; a large white swan (*chunguperedatei*) associated with the outsiders (*cojñone*) and the goods they brought to Ayoreo culture, especially iron and woven cloth. The sun approximates the notion of divinity, associated with the Guaraní concept of *Tumpa*, which the Jesuits introduced to the Ayoreóde in the missions of Chiquitos.[5]

The Ayoreóde divide known time in two parts: the ages of their ancestors and of living persons. The tension between nature and culture imbues both of these epochs, moving toward their separation and the emergence of human societies bound by a set cosmic order. Furthermore, the separate powers ascribed to different animals and mythic beings explain the social organization of the Ayoreóde in clans and provide a rationale for taboos and moral prohibitions that uphold a balance between counterposing forces. For example, evoking the cross (*curezei*), an enduring image of Christianity, can abate a dangerous flood, but spoken at the wrong time, it can bring on a period of drought. Spirituality emerges from the shamans' transgressive powers to move between nature and culture, wakefulness and sleep, life and death. Spiritual forces that shape character, maintain physical and mental states of well-being, and link individuals to their family lineages are often related to particular animals and plants, as well as mountains, rivers, and forests, in ways that connect human destinies with nature. The interpretations of Ayoreo culture presented here do not comprehend a uniform system of beliefs upheld by all members of the clans and settlements identified as Ayoreóde. Individual variations and inconsistencies in the meanings ascribed to religious knowledge render cosmologies a living, historical quality that develops over time in tandem with social relations in Ayoreo communities and in contact with the mestizo world (*cojñone*).[6]

The direct experience of the Ayoreóde in the colonial Jesuit missions of Chiquitos province was relatively brief, but its imprint on their worldview and historical memory provides a significant linkage with the past and a source of creative innovation in their culture. Jesuit missionary Juan Bautista de Zea first succeeded in visiting the Zamuco-speaking Ayoreo *rancherías* in 1717. Encouraged by the apparent approval of two of their caciques and the Indians' welcoming signs, Zea began his efforts to bring them into the Christian fold by raising a banner of the cross in a forest clearing and venerating a small image of the Virgin Mary. Accompanied by newly converted neophytes from the missions in San José and San Juan, Zea led the Zamucos in kneeling before the cross and intoning litanies to the Virgin Mary. Neither Zea nor the missionaries who followed him understood how the hunter-gatherers of the arid northern Chaco translated these powerful symbols of Catholicism, although the reference above to the protective invocation of *curezei* provides a clue to its symbolic adaptation.

Juan Bautista de Zea's peaceful contact and the *entrada* undertaken the following year by Jesuit Miguel de Yegros to bring the Ayoreóde out of their scattered *rancherías* and relocate them in the pueblo of San Ignacio together with Morotocos and Curacates foundered, in part, because of long-standing conflicts among these different groups and the fear that the Ayoreóde held of the Chiquitos peoples.[7] Warfare and the act of killing had both social and cosmic significance for the Ayoreóde. Warriors who slew another person or a jaguar acquired their powers; violent encounters were frequently woven into Ayoreo mythology and, historically, warfare marked relations among separate *rancherías*, especially with people they considered different or inferior to themselves—sedentary villagers, Europeans, and Africans.

The Ayoreo spiritual concepts of *ayipié* and *orégate*, only partially translated by the Christian notion of *soul* and collected through current ethnographic research, seem to be linked to the historical cosmologies of the Manasicas and other Chiquitano peoples.[8] The Jesuits who first made contact in the early eighteenth century with the *rancherías* that extended northward from the mission of San Xavier—known by different names and dialects, but identified collectively as the Manasicas—described a fairly elaborate religious cult of deities who appeared and spoke in public gatherings through the shamans. Manasica gods and their priests occupied hierarchical positions of greater or lesser authority in accord with the power they commanded. The most important gods had specific names and genders, both masculine and feminine; a lesser order of divine spirits, the *isituús*, guarded the rivers and lagoons and kept them stocked with fish. Their good will was assured with offerings of fish and the incenselike smoke of burning tobacco. Shamans vied for prominence by demonstrating their ability to communicate with the gods: highest among them was the *Mapono*, who flew with the gods in their earthly visitations and transmitted their prophecies to the people. The *Mapono* transported the souls of the dead (*Oquipaú*) to a kind of heavenly paradise, traversing figurative landscapes of mountains, dense forests, swiftly flowing streams, lagoons, and swamps that mirrored and augmented the physical environment of Manasica villages.[9]

Jesuit representations of Manasica rituals and cosmic beliefs suggest reciprocal processes of transculturation in the spiritual encounters of native and Catholic religious practices. Native peoples of the Chiquitanía were not formally settled in mission pueblos until the turn of the eighteenth century, yet many of them gained familiarity with Christian symbols through Jesuit ministries to the Indians taken into *encomienda* service in the Spanish villas of Santa Cruz de la Sierra and San Lorenzo de la Barranca, begun in 1587, or in the Mercedarian missions of Buenavista and Porongo established west of Santa Cruz.[10] They may have begun to recognize new deities that, in some ways, resembled Catholic images of the Trinity and the saints. Jesuit writers, for their part, interpreted what they observed and the stories of native religious practices passed on from one missionary to another in terms of their own theology. The priests who labored in the Chiquitano *reducciones* brought with them intellectual traditions steeped in medieval scholasticism and the Counterreformation. Their faith in the efficacy of Christian conversion was strengthened by their conviction that Saint Thomas the Apostle had visited native peoples of the Americas some time in the ancient past, bringing them word of the gospel of salvation.[11] Conversely, they believed as firmly in the devil as they did in God, personifying as demons the forces of evil that obstructed their evangelizing mission, whether in the severity of the climate and terrain or in the hostility of native shamans. Moreover, the Jesuits sought martyrdom, portraying the physical dangers to which they were exposed and the murderous attacks that at times ended their lives as a blessed pathway to sainthood.[12]

Through the confluence of their own intellectual and religious traditions with native rituals, only dimly understood, Jesuit narratives arranged indigenous deities in gendered hierarchies that resembled Christian attributes. They portrayed the shamans—*maponos* and their assistants and apprentices—in a ranked order that paralleled the civil authorities, whom they recognized as caciques. Father Lucas Caballero's account rendered the four principal gods revered by the Manasicas as a kind of holy family: father, mother, son, and spirit. When the *Mapono* returned from his exhausting flights into the spirit world, he descended to the waiting *ranchería* in the arms of *Quipoci*, the maternal deity. Her melodious songs were echoed by the women, who danced and praised her victories in battle. *Quipoci*, like the Christian Mary, interceded on behalf of humans before the fury of her male counterparts, "who afflicted them with illnesses, losses, and defeats."[13]

Mapono, the most knowledgeable and powerful of the shamans, was distinguished from the cacique, although both leaders received respect and food offerings of game and harvested crops. Secular and spiritual authorities merged during religious celebrations, when they shared the same space. Manasica *rancherías* did not have permanent structures set aside as temples; instead, the cacique set aside space in his home, curtained off with palm mats woven for each religious festival. Opening ceremonies sanctified the room for the gods' appearance: an elderly woman, considered devout and pious, approached the cacique, bowed to him, and lightly injured him with a sharp stone. She moved through the room on her knees, with audible sighs of devotion, followed by the *Mapono* who blessed the "tabernacle"; the festival then commenced with *chicha*, abundant food, music, and dance.[14]

Manasica shamans assured the continuation of their spiritual powers through the apprenticeship of young men, initiating their training before puberty ("before they grow a beard"). Juan Patrício Fernández reported that the shamans and their young assistants observed rigorous fasts, obeying the taboos against eating certain animals and fruits. Furthermore, before important religious festivals, the entire *ranchería* was obliged to fast, abstaining from eating meat, keeping silence, and employing all their time in weaving the mats used to dedicate a new temple. Perhaps in keeping with this practice, one of the Manasica communities, the Quiriquicas, on their peaceful reception of Lucas Caballero and acceptance of baptism, offered him their children as apprentices.[15]

Missionaries carried their labor of evangelization to the *rancherías* of the Manasicas and other Chiquitano peoples, seeking the support of newly converted neophytes. Jesuit priests like Zea, Caballero, and Fernández allied themselves with trusted caciques, relying on them to open trails in the dense undergrowth, guide them to new villages, and serve as a counterweight to the *maponos*. Missionary narratives recounted their progress and setbacks in the language of miracles, perceiving the presence of both God and the devil in the Indians' responses and in the hardships they endured. The Quiriquicas had initially opposed Caballero's entry into their *rancherías* with bows and arrows poised to shoot. As Fernández told the story, when the saintly missionary unfurled the standard of the Virgin Mary, the warriors found that their limbs were useless to release the arrows they had prepared against Caballero and the Christian Indians who had accompanied him.[16]

Violence followed the advances of the Christian banner, despite the missionaries' stories of a miraculously peaceful reception. Indian allies who accompanied them risked capture and death since many of the *rancherías* through which the missionaries passed were at war with one another. Frequent epidemics were recorded during these early *entradas*, decimating entire villages, which the Jesuits attributed to divine retribution against the shamans who had tried to harm them. Missionaries overturned native idols with force, ordering their destruction by fire and the razing of their temples, as occurred when Caballero followed up his advantage in the village of the Quiriquicas after their warriors had fled into the forest.[17] Missionaries, shamans, and caciques were engaged in a power struggle in parallel planes of symbolic intercession and political authority. The cross and the images of Mary and Saint Michael the archangel carried that struggle into numerous *rancherías*, where their symbolic force did not so much displace as reconfigure indigenous practices in song, dance, and feasting that appeased the fearful powers of the cosmos and reaffirmed the social bonds of community.

Sonora

Among the many indigenous groups known historically in northwestern Mexico, similar principles of shamanic intercession sustained different sets of beliefs, taboos, and ritual practices.[18] As discussed above for Chiquitos, religious traditions became reproduced within the fabric of social relations in heterogeneous ways, subject to change and innovation. Cosmologies grounded in received wisdom developed over time through the repeated enactment of rituals and storytelling. Our knowledge of these cultural processes is rich in ethnographic interpretations, but it is also fragmented—found partially in written texts, archeological remains, and oral traditions. The abundant literature for the O'odham, Cahitan, and Rarámuri peoples of the Sonoran Desert, upland river valleys, and cordilleras of the Sierra Madre Occidental illustrates the transcendent spiritual powers ascribed to shamans and particular animals and places that touch human souls both to heal and to harm; their religious content is laden with cosmic and historical meanings.[19]

Desert-dwelling peoples practiced religious rituals associated with hunting bighorn sheep, pronghorn antelope, and other large ungulate game from antiquity to recent historical times. Cremating the bones of felled animals and stockpiling rams' horns have left enduring shrines in the desert pavement of the Pinacate and in the archaeological excavations of the Hohokam villages of the Gila River valley, the site of late prehistoric extensive agricultural development with canal irrigation and ceremonial centers.[20] Hunter-gatherers sought their prey near *tinajas*, natural reservoirs or hillside springs, where animals and humans found precious water in their movements across the desert. Historic references to mounds of calcined (burned) bones and stacked horns come from the late seventeenth-century report of Juan Mateo Manje, who traveled through the Pimería Alta with the Jesuit Francisco Eusebio Kino, and from the journals of Juan Bautista de Anza, during his exploratory trek overland from Sonora to California (1774). Their explanations for these shrines resonate with O'odham oral traditions concerning hunting rituals that quieted the spirits of dead animals to prevent the occurrence of destructive winds and to bring rain. The sacred remains of animals—cremated bones, horns, and hides—were stored away from the villages, in protected places, often in rock crevices near the *tinajas*. Hunters, themselves, disposed of these remains, but the O'odham tell stories of religious specialists (shamans) who poured water into the sheep horns and performed rituals with singing and dancing to break a prolonged drought.[21]

Wind, rain, and animal spirits figured prominently in shamanic curing practices known historically through *o'odham himdag*, the people's way. O'odham theories of health and sickness comprehended two parallel traditions of sacramental rituals that at times shared symbolic imagery but remained distinct in their performances of public ceremonialism and private healing rites. *Sa:nto himdag*, the saint way, illustrated the enduring presence of Catholic missionization among the O'odham, centered in chapels, roadside crosses, and home altars. The saints were celebrated publicly and in family gatherings on their feast days, especially All Souls Day (approximating the Mexican Day of the Dead) and in the annual pilgrimage to Magdalena, Sonora, for the Feast of Saint Francis. *Jiawul himdag*, the devil way, fused Christian notions of good and evil with O'odham concepts of spiritual forces pictured in the cosmography of the Sonoran Desert. The devil way asserted its power in the context of particular forms of staying sickness caused by irreverence to cattle and horses, requiring the intervention of shamans for their correct diagnosis and cure. David I. López accounted for the strength of this tradition in the following words.

> Like it was at that time, the way that these traditions were done:
> devil way, spirit way, and other dangerous things, different ways,
> and that is where it started coming from, the elders were just listening
> about ourselves, where we came from, us Papago [O'odham],
> and from there it's our tradition, and it's our sickness, and it's our spirit
> respect.[22]

Devils, the spirits of cowboys who reside in hollow mountains, select the persons with whom they communicate either to inflict illness or to convey power and knowledge through dreams. Devil songs and their curative powers transcend the temporal boundaries of past and present and the cosmic boundaries that separate human communities in the finite "staying earth" (*ka:cim jewed*) from the spiritual world imagined as breath (*i:bui*), wind, and the powerful movements of horses, cattle, and cowboys.

> Great night horse, great night horse
> It came to me and blew inside of me, all over it massaged me.
> My great ash-colored horse, my great ash-colored horse
> From far away running, from far away song runs
> And disappears in the ground.[23]

These verses, like many of the devil songs, express a sensuous quality in the movement and beauty of the "great ash-colored horse" and in the carnal sensations of breath and touch, even as water moistens and runs into the ground. The devil way, or *himdag* more generally, implies a mode of knowledge and an aesthetic experience, both exhibited in the devil songs. Devil spirits impart their powers to shamans through sacramental acts of blowing and massaging, gestures repeated during diagnostic and curing sessions together with the recitation of devil songs, making the sign of the cross, and sprinkling holy water on the patient.[24]

Both the saint way and the devil way developed as transculturative processes, through the colonial missions, wage labor and peonage in mines and ranches, and the cultural importance of livestock as both property and the object of ceremonial intercession between humans and nature spirits. The two separate but interrelated paths of *o'odham himdag*—employing the imagery of Catholic saints, livestock and all their accoutrements (lariats, ropes, saddles, and bridles), and the masculine figure of the cowboy—provide a historical narrative for explaining long-term societal and economic changes. *Himdag* generates creative song-poetry for curative intervention and restorative rituals in the lives of individuals and communities. Its significance for the O'odham is similar to the role of shamans and the symbolic meanings noted above for the animal and bird spirits that inhabit the cosmos of the Ayoreóde and other Chiquitano peoples of lowland Bolivia.

Stories of migrations and kinship connections across ethnic and binational boundaries, forged largely during the twentieth century, compress the temporal and spatial histories of religious syncretism, selective adaptation, and tradition making that nurtured both shamanic and Catholic rituals. Anthropologists Donald Bahr and David Kozak assert that the Tohono O'odham "self-Christianized" during different phases of village consolidation and dispersal from 1825 to 1875, a period marked by prolonged drought, warfare with the Apaches, and the U.S. invasion of Mexico, ending in the establishment of a definitive boundary between the two countries. Each of these events, in turn, influenced O'odham foraging and pastoral practices, as they alternatively hunted and herded livestock. By the end of the nineteenth century, extensive commercial cattle operations and railway communications between Sonora and Arizona irreversibly altered the ecological and cultural bases of subsistence for the riverine and desert-dwelling O'odham.[25] Viewed historically, devil lore bears witness to the dramatic impact of these changes and the invention of traditions to assimilate them, drawing on a complex repertoire of cultural signs and meanings. The texts, images, and rituals that compose those traditions emerged (and were, no doubt, rearranged) during three centuries of Catholic and indigenous cultural production stemming from Hispanic conquest and evangelization. In the early twentieth century, Anglo-European Franciscans undertook another phase of missionization among the O'odham living in the Papago Indian Reservation of Arizona. Their endeavors at reinstating the formalities of the Catholic Mass and building village chapels followed, and necessarily adapted to, the creative traditions of *sa:nto himdag*. The O'odham saint way and devil way allowed for a plurality of practices and beliefs through the repeated enactment of rituals, processions, and feast days that varied over time and space.[26]

Cahitan Spiritual Power

Missionaries and shamans confronted one another from the initial Jesuit advances into northwestern Mexico of the early seventeenth century. In their evangelizing zeal, the Jesuits emulated the first apostles as sacramental priests by baptizing, preaching the gospel of salvation, and following the biblical injunction to cast out demons and cure the possessed. As did Juan Patricio Fernández in the *Relación historial* for the province of Chiquitos, Andrés Pérez de Ribas, in his monumental *Triunfos de nuestra sante fé*, portrayed his order's mission in terms of a battle between good and evil, in which the enemy consisted of the devil and his earthly minions, the shamans. Pérez de Ribas's narrative of the Cahitan missions, in this preliminary stage of resettlement and indoctrination, is richly punctuated with episodes of the devil's treachery and the ultimate triumph of Christian symbols against the fearful power of Satan.

Pérez de Ribas denounced shamans as sorcerers, false healers, and deceitful preachers, even as he minimized the dangers of idolatry among the newly converted Cahitan Christians. He dismissed the shamans' knowledge as a pact with the devil, attributing to these "demonic healers" the transgressive faculty of moving between the realms of nature and human sensibility and accusing them of instigating rebellions that threatened the newly established missions.

> The devil frequently spoke to the Indians when they were gentiles, appearing to them in the form of animals, fish, or serpents. . . . The Indians greatly respected and feared him whenever he appeared, and as a title of respect they called him "Grandfather." They did so without distinguishing whether he was creature or creator. . . . When the devil sees that the light of the Gospel and the catechism are undoing the healers' deceitful tricks, diminishing their authority and the people's interest in them as healers, and putting an end to the Indians' vices, then the enemy of humankind marshals all his efforts through the sorcerers to persuade the pueblos to rebel, to burn the churches, and to return to the montes to live as they please.[27]

In this passage, *gentiles* referred to non-Christian Indians, who had not been baptized. *Montes* indicated uncultivated scrub forest and, in the Jesuits' cosmology, a dangerous place, the realm of the devil.[28] Missionaries countered these forces of evil with their rites to save souls, heal illnesses, and bring rains. During a severe drought that threatened the autumnal harvest in the province of Sinaloa, Martín Pérez urged his neophytes to pray, whip themselves, and process with the image of the Virgin Mary. According to Pérez de Ribas, they followed his advice and, after three days, the skies filled with clouds and delivered rain to the parched fields. This story was followed by the confession of a woman who admitted that she had used her powers of sorcery to disperse the clouds and prolong the drought. Christian Indians brought her to the church where she was publicly questioned and punished.[29] The Jesuits' account of these events suggests the ways that missionaries inserted themselves into village ritual practices of social control during this early phase of *reducción*.

Pérez de Ribas was equally zealous in extolling cases of repentant shamans. In recounting the successful mission undertaken by Pedro de Velasco to the mountainous peoples east of the Villa of San Felipe (Sinaloa) in 1607, he underscored "a very singular case" that illustrated "the virtue of the holy Cross." A baptized shaman, who had vowed to renounce his former pacts with the devil, found himself besieged by spirits who nearly succeeded in persuading him "to return to his diabolical arts." At Father Velasco's urging, the shaman repeated his renunciation of Satan and placed crosses in his home—a talisman that drove the devil away.[30] Turning to a more complicated incident, Pérez de Ribas brought together themes of human illness, demonic visitations, and idolatry in his report on the Guasave mission along the coastal delta of the Sinaloa River, under the care of Alberto Clerici.

On the eve of the feast day for Saint Ignatius (31 July), Father Clerici was advised by a native catechist that the devil was preaching to a baptized woman, who lay sick in her own home. The catechist was convinced that the culprit was not a native sorcerer, but a devil, because he spoke with superhuman skill and oratory. This male intruder treated the woman violently, striking her and threatening her with death if she did not recognize him as her lord and follow him "into the *monte,* to her former dwelling places, where each person lived as he wished, instead of here [in the mission], where the priest deceived them with his inventions." Father Clerici visited the woman several times, eliciting her confession with the aid of a crucifix, holy water, and a medal of Saint Ignatius; he exorcized the devil from her room, "having donned a surplice and bearing a lit candle in one hand and the Book of Exorcisms in the other."

Clerici's repeated questioning of the woman brought to light a history of idol worship, in which her father had been the "sacristan" of a stone image that represented a hooded figure "covered in rich plumage" who had led her people in war and, following her father's death, had commanded her to accept him as her protector. Pérez de Ribas framed this story in spatial and temporal terms that distanced the woman's pre-Christian past from her present condition of baptismal confession. The hooded figure had visited her "forty years earlier," when she lived among her ancestors in the *monte*, some twenty-six leagues (over one hundred kilometers, or sixty miles) away from the Guasave *reducción*. Her exorcism was complete when the idol was found and burned, and the ashes thrown into the river. "God was glorified, and from then on, this woman was free."[31]

When we compare Jesuit accounts of their moral battles with the forces of evil in the evangelizing mission with the *o'odham himdag* of more recent times, we note layers of ambiguity in the imagery of devilry as it is revealed in texts and images of different historical moments. The O'odham devil way has deep roots in medieval Christianity, as well as in the cosmographies of the Sonoran Desert, traditions reworked and entwined in the course of missionization. To be sure, the devils who inhabit hollowed mountains and move in a dream world between living communities and the ghostly afterlife of death are somewhat more playful and less terrifying than the devils who threatened the Jesuits' labor of salvation in the colonial missions of Sinaloa and Sonora. Nevertheless, O'odham cowboy devils exercise dangerous powers: they select their human victims and apprentices, they cause debilitating sicknesses, and they alone can teach shamans their curative songs. The O'odham saint way and the devil way both draw on Christian rituals, using the imagery of crosses, holy water, and patron saints to ward off disease and misfortune.

Gender adds a further note of ambiguity to the layered meanings of shamanism in the spiritual worlds of northwestern Mexico. The episodes I have summarized here from Pérez de Ribas's numerous accounts of priestly encounters with devils and shamans portray women, at different times, as both agents and victims of destructive powers. In one instance, a woman is accused of harming a village (representing numerous *rancherías*) because, in her anger, she had dispersed the clouds and prolonged a drought that threatened the livelihood of Indians and Spaniards alike in early colonial Sinaloa. In the second instance, a woman fell victim to demonic visitations with violent overtones of patriarchy and domination. The devil who tormented her and (by inference) caused her sickness was gendered male; he claimed his right of place in her psyche as a father figure even as he struck her and demanded that she follow him or face death.

The O'odham devil way, centered in the imagery of livestock herding, is predominantly—but not exclusively—masculine. It is men who dream and learn the songs to cure staying sicknesses, but devil mountains comprise two chambers for "cowboys" and "cowgirls," the afterlife abode for the deceased ancestors of living O'odham villages.[32] Three-quarters of a century ago, Chona, a Tohono O'odham woman who told her life story to Ruth Underhill, spoke in different ways about rituals, curing, and power. Chona confirmed that women had restricted roles in the games and rainmaking ceremonies so important to her people's culture and that, furthermore, they were "not supposed" to dream or have shamanic knowledge. She, however, learned songs from several medicine men in her family and, in different episodes, had dreams and possessed crystals and feathers that gave her curative powers. Before she reached puberty, at her father's bidding, a medicine man sucked the crystals from her breast and put them safely inside a saguaro. Even as the shaman removed her power from Chona, he told her that the crystals were a gift: she could have been a medicine woman.[33]

The elusive gifts of the shaman, with powers that are both admired and feared among the indigenous peoples of Sonora and Chiquitos, may imply further dimensions of sexual and gendered ambiguity. Through their transgressive journeys into hidden spiritual realms and their authority over apprentices, it may be inferred that shamans performed homoerotic acts as part of their privileges and rites, occupying a third sexual space that had an accepted place in native cultures. There is no direct evidence for this, however, in the ethnographic and historical accounts that support the discussion of priests and shamans developed here.[34]

Spiritual Landscapes and Sacred Architecture

The layered meanings of gender, place, and time that emerge in the foregoing histories of Christian and shamanic observances have created spiritual landscapes inseparable from the geographies of northwestern Mexico and eastern Bolivia. Sonoran and Chiquitano peoples anchored their cosmologies in specific reference points of their natural environments: deserts, mountains, springs, forests, lagoons, and streams. Holiness resided in wild places, beyond settled villages and cultivated fields, in mountain ranges, caves, and waterways where humans encountered realms of spiritual power.

Sonora

The concept of wilderness understood by the contemporary O'odham, *doajkam*, signifies health, wholeness, and the renewing life. It is associated with specific sacred sites and with the desert, *tohono*, a "bright and shining place . . . of songs," which the Tohono O'odham claim as their identity. Yet as we saw in chapter 1, the quality of wilderness is sculpted and nurtured by humans, who create sacred spaces through rituals, hunting taboos, and cultivation—preserving seeds, pruning, and transplanting selected plants.[35] Quitobac, a spring-fed lagoon in the Altar Desert, becomes sacred ground through the annual celebration of *wi:gita*, a ritual of dance, song, and storytelling that brings together O'odham families from Arizona and Sonora. O'odham stories give mythical weight to Baboquivari mountain, home to their man-god creator *I'itoi*, the Gulf of California, and the Colorado River, as well as to a number of specific sites like the Children's Shrine, the Witch's Cave, and Quitovac, where a mythical O'odham water beast, *Ne:big*, emerged periodically from the lagoon, unleashing dangerous winds that sucked humans into the belly of the monster.[36]

The sacred quality of wilderness embodied in the O'odham concept of *doajkam* is mirrored in the Cahitan notion of *huya aniya*, which may be translated as "forest world," or, in Spanish, *monte*. The Yaquis of southern Sonora associate *huya aniya* with the Sierra Bacatete, a mountain range that marks the eastern boundary of their traditional territory. Yaquis took refuge in the sierra from Spanish and Mexican military forces on innumerable occasions, from the sixteenth to the twentieth centuries. Moreover, the Bacatetes—the Tall Cane-Reed Mountains—harbored a profound sense of power, arising from the geology of uplifted earth and desert plain, and expressed in Christian terms as the "place where Jesus walked."[37]

On the eastern slopes of the Sierra Madre Occidental, the Rarámuri (Tarahumara) today ascribe a number of different meanings to wilderness, *kawichí*, implying both renewal and potential danger. Although *kawichí* can refer simply to the earth or specifically to one's homeland, "the wilds" are areas outside the moral center of church and pueblo. Rarámuri cosmology associates uncleared forests and deep bodies of water with the devil, distinguishing them from cultivated lands that link the people to their ceremonial sites and to God. The devil's allies, who threaten the Rarámuri with illness and death, live in the underworld and beneath streams and lakes; they can appear as water monsters, rainbows, whirlwinds, or as coyotes, foxes, and human strangers.[38]

O'odham and Rarámuri spiritual geographies bring together Christian and native symbols of holiness and power. Their conceptual mapping of sacred space links the familiar landmarks of their physical surroundings to the mission pueblo and to the abstract notion of the divine Trinity: a god in human form revealed in three persons.

The solar journey from east to west marks the horizontal axis of O'odham cosmology, centered in the "staying earth," where all humans and the natural world exist and, moreover, devils and devil owls make their homes in earthly mountains. Sunset place marks the land of spirit animals and some devils; its geographical referent is the Gulf of California, a site of pilgrimages for

BOX 2 O'odham Cosmology

	da:m kacim (up-above lying)	
huduñig (sunset place)	*ka:cim jewed* (staying earth)	*si'al ig weco* (beyond the eastern horizon)
	mehi weco (fire below)	

Source: David L. Kozak and David I. López, *Devil Sickness and Devil Songs: Tohono O'odham Poetics* (Washington: Smithsonian Institution Press, 1999), 65.

shamans and for gathering salt. The O'odham image of beyond the eastern horizon describes a verdant landscape with abundant rainfall, wildlife, and cultivated fields. This is the destination of the O'odham dead, where they find relatives and friends and live in pleasure. Deceased O'odham spirits may return to earth as owls or devils. "Up-above lying" may be interpreted as the Christian heaven, the dwelling place of God, Christ, Mary, the saints, and worthy non-Indians in accord with *sa:nto himdag*.[39]

The center of the Rarámuri universe, as in O'odham cosmography, is the earth, where all humans—Rarámuri and Chabóchi ("bearded" non-Indians)—live. The realms of water people and plant people, who inhabit the earthly plane, represent danger for the Rarámuri. These are malevolent beings who threaten the souls of living Rarámuri, especially during sleep and through their dreams. People of the edge are associated with particular species of plants, both helpful and harmful to Rarámuri, and with the animals who serve the sorcerers in their evil deeds. The devil and his wife dominate the underworld, with their helpers—"soldiers," deceased Chabóchi—who bring diseases and capture Rarámuri souls. Heaven is peopled with the issue of the holy union of God and his wife: Our Father and Our Mother. Adapting Christian imagery and vocabulary, the Rarámuri refer to God's sons as *sukrísto* (Spanish, *Jesucristo*) and to God's daughters as *sánti* (saints). They associate *sukrísto* with metal crucifixes and *sánti* with medallions, icons that confer protection on their owners. God's helpers, comprised of the souls of deceased Rarámuri,

BOX 3 Rarámuri Cosmology

Heaven
Our Father and Our Mother
Children of Our Father and Our Mother (*sukrísto, sánti*)
God's helpers (Rarámuri souls)

Earth

water people	humans	people of the edge
water monsters	Rarámuri and outsiders	plant people
rainbows	(Chabóchi)	

Underworld
The devil and the devil's wife
The devil's helpers

Source: William L. Merrill, *Rarámuri Souls: Knowledge and Social Process in Northern Mexico* (Washington: Smithsonian Institution Press, 1988), 74.

serve Our Father and Our Mother, bringing them food, performing tasks, and serving as soldiers against the devil's allies.[40]

We see the common influence of Christianity in both these cosmologies, expressed in the struggle between good and evil, the personification of God, and references to the saints. The Rarámuri envision the opposing forces of good and evil in the soldiers of God and the devil, while the O'odham recognize two parallel spiritual powers: the saint way and the devil way. The O'odham and Rarámuri interpretations of the Trinity further humanized the god of Judeo-Christian traditions. In the Rarámuri vision, Our Father and Our Mother procreated sons and daughters with beneficent, protective powers to guard faithful humans from disease and sorcery. Their image of heaven restored sexuality to the notion of the divine by placing the conjugal couple at the center of the spiritual universe and portraying these male and female deities as the Sun and the Moon.[41] The saints play an important role in both interpretations of the Christian gospel, rooted in the colonial mission. We may recall that beginning with the earliest evangelizing expeditions, missionaries presented crucifixes, banners of the Virgin Mary, rosaries, and saints' images as cultural talismans for protection against devils and sorcerers. Furthermore, they distributed crosses and medallions in each shipment of trade goods that constituted the moral economy of reciprocal payment for the Indians' labor. Thus these objects became part of the cosmic vision of the Indians who entered the missions.

Rarámuri and O'odham attention to particular species of plants and animals, endowing them with spiritual power, placed the Christian cosmology of heaven, earth, and underworld in landscapes with locally specific meanings. Plant people of the Rarámuri earthly realm refer to three species of cacti and a bulrush, each associated with ritual specialists who can cure diseases caused by the souls of these plants. Water people and water monsters are associated with actual streams and springs and, in the Rarámuri imaginary, with the dike that surrounds the island earth.[42] O'odham creation myths honored the creosote bush (greasewood; *Larrea tridentata*; *shegai* in O'odham), a common plant with ancient roots in the Sonoran Desert, as the first plant to emerge from the soil that Earth Maker shaped. Creosote produces chemically potent resins from which O'odham and Cunca'ac peoples have found many curative powers. The plant lives in symbiotic relationship with an insect, *Tarcardiella larrea*, that generates a lac casing from creosote resins. According to the legend, Earth Maker sang and pounded the insect lac into the earth's crust.[43] It is noteworthy that two seemingly humble species should attain primary significance in O'odham stories of the cosmos.

Chiquitos

A number of striking symbolic points of reference emerge in the images that marked spiritual places among the O'odham, Rarámuri, and Cahitan peoples of northwestern Mexico and the Ayoreóde and Chiquitos peoples of lowland Bolivia. Comparing these similarities is not intended to conflate different historical traditions and systems of meaning that, as we have seen, developed in distinct geographical settings. Rather, underscoring common themes in these religious practices and cosmologies serves to integrate this chapter and to connect spiritual landscapes in nature with the sacred architecture of built environments intended for worship, religious discipline, and indoctrination.

The sun and the moon figure prominently in the mythologies and cosmographies of all these peoples (perhaps as a universal principle), but especially among the Rarámuri and Ayoreóde, these heavenly bodies appear as named deities associated with procreative powers. Watercourses represent natural and spiritual power in both regions, at times personified or portrayed as mythic animals. We have seen above that the O'odham revere desert *tinajas*, water catchments or reservoirs hidden in rocky hillsides, and that the Rarámuri recognize a separate realm of water people on earth. Sonoran folklore of indigenous and mestizo sources recounts numerous stories of the *corúa*, a snake that lives in springs and streams, its sinuous form serving to protect the flow of water. Generally they are harmless, but should they be killed, the spring will dry up.[44]

For the Chiquitos, spirits of the natural world known as *jichis* have different names in relation to water, hills, rain forests, and savannas. Like the Sonoran *corúa*, *jichis* are visualized as serpents, but they are further conceptualized as the masters of each realm of nature: animal, vegetable, and mineral. *Henaxíx-tí* represents the spirits of all animals (except fish, who belong to the *jichi-tuúx* of water), and *toíx* is the spirit of the *chacos* and their produce, who causes the crops to grow and their fruit to mature. Humans must respect the separate powers and territories of each of these beings if they wish to avoid suffering illness or death and in order to protect their gardens and assure plentiful game for hunting and fishing. Various taboos that restrict human activities in work and play—crafts like weaving and ceramics, bathing in natural springs, and sexual intercourse—to particular times and places in accordance with the seasons of planting and harvesting highlight the central place of the *chaco* in Chiquitano lifeways and in the cultures of meaning the people have created around the spatial rhythms of garden and forest. Men and women are enjoined to abstain from sex before sowing and harvesting, but couples take their lovemaking to their *chacos* after the harvest, thus linking their sexuality with the fertility of the land.[45]

In addition to the *jichis*, Chiquitanos acknowledge and fear the presence of humanoid dwarfs known as *jatokaáx*. Described as both comical and dangerous, the *jatokaáx* are held responsible for the disappearance of children; they appear seductively to adolescent boys and girls, promising riches and enticing their intended victims to follow them to the savannas (pampas) outside the villages. The sexual innuendos of the seductive powers ascribed to these not-quite human beings, associated with merchandise originating outside Chiquitano villages, may be linked to instances of rape that occurred between the propertied *estancieros*, and other non-Indian ranchers and cowhands, and Chiquitana women, gendered violence that occurred at the shallow lagoons where women and girls bathe and wash their clothes.[46] The maleficent power of the *jatokaáx* bears some resemblance to the devils who afflict the O'odham and the Rarámuri with illness and misfortune, and whose songs prescribe the path to recovery from staying sickness.

The overriding importance of shamanic healing rituals is evident among all the Sonoran and Chiquitoan peoples. Shamans' transgressive powers to heal or inflict illness are often associated with particular plants, animals, and birds. Both men and women may be recognized as healers or feared as sorcerers, but most often it is men who assume the position of shaman in local communities.[47] The references to shamans as *mapono*, in the Jesuit sources noted above, may correspond to two categories of healers and sorcerers in present-day religious practices—the one (*cheeserúx*) honored and revered, the other (*oboíx*) feared and reviled—who personify their moral universe of good and evil. The beneficent or maleficent powers ascribed to individual shamans are defined, in part, by the networks of kinship relations that distinguish different families and create divisions in the communities. Through massaging, blowing, and sucking, healers remove the objects that cause pain, illness, and weakness believed to have been inflicted by sorcerers. Both the healing capabilities and the destructive powers of shamans are associated with the Chiquitanos' complex and detailed ethnobotanical knowledge of herbal remedies and poisons.[48]

The Chiquitano universe describes a horizontal plane of interlocking earths and a vertical depth of layered heavens, distinguished by the humans and humanlike beings that live in them and by the type and place of death. Their cosmography of concentric spheres arranged hierarchically comprehends different habitats and the spirits that control them within the material landscapes and the imagined spaces of their world.

The structures and artifacts of sacred space correspond in different ways to native concepts of spiritual places and to the Christian architectural precepts of godliness. We may recall that Manasica religious practices took place in a room so designated in the house of the village cacique by erecting temporary walls of woven palm mats. Chiquitano healing rituals observed historically and in recent times have generally been performed in the homes or on the patios of the shaman or the patient. Turning to northern Mexico, Rarámuri curing ceremonies centered on the hallucinogenic properties of the peyote cactus occurred on household patios; their attendance and observation was restricted to the direct participants in the healing ritual, who consumed peyote and traveled in its dreams. The O'odham *wi:gita* and Cahitan religious dances for the deer and ritual clown (*pascola*) took place under ramadas or in earthen,

BOX 4 Chiquitano Cosmology

seven layers of heaven, inhabited by:
Jesus and his grandfather, *Tupax*
persons who died in the village
persons who drowned in the rivers or lagoons
persons who died in the forest

seven interlocking earths
small humanoid beings

earth
spaces and habitats: *chaco*, forest, village
spirits of the natural realm (*jichis*)
humanoid dwarfs (*jatokaáx*)

Source: Riester and Fischermann, *En busca de la loma santa*, p. 152.

open-air plazas that had been swept clean and marked off for the purpose. These examples from both Sonora and Chiquitos suggest that the boundary between intimate, familial rituals and public ceremonies was less clearly defined for indigenous sacred traditions than for Roman Catholic rites of worship. Similarly, the demarcation between constructed space within doors and the natural environment out of doors was more sharply accentuated in the Judeo-Christian notion of temple than in native practices.[49]

Early continuities between indigenous and Catholic religious sites, when mission chapels were constructed as partially roofed ramadas, gave way to the designation of formal sanctuaries, built according to Catholic architectural canons, and of *colegios*, *conventos* (convents), plazas, and cemeteries as sacred spaces with particular functions and meanings. Catholic tenets of holiness insisted on height, prescribed in the dimensions of mission churches, in striking contrast to depth and intimacy with the earth, principles that associated caves with spiritual power among many Sonoran peoples. Mission compounds were distinct from shamanic connections to the forest and desert in their spatial design and physical construction, external boundaries, and central plazas marked by the Christian emblem of the cross. The pueblos created a cultivated landscape of fields, gardens, and indoor dwellings in contrast to the wilderness of rain forest and *monte*.

The *reducciones* established town sites that encompassed the church, an adjoining convent (missionaries' residence), a *colegio* for instruction and the deliberations of the *cabildo*, and additional workrooms and storage areas for granaries, merchandise, and mission produce. Sonoran missions, often founded in preexisting native communities, maintained the dimensions of a village in their design and size. Located on mesas or plateaus overlooking streams and planted fields, mission pueblos of Sonora did not conform to the Spanish grid pattern of colonial cities, but adapted to the topography—mindful of the conditions required for agriculture and stock raising—and to the need for security against raiding. The most prominent buildings in the pueblo were the church and convent, surrounded by the houses of Indians who had relocated to the missions or who, as frequently occurred, visited seasonally from the outlying *rancherías*. Family dwellings conformed to native building techniques, combining brush huts, mud-covered wattle, and, in the eastern Sonoran highlands, stone and earthen walls. As the missions consolidated into permanent towns, houses constructed with adobe bricks gradually replaced the movable huts of brush and reed matting. Francisco Barbastro, writing from Tubutama in 1793, noted with pride that "straw houses have been removed from this Pimería Alta, thus the Indians live in decency . . . in adobe houses with doors, and many of them have keys with locks, others have wooden locks in the form of a square bolt, which they know how to make."[50] Barbastro's reference to locks implies that the Christian O'odham living in the missions had developed the need for guarding their possessions and adapted European techniques for securing the interior spaces of their homes.

Missionaries rendered each village a holistic Christian space that was connected, in turn, to a network of villages and, metaphorically, to the authorial presence of God and king. Church design was nearly always cruciform, reproducing in its architecture the passion of Christ. Wooden beams, sometimes cradled in ornate *zapatas* (brackets for beams), supported the roofs, while the interior and exterior walls were finished in stucco or mortar, providing surfaces for sculpted and painted decoration. Their religious imagery, crafted for the most part by native artisans, created visual icons that illustrated Christian doctrine. Altars and niches, adorned with painted *retablos* (altar ornaments) and statues, interpreted liturgical and biblical themes, enhanced by mirrors, candles, and the contrasting effects of sunlight and shadows. Sonoran mission

Facade of the mission church at Opodepe, Sonora. Photograph by the author, 1982.

churches, in theory and intent, conformed to canonical precepts for Roman Catholic houses of worship; in practice, however, they proved heterodox, built by stages over long periods of time and, more often than not, in several locations. Their material construction, aesthetic qualities, and symbolic meanings varied according to the religious philosophies of the Jesuit and Franciscan missionaries who oversaw the design and execution of the churches, as well as the cultural traditions of the Indian neophytes who built and maintained them.[51]

Sonoran Christians interpreted the spatial dimensions of Roman Catholic houses of worship through their conceptions of spiritual power. From the colonial foundations of church and mission, Indian and mestizo villagers developed postcolonial religious practices during the nineteenth and twentieth centuries that informed their design and use of ceremonial spaces and their cosmographies, as in the staying earth and up-above lying (box 2). O'odham communal structures, round brush houses placed in the center of their *rancherías*, served to shelter the council meetings of male elders and to store

Bell tower of mission church at Onavas, Sonora. Photograph by the author, 1979.

Facade of mission church at Onavas, Sonora. Photograph by the author, 1979.

ceremonial masks, pots, and other implements used for *nawait* (ceremonial fermenting and drinking of wine) and *wi:gita* rituals. Their simplicity and centrality to Tohono O'odham communities provided enduring principles for building and maintaining chapels on the Papago Indian Reservation of Arizona in the adaptive tradition that James Griffith has called "folk Catholicism."[52]

In this blending of modernity and the saint way (*sa:nto himdag*), O'odham chapels distinguish themselves from Roman Catholic churches in their physical proportions and uses. If the latter provide shelter for the regular priestly celebration of the Mass and the ecclesiastical administration of church records, as well as house the sacred images of the Christian faith, the former are principally associated with the annual feast days, including religious processions, dances, and a festive communal meal. O'odham chapels, while conforming outwardly to the architecture of Catholic churches, constitute ceremonial spaces that comprise the building itself and an outdoor fenced atrium with a pedestal on which to place a large decorated cross, a dance floor, and a sheltered kitchen, all essential for feast days. Paper flowers, arches (rainbows, allusive to rain), and saints' images adorn the interior of these small, low-ceilinged chapels, where they are also stored, retouched, and moved in preparation for each annual or biennial festival. In addition to these special occasions, O'odham Christians enter their chapels in the evenings to recite the rosary led by lay specialists, alternating hymns and prayers.[53]

Native adaptations of Christian imagery have created similarly powerful cultural translations in Chiquitos, but in distinct architectural spaces of the sacred. Jesuit *reducciones* of Chiquitos established ten mission towns, in which the Indians' houses arranged in linear rows occupied three sides of the central plaza, while the church, the *colegio*, the cemetery, the gardens, and workshops stood together on the fourth side. Two intersecting axes, passing through the central plaza, established the geometrical design for these missions and designated the obligatory approach from the small chapel of Bethany on the edge of the village—symbolizing Jesus's point of departure for his triumphal entry into Jerusalem on Palm Sunday—to its spiritual center marked by a large wooden cross and four palm trees planted in the middle of the plaza.[54] The first phase of Jesuit church construction, dating from the closing years of the seventeenth century, followed indigenous building methods and materials. Early mission chapels comprised a singular, undivided space, its structure formed by wooden pillars that supported a pitched roof covered in woven palm leaves and enclosed with walls fabricated of woven mats and reeds, reminiscent of the ceremonial spaces secluded in the caciques' homes for the spiritual visitations of the *Mapono*.

By the mid-eighteenth century, the consolidation of the missions in relatively permanent locations and the development of their economy afforded the material wealth and labor needed to undertake formal church construction. Jesuit missionaries, European lay brethren, and native master builders created artisanal structures of unusual beauty and symbolic power by enhancing native building techniques and converting them into a distinctive architectural style. The single galleries of wattle and mud that characterized the early phase of the *reducciones* became spacious naves, subdivided into three parts by rows of wooden arches and covered by steeply inclined roofs that overhung the walls, thus creating shaded verandas that protected the interior of the churches from rain and the intense heat of their tropical setting.[55]

In contrast to the missions of Sonora, where the principles of baroque architecture were adapted differently for each of the churches, the temples of Chiquitos followed a nearly uniform pattern of construction. In six of the seven standing churches of the Chiquitanía, the basic building materials comprised enormous carved wooden pillars of hardwood that supported their massive roofs, their walls enclosed with adobe; palm-and-thatch roofs were gradually replaced with fired clay tiles. The striking similarity in their design is due, most likely, to the lack of stone and lime for firing brick in most of the province, thus reinforcing the use of timber and earth, and to the intense period of church construction, 1747–67, in which the singular influence of a few leading missionaries developed the artistic and architectural styles of Chiquitos. While using space in nearly identical ways, each church was distinguished by its decoration and the placement of particular elements, for example, the bell tower or the solar clock. Native artists embellished the walls with floral designs and images of the saints, using vegetable and mineral dyes. Indigenous and Creole artisans carved wooden statues of the saints, crucifixes, and intricately decorated pulpits that, together with the sanctuary and the two side altars, dominated the main nave of the church.[56]

The architecture of the sacred extended beyond the church and *colegio* to encompass the entire mission through the processional spaces marked by large wooden crosses. It was the movement of people along these routes through residential streets in each *parcialidad* and along the main axis leading from the Bethany chapel to the central plaza that sanctified the pathways they described, integrating the town in this ritual act. The *capillas de posa* (chapels along

Mission San Miguel, in the province of Velasco, Chiquitos. Photograph by the author, 1997.

Mission San Miguel, detail from the sacristy. Photograph by the author, 1997.

Baptistry, mission San Xavier, in the province of Ñuflo de Chávez, Chiquitos. Photograph by Xicoténcatl Murrieta, 1999.

Detail of baptistry, mission San Xavier. Photograph by Xicoténcatl Murrieta, 1999.

pilgrimage routes) on the four corners of the plaza and along the exterior routes, customary places to pause or where two processional columns met, did not constitute permanent structures; rather, these spaces were designated for religious festivals with crosses or arches made of woven palm branches.[57] Processions evoked both solemnity and festivity by announcing feast days and thus marking the cyclical recurrence of special events in the Catholic liturgical calendar. Their enactment set apart celebrations in space and time, joining Iberian medieval traditions of penitential procession with indigenous rituals of dance and song that preceded shamanic healing ceremonies.

Feast days brought together different kinship networks and *rancherías* in an affirmation of communal identity under the auspices of Catholic ritual. Special moments of liturgical space-time occurred during Holy Week, Corpus Christi, and the patron saint's day of each mission, bringing the renewal of spiritual power in combined ceremonies of the sacred and the profane. In historical times and in the present, dances of European origin, like the *saráo*, and masked performers with particular connotations in different pueblos—for example, the clowns (*bufones*) or grandfathers (*abuelos*) who perform in San Rafael, San Juan, Santiago, and San José, or the birdlike *yarituses* who play a prominent role in the feast of Saint Peter and Saint Paul in San Xavier—recreate living cultural traditions.[58] The *pozoka*, *cabildo* members who visit neighboring pueblos to honor their feast days, are integral to the theatricality and hospitality that mark religious holidays. On the eve of the feast of San Ignacio (26 July), as many as twenty-four *chininises* (ritual dancers) leave the pueblo to await the visiting *pozoka* from San Miguel, San Rafael, and Santa Ana in the village cemetery; together, they pay formal visits to the parish priest, the prefect, and the *corregidor* and, in turn, are recognized by the ecclesiastical and civil authorities of the pueblo. The *pozoka* of San Ignacio return their visits to these neighboring pueblos and travel as far as San José for their respective saints' festivals.[59]

Comparison with the religious festivals of the O'odham, Cahitan, and Rarámuri communities in northern Mexico brings the overlapping realms of the sacred and the profane more sharply into focus, revealing layered meanings. Each of these groups, whose cultures were steeped in the Catholic missions, developed ceremonies that illustrated the Christian story and bestowed legitimacy on the annual election of *cabildo* officers, at the same time that they strengthened linkages to their natural environment and to the spiritual forces of their cosmos. The Feast of the Holy Trinity, coinciding with the onset of summer rains, the pageantry of Lent, and the ritual battles of Holy Week, became the occasions for theatrical episodes of spiritual power in the liturgical cycles of religious observance. They combined dances of both Spanish and indigenous origins, especially the *matachines* (ritual dancers) whose practiced movements sanctified the spaces in which they moved, and the masked performers—the sacred clown, the deer, the *fariseos* (pharisees) and their demonic assistants, the *chapayeka*. These rituals opened transcultural spaces for enacting particular identities and created a distinctive poetics for telling their histories.[60]

Language, Moral Suasion, and Translated Meanings

Language provided the necessary vehicle for the adaptation of Christian concepts integral to the imagery of religious festivals and the creation of sacred spaces. The richly varied symbolism and complex cosmologies that developed among the native peoples of both Sonora and Chiquitos through the mission experience attest to the parallel processes of translation from Latin and Spanish to the many indigenous languages spoken in each region and of transposed meanings across different cultural settings. To understand these processes and their centrality to the designation and renewal of spiritual landscapes, it is important to consider the historical production of a lingua franca, or a comprehensive colonial language, the segregation of gendered spaces and idioms in the rites of worship, and the communication of knowledge through the recitation of the catechism, the rosary, and sermons.

The codification of indigenous languages, as we have seen in chapter 4, formed an integral part of the colonial project. Missionaries compiled grammars and vocabularies that served as codes of communication to maintain discipline over the native families resident in the pueblos, whom they considered to be neophytes, people newly and only partially integrated into the Christian faith. Pima-Tepehuán, Opata, and Eudeve, in Sonora, like Chiquitos and Guaraní in Bolivia and Paraguay, became enduring general languages in the mission provinces. Jesuits intended them as the means of conversion and control; significantly, however, they functioned as a dual pedagogy through which the missionaries struggled (in their own words) to learn comprehensive languages through which to communicate the orthodox message of Christian evangelization.[61] Francisco Barbastro, reporting on his experiences in both the

Pimería Alta and the Opatería of Sonora, wrote candidly that the continual migration of Tohono O'odham into the Pima missions represented a constant flow of neophytes who stood at different stages of instruction. Barbastro explained further that the Indians heard the Christian message in Castilian but that native teachers taught from manuals written in their own languages.[62] Native catechists and *cabildo* officers turned these general languages into formal modes of ritual expression and political representation in the presence of colonial authorities.

That missionaries were students as well as teachers is illustrated pointedly by the working lists of words and phrases that circulated among the missions, as well as by the formal *artes*—vocabularies and grammars—approved for publication by the religious orders.[63] Barbastro compiled a manual of sermons, prayers, confessionals, and word lists in the Hehíre and Tegüima dialects of the Opata language during his last assignment in Sonora, when he was transferred from Tubutama, in the Pimería Alta, to Aconchi, in the Sonora river valley.[64] The practical nature of Barbastro's workbook, in which some of the Spanish text appears with Opata translations written above individual words, suggests that it was used by indigenous doctrinal teachers who served as religious and disciplinary intermediaries between the missionaries and the wider community of baptized Indians. Missionaries depended on native interpreters for these plural translations from written text to oral recitation, and from Spanish and Latin to the native tongue.

Two of Barbastro's sermons copied into the manual illustrate the process of transculturation through which both Jesuits and Franciscans assembled words and images in native tongues in order to transpose Catholic doctrine into local surroundings. Barbastro's sermon for the Feast of the Conception of the Virgin Mary enjoined his parishioners of Aconchi to pray to her "in your language with all your heart," as we saw in the opening epigraph for this chapter. The words he directed the Opatas to recite contrasted their sinfulness with the purity of Mary and personalized the dual concepts of moral transgression and forgiveness.

> I am very happy because God has freed you [Mary] from original sin and he made you holy and beautiful; I am bad and I had original sin, but you were always very good and you never had sin. For this reason, have pity on me and teach me all good things, help me to confess well, to separate myself from bad people, and to leave all my sins forever. Holy Mary, mother of my Lord Jesus

> Christ, I love you with all my heart, and I give you my soul and all that I have, because you are my mother and my guardian [*patrona*]. Take good care of me, my mother, have pity on me, and ask God for me to pardon all my sins. Take my soul to heaven to see you and to make a great feast for you there in heaven with the angels and saints.[65]

This mode of second-person address placed the notion of sin at the center of Christian beliefs and procedures, comparable to the patterned questions and responses that structured confessional manuals in nearly all colonial mission fields. Following Catholic tradition, it gendered the attributes of judgment and merciful intercession ascribed to God and to Mary. Barbastro's prayer wove into the language of atonement references to the rewards of forgiveness in a festival to honor the Virgin Mary that would bring together the earthly communities of Opata villagers with the angels and saints of heaven.

Barbastro's Christmas sermon, dated in 1792, placed the birth of Christ in the spiritual geography of Sonora. In this Christmas story, Mary gave birth to Jesus in a cave, a holy place in the Sonoran Indians' cosmology, where the shepherds came to adore him. Mary and Joseph swept and cleaned the cave and, because they were hungry and cold, Joseph made a fire, then they cooked and ate some food. Joseph retired to a corner of the cave to rest and pray, while Mary prepared to receive her son. She knelt and, assisted by the angels, gave birth at midnight without pain or travail. Mary and Joseph joyfully worshipped their son; then she took him into her arms, wrapped him in the cloths she had prepared, and nursed him "with the virginal milk which the Lord put in her breasts." The child God suffered cold and wept for our sins, "because in that time it was very cold," until an ox and a mule came running to the cave and warmed him with their breath. Barbastro's adaptation of the biblical text responded to the Opatas' fears of hunger, cold, and death in childbirth. His sermon wove together the ethereal and earthy elements of Christ's birth: Mary was holy and a virgin, but she squatted to give birth, and Jesus came out of her body (*entrañas*) and suckled at her breast.[66]

These exercises in doctrinal translation led to multiple interpretations of the Gospel and its implications for moral behavior. Scholars of colonial histories and literatures have pointed out both partial and missed translations of sacred texts, which occurred in different imperial settings of America and Asia. James Lockhart has written extensively on the "double mistaken identities" of Nahua nobility and Spanish friars who thought they recognized each other in their own cultural conventions; Alfredo López Austin, Louise Burkhart, and Serge Gruzinski have shown eloquently the transferred meanings of key religious concepts like *sin* and *redemption* in the Nahuatl texts that circulated in central Mesoamerica; Sabine MacCormack analyzed the layered symbols of imagery and text that accompanied evangelization in the Andes.[67] Vicente Rafael has scrutinized the translation process itself in the Spanish Philippines as a complex instrument of power and submission, on the one hand and, on the other, of interpretation and resistance. Rafael argued that the rendering of redemption as indebtedness, with its connotations of reciprocity and deferred payment, apposite to the notion of shame in Tagalog religious manuals, produced confusion and "shock" that altered the intended meaning of Catholic doctrine. His analysis of Tagalog texts concluded that native listeners—the intended neophytes of Catholic missionaries—fragmented Spanish phrases into phonetic bytes that were then reworked into different compositions. Furthermore, he argued that this adaptive translation set a boundary between Tagalog and Spanish, affording some protection from the hierarchies of power imposed by the colonial order.[68]

The intriguing implications of Rafael's research offer significant points of comparison and contrast with the two provinces that comprise our story. When examining the languages of moral suasion and religious belief conveyed in confessional texts and their transformation in the indigenous cosmologies of both regions, as we have seen above, their significance is sharpened through association with the natural environments and sacred spaces of their worlds. Visual and textual elements like the devils of O'odham and Rarámuri lore or the serpentine *jichis* of Chiquitos (and the *corúa* of Sonora) were lifted from Christian sermons and native traditions and reworked in new combinations, in ways that are analogous to the metonymic arrangement of speech particles in colonial Tagalog texts. But when we juxtapose words and images with cultural landscapes, as these were carved into the Sonoran Desert and the Chiquitano rain forest, we find that colonized peoples appropriated whole stories and wove them into composite narratives. Their moral universe transcended individual units of speech and writing, inserting the Christian transcript into spatial conceptualizations of cosmic forces that stood for good and evil and generated the contingencies of life and death.

The cosmologies of Sonoran and Chiquitano peoples rested on parallel masculine and feminine sources of power; accordingly, religious languages and ritual spaces were gendered. Opata discourses included separate vocabularies for men and women, which were preserved partially in the terminology relating to kinship. Gendered speech emerged as the central principle in the evolution of Chiquitos as the general language of the missions. Distinct phrases were spoken by men and women, according to the speaker and to the objects or persons of reference, and codified in the written manuals produced by the missionaries with their Indian assistants. Women's speech constituted the basic language, employed by them and by men when they referred to women, animals, and inanimate objects; when men referred to God, to men, or to devils, they used a more elaborate vocabulary, with different words and phraseology. Julián Knogler emphasized the importance of preaching in both the masculine and feminine idioms, lest their sermons create confusion among the neophytes and make the missionaries appear ignorant of their languages.[69] Gendered divisions that distinguished speech also governed the spatial distribution of worshippers during the daily rites of Mass, catechism, and rosary and on special feast days. Young men and boys knelt before the altar close to the communion rail, followed by the men who stood in rows that extended as far back as the center of the nave. Behind them knelt the young girls, while adult women comprised the final group to enter the church.[70]

Men assumed proprietorship of oral preaching, in both northern Mexico and eastern Bolivia, chanting sermons in their languages with references to Latin and Castilian. Recitation of devotional verses and moral exhortations by *cabildo* officers or other leading male figures in the community, usually at the atrium of the church on patron saint festivals or other religious feast days, had its roots in both aboriginal and missionary modes of governance and spiritual observance. Native sermons were linked to Catholic traditions of doctrinal teaching and procession; their content alluded to the saints, and the context of their delivery followed the Christian liturgical calendar. Before missionization, however, oratory was a hallmark of the caciques' authority to command allegiance and labor from their kinspeople and villagers. Eloquent speech enforced the masculinity of cacique status, as Knogler described it among the Chiquitanos: "If someone wants to praise himself and let it be known that he is better that the others, then he will say, 'I am a man in all the extension of the word, or I am a whole man.'" The missionaries recognized the caciques' "sense of honor" and elevated their position in the pueblos with ceremonial dress for religious festivals, places of honor to sit in the church, and the canes of office that they carried to all meetings.[71]

The language of religious convocation and reference that infused Chiquitano sermons constituted the rhetoric of political discourse used by *cabildo* officers when they addressed colonial authorities. A few select documents transcribed into the petitions presented to Spanish governors and the investigations of local disputes and tumults, in both Castilian and Chiquitano, illustrate the rhythms and themes of sermonizing that developed as a distinctive tradition in the missions. In the course of the 1790 tumult in San Ignacio on the Eve of the Feast of Corpus Christi, council officials addressed a letter to Antonio Carvajal, governor of Chiquitos, that began with a blessing: "Praised be Jesus Christ, Amen."[72] Twenty years later, the *cabildo* officers of San Rafael intervened in a complicated power struggle between the administrator of their mission and the governor of Chiquitos, Juan Bautista de Altolaguirre, who was violently dispossessed of his office and held prisoner in the mission during January 1811. Altolaguirre mounted his defense before the *audiencia*, beginning with the Indians' written statement. The judges of San Rafael, who supported the governor and denounced the administrator's abusive practices, prefaced their statement with a religious invocation that paralleled the prayer that appears at the beginning of this chapter. They transposed the formal language of the sermons into a written salutation to establish the legitimacy of their testimony and present their own grievances.

> Praised be our Lord Jesus Christ. How are you, lord governor Don Juan Bautista? May God reward you, lord, thanks be to God who has given you life in all your journeys, thanks be to the Virgin Holy Mary, who with great effort has implored the grace of our God for you, our judge. Lord governor, for a long time we Chiquitos in this land have waited for you, our judge; you have come because God wanted it so, and because the King Don Fernando VII has ordered you to look out for us.[73]

The Indians accused Miguel Hurtado, the administrator of San Rafael, of subjecting men and women to flogging, denying them adequate rations of meat, stealing mission produce that "belonged to the King," and of dissipating the livestock and harvested crops: "Everything is lost in his hands." Their language wove together apt allusions to the divine and earthly majesties that ruled the Spanish empire and reminded the governor of his duties to the Indians who had long awaited him.

The recitation of sermons together with the maintenance of *cabildo* offices and the observance of religious festivals sustain enduring traditions of cultural identity to the present day. In the Chiquitanía, ritual language and faith have fused into a collection of texts, conserved through oral declamation and in written manuscripts, which are carefully maintained and passed on from generation to generation. The chanted verses repeated in the sermons seem stilted and removed from the cadences of everyday spoken Chiquitos, converted into the privileged knowledge of male religious leaders. Nevertheless, their performance serves to reaffirm the bonds of community and to demonstrate the integrity of their culture to the wider Bolivian society in which the Chiquitanos are immersed.[74]

Distinctive traditions of chanting sermons and memorizing sacred texts appear with equal force and for reasons of cultural reaffirmation among the Rarámuri and O'odham of northern Mexico. As I noted above, the O'odham devil way comprises collections of song cycles, whose memorization and performance is the skilled training and exclusive knowledge of the shamans, who heal devil sickness, and the spiritual leaders who chant in the annual celebrations of *wi:gita*.[75] Rarámuri villagers place special importance on the gifts of oratory because they are convinced that eloquent speech is a sign of good thinking, which is necessary for good behavior. Rarámuri sermons occur outside the church before a formal assembly of the parishioners after Mass or the recitation of prayers, delivered by village officers, to convey religious beliefs and exhort the people to practice moral behavior during church services and drinking ceremonies.

> Follow the path of Our Father and the path of Our Mother. May you encounter old age a long way away. Grasp the staff of Our Father and also the staff of Our Mother, the flower of Our Father and the flower of Our Mother. In this way each of you who is standing here will have strength. Do not be discouraged. Do Our Father and Mother become discouraged as they unfailingly provide light so that we can go around contentedly?[76]

The sermon from which this verse is taken elaborates on the day-to-day relations among the Rarámuri and, by extension, the fulfillment of the people's duties to their deities, Our Father and Our Mother, the Sun and the Moon. Portions of the sermon give separate admonishments to men and women concerning their conduct within the spatial landscape described by the church and village center. William Merrill has contrasted the reductive and rhetorical quality of the imagery and ideas expressed in the sermons with the complex and varied cosmologies—representing different spiritual forces and earthly planes—revealed to him gradually during spontaneous conversations and drinking ceremonies (see box 3 above). Merrill concluded that Rarámuri sermons served not so much to communicate theoretical knowledge about the cosmos, but "to reiterate time-proven advice for the proper conduct of life."[77]

The sermons intoned by village leaders are central to the religious traditions generated in the colonial missions over two centuries. Each of the three groups under discussion here, however, interprets the importance of the sermons in different ways. The central feature of Chiquitano sermons is the language itself, emblematic of ethnic distinctiveness and cultural integrity. Among the O'odham, the prescribed sequence of song cycles holds curative powers, while for the Rarámuri, sermons comprise a mode of formal counsel and a reaffirmation of correct social and political relations within the community. Common to all three traditions are the gendered settings in which they take place and their association with the *cabildo*: the oratorical skills of sermonizing and the honor of election to the *cabildo* are reserved for men.

The formal hierarchies of ranked offices associated historically with the *cabildos* and liturgical ceremonies were masculine even as attendance at religious services was organized by gender and age, separating men, women, and children. Behind the formal edifice of political officeholding and ritual declamation, however, women contributed in fundamental ways to cultural production and public life in the missions as producers of fermented corn drinks: *chicha* in lowland Bolivia and *tesgüino* in northwestern Mexico. *Tesgüino* and *chicha*, in their respective domains, served as libations to enliven a good party, but they signified much more in the indigenous societies of these two frontiers. Fermented brews supplied the necessary ingredient for social conviviality; moreover, the rituals observed for their production and consumption offered conduits to spiritual power. Notwithstanding the missionaries' energetic efforts to curtail the use of *tesgüino* and *chicha* and the drunkenness that they condemned in village fiestas, it remained a centerpiece of Sonoran and Chiquitano cultural practices. In Chiquitos, caciques were expected to provide hospitality to visitors from neighboring villages, symbolized by the *pozokas*, who would later reciprocate in kind. A good cacique served abundant food and drink, and thus needed a wife who knew how to make good *chicha*.[78] Similarly, fiestas among the Sonoran and Rarámuri villagers accompanied by copious amounts of corn beer brought together extended families from different *rancherías*. These social occasions at the same time proved deeply religious, linked both to the Catholic liturgical calendar and to the Indians' moral universe.[79]

The ethical content and formal delivery of indigenous sermons, together with processions and festivals, illustrate the theatrical quality of religious performance. Choral and instrumental music figured prominently in the heterodox traditions of both regions, complementing the architectural and artistic aesthetic central to the missionaries' evangelizing project and the Indians' adaptation of Christian theology. In the missions Indians expanded their repertoire of instruments, whose manufacture and execution were closely linked to their physical environment, with European orchestral stringed and wind instruments. Formal choirs intoned polyphonic liturgical songs, often accompanied by instruments, which served to teach Catholic doctrine and dramatize the solemnity of religious holidays. Choirmasters occupied an important post in all the missions as musical directors who accompanied the Mass, led processions, and contributed to liturgical pageantry. In northern Mexico, especially in Sonora, Sinaloa, and Baja California, women and men sang in the mission choirs; in the Chiquitos missions, however, participation in the choir and physical access to the choir lofts in the churches remained reserved for men.[80]

The visual and sonorous standards of artistic performance that flourished in many Hispanic American provinces, including northern New Spain, Moxos, Chiquitos, and Paraguay, formed a baroque mentality in which music and art constituted integral parts of both daily life and stylized religious devotion. Baroque architecture and plastic art were thickly textured, the exuberant use of sculpture and color filling liturgical space. Music and art composed within the baroque sensibility aspired to express emotion and didactic content, as well as to convey formal beauty. Central to European and Creole conceptualizations of baroque aesthetics was the notion that life experiences paralleled a theatrical performance in which God assigned particular roles to individuals.[81] Comparable to the rhetoric of indigenous sermons as prescriptions for ethical behavior, baroque music and drama constituted a powerful pedagogical tool in the hands of missionaries and native religious leaders. Moreover, in the cultural and environmental settings of these mission frontiers, baroque forms proved varied and expansive, allowing for combined European and indigenous symbols with heterodox interpretations.

The Indians' representations of the Passion during Lent and Holy Week, documented historically and recorded in contemporary ethnographies, illustrate their comprehension of the fearful power of deities and the cosmic struggle between opposing forces for good and evil, health and illness, life and death. Indigenous musical traditions provided the aesthetic and symbolic foundations for theatrical creativity in the colonial missions. Even as the missionaries sought to displace the power of shamans, they turned to this ceremonial foundation in order to embellish the Catholic liturgy with dramatic arts. In addition to sung masses and vespers, musicians who served in the missionary orders adapted European baroque operatic forms to compose operas, oratorios, and musical dramas based on themes drawn from biblical stories and classical mythologies.[82] Through their participation in these performances, rendered partially or entirely in native languages, the Indians interpreted pointed messages of Christian faith and perseverance, ethical conduct and, in their terms, good speech and thought.

The scenarios created for these musical dramas were adapted to the natural environment, as the Jesuit Knogler described in some detail for the renditions of the conversions of San Eustaquio and San Francisco Xavier in the mission of Santa Ana, in Chiquitos. Each of these operas was staged in the forest, carving out a space for the public to view it, and the legend of San Eustaquio alluded to the vision of Christ appearing between the antlers of a deer, an image which Knogler found "particularly appropriate for the Indians who spend their entire lives hunting in the *monte*."[83] Two operatic scores preserved in the musical archive of Chiquitos, allegorical pieces dedicated to the exemplary lives of San Ignacio and San Francisco Xavier, founders of the Jesuit order, and fragments of a third, alluding to God as the just shepherd, show the determination of native council members to maintain the artistic and religious heritage of their past. Comparable to the written sermons recited on feast days, the musical texts for liturgical and theatrical performances became treasured cultural artifacts of Chiquitano faith and identity. Native scribes copied the musical notations from one manuscript to another, long after the departure of the Jesuits and their original composers, conserving them together with the sacred images for Mass and festival processions.[84]

The efficacy of Christian musical pageantry as a tool of evangelization could not obscure the violence of conquest, subordination, and religious indoctrination. Indigenous responses to the plethora of icons, texts, and dramatizations they encountered in the mission created contested meanings that fueled numerous instances of political and religious confrontation. These symbols of spiritual power inspired both reverence and rage, emerging during acts of rebellion that were motivated, at times, by the Indians' determination to protect sacred objects and spaces within the missions. The Manapecas of Concepción revolted against their priest in 1779, over his flogging one of their kinsmen and violating the sanctity of the *colegio* by keeping a mistress in his quarters. Uprisings occurred in the Chiquitano pueblos during liturgical festivals, as we have seen for San Ignacio in 1790 on the eve of Corpus Christi. Native rebels carried theatrical elements of ritual into their tumults, with drums, flutes, and inflammatory speeches, and repeatedly excoriated their priests with accusations of immoral behavior.[85]

Documented uprisings in Sonora exhibited similar signs of the inversion of religious rituals, as well as statements of political defiance and millenarian challenges to the colonial order. During the spring of 1737, several thousand Indians of northern and western Sonora joined a pilgrimage to a shrine in the Cerro Prieto, a desert range that long served Pima and Seri rebels as a refuge from Spanish troops. A prophet named Ariscibi, who led a cult to the god Moctezuma, had created a shrine with a wooden idol of Moctezuma, dressed in the robes of Christian saints, and decorated with ornaments taken from the mission of Belén. The prophet received his followers with the sign of the cross and performed rituals based on the Catholic Mass, incensed with tobacco smoke; he preached the coming of a new age in which Indians would rule over Spaniards, there would be no sickness, and the land would yield abundant food and water. The movement was violently suppressed when Juan Bautista de Anza assaulted the Cerro Prieto, dispersed the cult followers, and captured Ariscibi. Spanish authorities subjected the prophet to flogging and public execution by hanging, intended as exemplary punishment, but new rebellions shook the province with strong religious and political overtones.[86]

The Cahitan revolt of 1739–41, a major uprising of Yaquis, Mayos, and Pimas, inflamed the provinces of Ostimuri and Sonora. For over two years insurgents forced the abandonment of mines and haciendas and the mobilization of presidial troops. Rebel leaders presented formal grievances to the governor of Sonora and Sinaloa and carried their petition to the viceregal court in Mexico City. They demanded the removal of missionaries against whom they had specific complaints, the expulsion of Spanish and mixed-race *vecinos* from their pueblos, the termination of compulsory labor in the missions, and the freedom to sell their produce and work for wages outside the missions. The Indians articulated their demands in terms of political economy, but they expressed their anger in the theft of religious icons from churches in the Yaqui Valley. Armed warriors stole the vestments and sacred ornaments from the missionary's house in Belén. In a deliberate reversal of sacred space, they tore the altar cloths and the saints' robes, reworking them into shirts and pants worn by the *matachines* who danced with them in the pueblo of Ráhum on the eve of Corpus Christi. The symbolic defiance of this act of vandalism, serving as the opening volley of full-scale rebellion, dramatized the Indians' increasingly militant rejection of the missionaries' authority over them in religious observances and day-to-day governance.[87]

Conclusions

This chapter has developed three central arguments around the concepts of spirituality, ritual, and knowledge. First, the cultural systems of meaning that evolved historically in the colonial provinces of Sonora and Chiquitos imbued material landscapes with the religious dimension of spiritual power; the Indians' moral cosmologies of life and death, good and evil, were envisioned in terms of the natural environment in which they lived. Their religious concepts and practices revealed inclusive notions of the divine that informed both shamanic and Catholic rituals. Second, the languages, images, and spaces of sacredness entwined Christian and indigenous points of reference in richly colorful and heterodox tapestries of symbolic values and practices. It would be pointless to separate their content into pre-Christian and Christian categories, an exercise that would distort the complex cultural heritages of belief and ritual that comprise the Chiquitano and Sonoran religions. Third, the boundaries between the secular and the sacred, the intimate and the public, and indoor and outdoor spaces, were translucent and porous. The political and the religious were recognized as separate spheres, but each informed the other, as is revealed in the language of prayer and legal testimony and in the space-time of historical events—like tumults and local uprisings—and of liturgical acts—like the dances of the *matachines* or the theatrical processions of Holy Week and the festivals for patron saints.

We should not assume the antiquity of indigenous cosmologies. Moral space, symbolic references, and spiritual concepts developed through both religious and secular experiences of the colonial past and recent historical periods. Demonology, sorcery, and the spiritual forces of the natural world that figured so prominently in both regions—to be feared, respected, and appeased—evolved together with Christian imagery and were adapted to Catholic doctrine. The stories, creatures, and curing rituals associated with illness suggest prolonged and traumatic experiences with debilitating and fatal diseases, which seem to come from without and require spiritual intervention for their healing.[88] Native theories of health may arise from the long history of colonial epidemics and endemic disease punctuated by outbreaks of yellow fever, malaria, and cholera, which continued through the nineteenth and into the twentieth centuries.

We may be tempted to view the syncretic cosmologies of contemporary Indian communities like the O'odham, Rarámuri, and Chiquitanos as childlike renditions of Christianity with truncated and transposed names and hybrid deities. This, in my view, would be a narrow interpretation that misses the rich historical traditions they express. These are creative religious texts, presenting complex and continually changing language and imagery molded and shaped by colonialism, political struggle, territorial conflict, and the reassertion of ethnic identities. They teach us that the semiotic dimensions of cultural history remain inseparable from the material bases of everyday life: subsistence, the social bonds of community, and the dimensions of conflict. Furthermore, cultural "authenticity" finds its strongest expression in heterodoxy—in the sometimes disjointed appropriations of Catholic and shamanic rituals and beliefs that appear at different times as part of the historical process. Indigenous holistic concepts of power, community, and spiritual intervention informed native peoples' participation in the Bolivian and Mexican independence movements and their aftermath, when the violence and institutional revolutions of the postcolonial order again turned their worlds upside down. The political and cultural implications for human landscapes of the transition from colony to republic in both Chiquitos and Sonora constitute the organizing themes of the following two chapters.

CHAPTER 7

Postcolonial Landscapes: Transitions from Colony to Republic

Nineteenth-century developments brought irreversible changes to the Spanish and Portuguese colonies in their internal governance, their external relations with the major European powers, and the configuration of private property, communal territory, and national boundaries. The long, destructive wars for independence that erupted in New Spain, New Granada, and Río de la Plata in 1806–10 and continued for over a decade, produced major battles and a number of important experiments in constitutional government for Spanish Americans who aspired to political autonomy. Notwithstanding the dramatic legacy enshrined in master narratives of popular revolt in New Spain and the sweep of insurgent armies led by Simón Bolívar, Antonio José de Sucre, and José de San Martín through the Andes, the wars for independence more often than not took on the character of guerrilla struggles among armed combatants in particular locations, with a heavy toll on the civilian population.[1] Their impact was felt in the borderlands of northwestern Mexico and eastern Bolivia as an intensification of border disputes and a palpable decline of Spanish institutional control in finances and imperial defense.

The independence movements did not have their most decisive events in either region, but the formative decades of the early republican period for both Mexico and Bolivia transformed the political cultures and societies of Sonora and Chiquitos. As important as these changes proved for reworking the economies, the definitions of citizenship and personhood, and the legal and customary relations across gendered and ethnic boundaries among Creoles, Indians, *castas*, and blacks, their significance is comprehensible only in terms of the colonial foundations on which they rested. Despite the eclipse of the formal dominion of the Spanish empire in mainland North and South America, the weight of colonial political economy, territorial divisions, and cultures persisted in the nineteenth-century construction of the nation-state in Mexico and Bolivia.

The present chapter captures both the enduring continuities of colonial traditions and the momentum of change that marked the histories of Sonora and Chiquitos during this period. Its objectives are to place these postcolonial processes in the spatial and environmental contexts that guide the entire study, bringing together major themes from previous chapters that illustrate the tensions between colonial institutions and republican innovations. Proclamations and acts of protest or open rebellion drew on the values and symbols of Catholicism, monarchical authority, and the legal structures that had upheld the empire under both the Hapsburg and Bourbon regimes. Enlightenment principles, which established the equality of persons and the secularization of culture, were conveyed to the colonies through the Spanish Cortes of 1812 and the Constitution of Cádiz, a document created in the midst of the Iberian dynastic crises that followed the Napoleonic invasions of Spain and Portugal. For these reasons, this chapter highlights the contrasting chiaroscuro of past and future that marked the postcolonial present for nineteenth-century actors in the uneven transition from colony to republic.

The guiding question for this chapter centers on how we can perceive postcolonial landscapes through historical narratives. Because the kinds of documentary evidence available for Sonora and Chiquitos differ more strikingly than in earlier periods, their comparative histories do not follow a parallel course. Thus the narrative structure of this chapter builds on stories of material changes in the land for both regions, as these relate to the recorded historical experiences of Sonoran and Chiquitano peoples in three broad areas: political economy, society and culture, and territory. They underscore the changing economic roles of the missions, the different modes of Creole expansion in each region, and the degrees of ethnic blending that conditioned citizenship and community identity.

Political Economy

The experiment with textile manufacturing in Chiquitos brings together themes of material culture, technology, and gender. Colonial administrators focused their policies for commercial development on this tropical borderland with the aim of increasing exportable production. During the brief but tempestuous career of Joseph Sibilat in San José de Chiquitos, traditional weaving practices collided with modern standards of industrial output.

Textiles, Handicrafts, and Machines in Chiquitos

The 1801 arrival of the French master weaver Joseph Sibilat in San José de Chiquitos for a brief time turned that pueblo into a center for textile manufacture, producing waves throughout the province. Traveling from Buenos Aires to Jujuy, and from there to La Plata, Sibilat purchased tools in preparation for building mechanized looms in the Chiquitanía, charging his purchases and traveling expenses to the General Administration of Missions of the *audiencia*. Once established in San José, Sibilat was assigned an interpreter, women spinners, and six Indian laborers from each pueblo, whom he was to instruct in their operation, to help him build new spinning and weaving machines. In addition, the administrator of San José supplied Sibilat with raw cotton and wax of different qualities, as well as food rations and gifts with which pay the Indians who worked with him. For his own maintenance and comfort, Sibilat took his place in the refectory of the mission *colegio* alongside the priests and the administrator.

Salaried by colonial authorities under contract, Sibilat had undertaken the transformation of textile production in the province in accord with Spanish ambitions to make native handicrafts commercially competitive. It was his commission to "teach the Indians of the ten pueblos of Chiquitos, simplifying their labor and, to that end, build simple machines that would facilitate their work, and illuminating them with the most advantageous, useful, and basic knowledge to perfect their technique, improve their products, and enhance the quality of their *lienzos* in texture and colors, so that they will produce more and their manufactures will merit much higher prices."[2] Furthermore, Sibilat had promised to build a loom for weaving stockings, caps, gloves, pants, and other articles of clothing, as well as different sizes of looms for producing different cuts and qualities of cloth. He had even offered to introduce flax cultivation into the province in order to process the linen and produce *bretañas* (fine linens) as good as those made in France. Turning to wax, whose market value had surpassed that of cloth over the past two decades, Sibilat was expected to build a lathe and introduce European methods for whitening wax.

During the first year of Sibilat's residence in San José, complaints began to accumulate concerning his arrogant conduct and his failure to deliver what he had promised in his contract. The general administrator of the missions reported that despite an outlay of 2,300 pesos, the fruits of this investment waited to be seen. Few of the promised looms had been built, of these the cylindrical *calandria* for polishing cloth had broken, and the Indians had not learned the new techniques. Only the spinning wheels were in use, for they seemed to be readily adaptable to the Indians of this region. Sibilat, for his part, protested that to produce new and varied textiles and to train the Indians in the use of his machines would take time. Tensions between the administrator of San José, Ángel Castedo, and his foreign guest erupted into clamorous arguments, with verbal insults leading to Sibilat's swaggering threats, shotgun in hand, and his challenge to a duel. The episode ended abruptly with Castedo's retreat to the pueblo of San Miguel and Sibilat's stormy exit to Santa Cruz and, from there, to La Plata to collect his salary and reimbursement for expenses. At the instigation of Castedo and governor Miguel Fermín de Riglos, the *audiencia*'s investigation into the experiment in textile and wax manufacturing at San José led to some perceptive observations concerning gender, technology, and the commercial economy of Chiquitos.

Carpenters and metalsmiths recruited from San Rafael and Santa Ana to assemble Sibilat's looms and presses responded to Castedo's query about whether they arrived promptly to work. The men explained that they arrived early, but that "Sibilat, who at times was strolling through the pueblo and at other times shut up in his room, failed to lay out their tools until nine or ten o'clock and, for this reason, out of boredom they had left."[3] Even more sensitive were the discussions around the physical and cultural adjustment required of the women and young girls who learned to weave on the spinning wheels that Sibilat had built. The master craftsman had begun teaching married women to spin, but at the order of Governor Riglos, Castedo had removed the married women and replaced them with single women—a decision that had particularly angered Sibilat. The exchange of married for single women originated with the governor's sense of prudence, at the behest of the *cabildo* officers of San José. One witness testified that he had seen as many as thirty young girls spinning at the wheel. Juan Baca, a soldier stationed in Santiago who had been sent to San José to serve as an interpreter for Sibilat, elaborated further on the women's complaints about the machinery.

> He had heard the women say that it was uncomfortable to spin this way, [that they preferred to work] as they were accustomed. He also observed that they spent most of the day laughing and playing with the machine. [They claimed] that the wheel was hard work and that they tired from [the position of] their

> arms and from standing to spin with this new invention. They could work more easily and with less effort the way they were used to spinning.[4]

For his part, Governor Riglos wrote in no uncertain terms that the textiles produced in the Chiquitos pueblos were not the work of Sibilat, but to be credited to the hard work of their administrators and to his own efforts to develop the weaving crafts by importing model pieces of cloth (*muestras*) and teaching weavers in different pueblos to reproduce them. Riglos noted that "even what the master craftsman has produced for his own use is not up to the quality of the weavings from San Rafael, San Miguel, and San Ignacio." Furthermore, the governor had denied Sibilat's demand to have the authority to whip the Indians, "which, if I had permitted it, would surely have had bad results."[5] Castedo denounced Sibilat's arbitrary disassembly of twenty looms kept in the *colegio* to weave pieces of muslin of over one hundred *varas* in length. Sibilat's action had obligated Castedo to distribute thread to the weavers in the pueblo, rather than have them work in the *colegio*, thus incurring the risk of losing the cloth in a fire or through theft of the material.[6]

The controversy surrounding Sibilat's brief tenure at San José illustrates several key points of Bourbon economic policies, Spanish standards of technology linked to the commercial value of craft production, and Chiquitano concepts of their manual skills and the material worth of their products. The declining market value of regional muslin during the late colonial administration was probably due to increased output from highland *obrajes*, nonguild textile workshops that produced cheap cotton and woolen cloth for circulation in the Andean urban and mining centers. Colonial officials, ever wary of clandestine imports of British textiles, noted the discrepancy and took deliberate steps to enhance the variety and quality of lowland weaving, thus increasing its monetary value. Intendant governor Francisco de Viedma, whose jurisdiction included the province of Chiquitos, devised a policy of developing the frontier with an eye to increasing both the volume and value of its commercial output.[7] The investment in Joseph Sibilat, authorized by the *audiencia*, was clearly intended to advance the economic assets of the missions and modernize their production.

In the cultural, ecological, and economic context of the Chiquitanía, where craft manufactures were conditioned by the seasonal rhythms of cultivation, hunting, and fishing, the introduction of mechanized workshops met a number of obstacles. The intended role of indigenous spinners and weavers was that of laborers whose access to the market was mediated by non-Indian administrators and clerics. They were enjoined to produce more and better cloth, in less time, but their immediate rewards did not increase accordingly. In the absence of powerful incentives, on the one hand, or, on the other, means of coercion, Chiquitano women and men would have resisted mechanized production as alien to the bases of their gathering and weaving skills that, in turn, were closely linked to their gardens and the forest.

It is not true, however, that Chiquitos Indians were hostile to new technologies on principle. It will be remembered that the history of European incursions into the eastern lowlands of Bolivia began with the natives' avid interest in the iron and metal tools the Spaniards brought with them. But these forest dwellers sought European implements and the technical advances they implied on their own terms, as evidenced in the recurring conflicts over *encomienda* and the Indians' flight from enslavement. As we have seen in previous chapters, Chiquitanos negotiated the conditions of their labor and remuneration in goods, thus molding the political economy of the missions even as they interpreted the aesthetic and symbolic iconography of Christianity in their own idiom. Their limited access to the market, both inside and outside the mission communities, in the light of the minimal material rewards they received, provided little incentive to increase their output of surpluses. Moreover, native crafts involved much more than turning out skeins of thread and bolts of cloth. Weavers' skills encompassed all phases of producing fiber: women planted, harvested, cleaned, spun, and dyed cotton thread; men and women tied and wove pieces of fabric that were at once utilitarian and artistic. The aesthetic and spiritual qualities of spinning, dyeing, and weaving transcended the market, linking the day-to-day tasks of mission life with the communities of pueblo and *ranchería* and the renewing power of the forest.[8]

The Corporate Mission Economy under Bolivian Rule

The republican administration of Bolivia maintained the colonial structures of mission economy in Chiquitos for over half a century after independence. Correspondence and ledger sheets originating in the prefecture and the bishopric of Santa Cruz attest to the continuation of accounting systems intended to control the flow of goods produced in the missions through a central receiving station (*receptoría*) in Santa Cruz, and the redistribution of merchandise to the priests and administrators stationed in the missions and to the

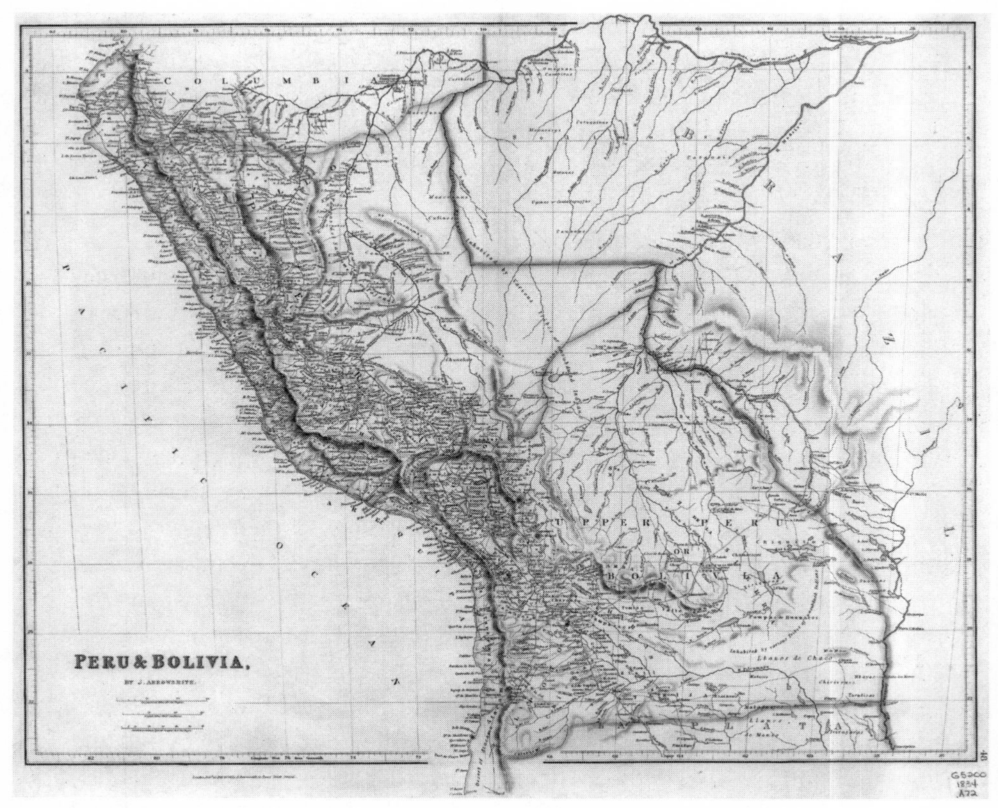

Peru and Bolivia, by J. Arrowsmith, 1834. Chiquitos is characterized by lagoons, placed within the department of Santa Cruz. The missions of Concepción and San Juan Bautista are named. The legend beneath Pampas de Huanacos reads: "Inhabited by tribes of unsubdued Indians." Courtesy of Map and Geography Information Center, Centennial Library, University of New Mexico.

Indians of each pueblo. Most of these goods repeated those listed in the shipping bills and inventories familiar to us from the colonial regime. Indigenous workers produced wax, raw cotton, thread, and woven cloth. Mission textiles now included tablecloths, napkins, stockings, and handkerchiefs in addition to burlap, uniforms, and tents destined for the Bolivian military. San Miguel had become a center for textile production by the 1830s, despite the late colonial experiment with mechanization in San José. Innovations included leather goods from tanned cattle hides for shoemaking, riding gear for horses and mules, and traveling pouches. In return, mission laborers received shipments of iron and steel for metallurgy, knives, needles, combs, imported cloth (*bayeta*), and wooden rosaries (of their own manufacture). In the eastern and southern pueblos—especially San Juan, Santiago, and San José—salt became a commodity of some importance during this period; it was traded within the province as a medium of payment at two pesos the *arroba* and exported from the region in exchange for cattle and metal tools.[9]

Three basic conditions that defined the nature of this commerce limited the economy of Chiquitos. First, all exchanges took place as barter, trading goods for goods, since very little coined money circulated in the province. This held true for remunerating the Indians' labor, paying the salaries of priests, administrators, and other officers, and for establishing equivalent prices for the pueblos, the city of Santa Cruz, and the wider commercial circuits that connected Santa Cruz to the Andean highlands. Second, price equivalents for the goods exchanged were established officially by the governors of Chiquitos.[10] Third, the subventions that had supported the salaries of clerics and administrators in the Chiquitos missions after the expulsion of the Jesuits, from 1768 to the end of colonial rule, against future repayment in mission products, were nearly always in arrears and paid in kind. After independence, financial support from the central administration became illusory, following the eclipse of the *audiencia* and its colony-wide oversight of the missions. The very goods produced in the missions—cloth, wax, and salt—circulated as currency for payment of debts, satisfaction of unpaid salaries, and loans among different individuals and pueblos. Equally serious were lapses in the delivery of *fomento*, the merchandise distributed to the Indians, that sustained rhythms of production in the pueblos. In January of 1829, the governor of Chiquitos, Gil Antonio Toledo, called a meeting of the mission administrators in Concepción to solicit their collaboration in a subscription of small loans from each pueblo in order to buy goods in Santa Cruz and redistribute them through the pueblos, with the objective of keeping the economy of the province from coming to a standstill. Without the incentive of *fomento*, Governor Toledo warned, there would be no production and no funds to pay the salaries of administrators and priests.[11] Two years later, his successor, Marcelino de la Peña, acceded to the clamorous demands of these clerical and lay employees for overdue salaries, paying them in mission textiles "until there was not one piece of cloth left in the warehouse."[12]

Revenues generated in the missions of Chiquitos and Moxos, the neighboring province to the northwest, figured substantially in the balance sheets of the prefecture of Santa Cruz during the first two decades of the republic. The 1830 general statement of revenues owed and collected from different categories of taxes, fees, fines, and duties showed that personal contributions, tithes, and products from the missions comprised the three most important sources of income. Together these three categories represented 54 percent of all revenues owed to the prefecture and 66 percent of the taxes collected.[13]

The personal contribution (*contribución personal*) constituted a tax on all non-Indian property holders, established by a national decree of December 1825 and implemented with accompanying legislation in January of the following year. It was intended to be calculated on the basis of a cadastral survey of properties and a census of persons, distinguishing between rural and urban properties, commercial warehouses and retail shops, artisans and apprentices.[14] Men and women of the Chiquitos pueblos, who did not hold property in their own right, were not subject to the personal tax, but they contributed through their labor to the revenues remitted to the prefecture from the missions. Chiquitanos herded livestock, quarried and loaded salt, gathered and processed wax, and spun and wove cloth—as they had done under the colonial regime—and, furthermore, they were recruited to cut roads through the forest and, at times, to drive the mule trains that carried mail and goods from the pueblos to Santa Cruz. The status of lowland indigenous peoples in relation to the Bolivian state contrasted significantly with that of tribute-paying Indian communities of the highlands. As in Chiquitos, the highlanders' labor supported many public services such as building and maintaining roads, delivering the mail, and operating roadside hostels for travelers (*tambos*). Their semiannual tribute payments, however, provided them with leverage to secure their communal possession of *ayllu* lands in a continuation of colonial practices.[15]

Colonial authorities from the *audiencia* in La Plata to the viceroyalty in Buenos Aires had frequently discussed the issue of tribute payment by the

TABLE 11 Tax Revenues for the Prefecture of Santa Cruz, 1830

	Amounts Owed	*Collected (pesos)*
Personal contribution	18,361	15,975
Tithes	15,247	13,152
Products of the missions	12,810	12,810
Subtotals	46,418	41,937
Total revenues	84,589	63,374

Source: Universidad Gabriel René Moreno, Museo de Historia, Fondo Prefectural 1830, vol. 1, exp. 10. Estado General del Tesoro Público de Santa Cruz, January–December 1830.

Indians of the Moxos and Chiquitos missions, during both the Jesuit and post-Jesuit periods. Individual tribute assessments from heads of household were never imposed in either province; rather, the missions collectively paid a modest tribute to the crown.[16] In response to Bourbon free trade policies, coupled with renewed emphasis on collecting tribute from Indians as individual subjects of the crown, the governor of Chiquitos, Josef Ramón de la Roca, in 1811 advised the bishop of Santa Cruz, Francisco Xavier Adazábal, concerning the potential benefits and risks of opening the province to contractual labor and commerce among Chiquitanos and Cruceños. Roca's optimistic view of the fertility of mission lands, the abundance of mission herds, and the skills of indigenous artisans led him to assert that "it should be easy for the Indians to pay tribute," but he warned against charging them ecclesiastical fees. Roca cautioned that Chiquitanos were naive and easily deceived in calculating the fair price of their products and, furthermore, accustomed to working "in common, all for all . . . under the direction of the heads of their guilds." If the new directives were taken literally, conferring "absolute liberty to work for themselves," the Indians could fall into destitution, idleness, and hunger and turn to robbery. Thus Roca concluded that free trade should be moderated in the Chiquitanía, maintaining collective organization in the pueblos and the leadership of the caciques.[17]

A closer view of two Chiquitos pueblos from the detailed censuses and inventories of buildings, *chacras*, and goods in storage taken in San Xavier (1823) and Santa Ana (1828) illustrates the economy in which Indians and Creoles lived during the first decade of the Bolivian Republic (table 12). San Xavier, the first of the Jesuit missions founded in Chiquitos (in 1691), constituted the westernmost pueblo and the one closest to the urban market of

TABLE 12 Censuses and Inventories for San Xavier and Santa Ana

	San Xavier (1823)	*Santa Ana (1828)*
	Population	
Families	518 (*x2*)	520 (*x2*)
Widowers	81	20
Widows	68	102
Adolescents	255	133
Children	635	301
Total	2,075	1,596
	One-Room Houses	
	344	139
	Cattle	
	858	620
	Horses and Mules	
	45	91
	Chacras (*almudes* planted with cotton, maize, beans, manioc, bananas, sugarcane)	
	44	25

Source: Universidad Gabriel René Moreno, Museo de Historia, Fondo Prefectural, vol. 1, exp. 9, 1829.

Santa Cruz. Santa Ana, one of the last Jesuit missions, founded in 1755, marked the eastern boundary of the province. Each pueblo presented detailed lists of merchandise and mission produce stored in the warehouses, cattle of different classes, and the tools in their workshops for metallurgy, carpentry, shoemaking, weaving, and carving rosaries. San Xavier reported thirty looms and 978 *varas* of cloth, with forty skilled journeymen; eight years later, in 1831, Alcides d'Orbigny observed forty looms in operation there.[18] Santa Ana, by way of contrast, counted only one loom and had no woven cloth in storage; rather, its surpluses were concentrated in salt and leather goods of different kinds.

The reports submitted for these two missions suggest intriguing differences in their demographic composition and stability for the early republican period, although they are separated by five years.

If indeed the total number of persons (*almas*, or souls) noted for San Xavier—2,075—and for Santa Ana—1,596—corresponds to comparable proportions for both pueblos during the colonial period, the structure of their populations suggests critical losses in certain age groups and by gender for Santa Ana. Both pueblos reported almost the same number of households (518 to 520), but Santa Ana's population of youth and children was considerably lower than that of San Javier. These differences may be expressed numerically in population ratios: the family-to-population ratio of San Xavier was 4.0, while it was 3.1 for Santa Ana. Summing adolescents (*solteros* and *solteras*) and children (*párvulos* and *párvulas*), the child-to-family ratio of San Xavier was 1.72 and 0.83 for Santa Ana. In addition, the high number of widows in Santa Ana (102) seems striking, and among adolescents and children, young girls were outnumbered by young boys in Santa Ana with a difference of thirty individuals in each category; in San Xavier, however, their numbers were relatively balanced. Finally, the one-room houses reported for Indian families living in San Xavier (344) were more than double those of Santa Ana (139), and the agricultural plots for the former mission (44) were considerably more than in the latter (25).[19]

Governor Marcelino de la Peña's 1831 reports to the prefect provide telling information on the climatic and social crises that conditioned life in the pueblos, helping to explain the profile of Santa Ana described in the census of 1828. The entire province had suffered droughts during those three years, with especially severe effects on the cotton harvest and textile production. An outbreak of chicken pox in San Xavier in 1825 nearly halved the population of that mission, largely due, in Alcides d'Orbigny's opinion, to lack of good medical care.[20] Fires in Santa Ana, in 1826, and in Concepción, in 1830, destroyed significant portions of their *colegios* and of the Indians' houses. In Concepción, the fire was followed by an epidemic, which Peña assuaged by establishing a hospital in the pueblo where over eighty victims were treated with food and "simple medicines" that saved most of them from dying. Peña expressed special concern for the widespread hunger and illness that he had observed throughout the province.

> The pueblos of this province should be wealthy and populous, but on the contrary, each day we see its inhabitants die off. Experience tells me that the principal cause of this ruin is not so much the gravity of accidents and epidemics as it is the general lack of resources, food, and medicines that plagues these unhappy souls.

> There is not one pueblo where we do not find a growing number of the sick, abandoned in their little huts, with no one to take pity on them or help them. Thus the lack of food is the worst hurricane that threatens to finish off these Indians.[21]

Governor Peña's frank observations revealed the underlying poverty of the missions and their vulnerability to climatic vagaries of drought and flood, despite their monitored production and trade in agricultural and craft products that led to the appearance of prosperity. Chiquitanos did not have direct access to the market, even though the sheltered conditions of their labor had prevailed—as prescribed by Governor Josef Ramón de la Roca in 1811—and two decades later the dangers of hunger and destitution to which he had alerted Bishop Adazábal seemed to erode the economic foundations of the Chiquitos missions. The range of products leaving the missions had diversified, but the commercial networks that supplied them were reduced substantially to a regional web centered on Santa Cruz. Prefects and provincial governors exchanged a number of proposals during the middle third of the nineteenth century intended to invigorate the commercial economy of Chiquitos across the Amazonian-Andean ecological frontier and to establish transnational spheres linking the province to Brazil and Paraguay. They proposed enriching the Chiquitos economy with plantations of vanilla and cacao, especially in the marshy environs of Concepción, but these projects were scuttled when, in 1847, the governor undertook emergency measures to trade cattle for corn in Casalbasco, Brazil, at the height of a four-year drought.[22]

The skewed demographic profile of Santa Ana in 1828 was emblematic of population fluctuations experienced in a number of pueblos, suggesting high rates of mortality and morbidity, as well as the Indians' movement from one mission to another, their migration to the city, and their flight to the forest in search of subsistence. The province-wide census of 1842 for all ten Chiquitos pueblos places this comparison of San Xavier and Santa Ana in wider spatial and temporal perspective (table 13). By the latter date, San Ignacio had emerged as the largest pueblo, followed by Concepción and San Miguel, then Santiago, San Xavier, and Santa Ana (of nearly equal size), and finally Santo Corazón and San Juan. The total indigenous population recorded in the missions increased gradually from 16,900 to 18,300, from 1835 to 1842 (table 16).

The economic and demographic data analyzed here suggests that the continuation of the corporate economy in the Chiquitos missions under republican

TABLE 13 Census for the Province of Chiquitos, 1842

Pueblos	*Population (number of souls)*
San Javier	1,391
Concepción	2,059
San Miguel	2,848
San Ignacio	3,839
Santa Ana	1,345
San Rafael	1,252
San José	2,132
San Juan	990
Santiago	1,442
Santo Corazón	1,093
Total	18,391
Signed: San Ignacio, 30 May 1842, Juan Felipe Baca.	

Source: Universidad Gabriel René Moreno, Museo de Historia, Fondo Prefectural, vol. 31.

rule—as described by d'Orbigny and repeated in later histories—was more apparent than real. Despite the disciplined production of mission commodities and the combined clerical and governmental bureaucracies put in place to manage their sale and distribution, the erosion of native subsistence under the burdens of this commercial economy that were already discernible in the late colonial period had taken its toll in the basic living conditions of the Chiquitanos. Governor Peña's reports of hunger and disease in what "should [have been] a wealthy and populous province" dramatized this situation.

Privatization and the Waning Mission Economy of Sonora

Heartfelt warnings of impoverishment, similar to those recorded for Chiquitos, reverberated through the early nineteenth-century documentation for Sonora, but they described different modes of economic change tempered by environmental and political crises. Late colonial policies restructured the corporate economy of the Sonoran missions under the post-Jesuit regime, when secularization converted missions into tithe-bearing parishes and weakened the clerical administration of their economy. During this same period a growing mixed population of *vecinos*, including middling ranchers and large landholders, fueled the expansion of commercial markets centered in the mines and military presidios that had spread throughout the province. These processes accelerated following Mexican independence, leading to the impoverishment of the *común* and the dispersion of indigenous subsistence strategies through contractual labor, peonage, and—in a few cases—entrepreneurship.

At the turn of the nineteenth century, only the missions of the Pimería Alta, in the northernmost portion of the province of Sonora, remained under the formal tutelage of a missionary order. The Franciscan Colegio de la Santa Cruz de Querétaro administered these missions, remitting periodic censuses and inventories to the intendants of Sonora and the viceroys of New Spain. Two of these reports, dating from 1818 and 1819, provide a frank assessment of the economic patrimony and the spiritual health of these missions peopled by different groups of O'odham cultivators and gatherers. Covering sacramental events, population counts, and standing inventories of crops and herds, the data compiled by missionaries in the field conveyed a picture of sparsely settled communities, and their marginal notations warned of periodic droughts with losses of harvests and livestock.[23] The earlier of the two, representing 1818, is summarized in tables 14 and 15.

Population data recorded for Indians, Spaniards, and *castas* confirms both the mission status of the pueblos of the Pimería Alta and the racially and culturally mixed communities they had become on the eve of Mexican independence. The friars noted baptisms for Spaniards, *castas*, and Indians; among the Indians, they distinguished between those born and living in the pueblo and the *gentiles*, adults who had not yet been converted to Christianity. Each pueblo had received at least one *gentil* neophyte during the year recorded in this census, and the three westernmost villages of Caborca, Bisanig, and Pitic had baptized the greatest number of *gentiles*, presumably Tohono O'odham or Hiach'ed O'odham who migrated seasonally from the desert. All the pueblos recorded the marriages and burials of both Indians and non-Indians; deaths occurring among Indians, compared with births and total population, were consistently higher than among Spaniards and mestizos. San Xavier del Bac, although supporting the second-largest indigenous population after Caborca, showed a dangerously high number of deaths among the Indians: fifty-one burials to twenty-nine baptisms. These numbers may reflect a smallpox epidemic that had begun in San Ignacio in 1816 and spread northward to Tumacácori and Bac. A decade later, a severe outbreak of measles swept through the region.[24] The native population recorded in the census clustered in the missions of Caborca, Bac, Oquitoa, Tumacácori, Cocóspera, and their respective *visitas*; non-Indians were dominant in San Ignacio and Magdalena, and they outnumbered the Indians in Tubutama, Oquitoa, and Sáric. Comparing total populations across the eight missions, that of Spaniards and *castas* together was three-quarters larger than that of the Indians.[25]

Neither Indians nor non-Indians, however, lived entirely in the pueblos. José Pérez designated the column for total population in the missions as *existentes*, which I have translated as "souls," indicating those who "existed" in the pueblos from one year to the next. A significant but uncounted number of inhabitants in the Pimería Alta lived and worked in desert *rancherías* and in the livestock ranches that were the property of Sonoran Creole families. A number of the Franciscans assigned to the northern Pima missions served the spiritual needs of numerous ranches spread through an extended radius from their pueblos. Vicente Giner, stationed in Sáric, was in charge of the ranches of Agua Caliente and Alamo. Narciso Gutiérrez of Tumacácori, in the middle Santa Cruz drainage, served as chaplain at the nearby presidio of Tubac. Francisco Solano García, assigned to Cocóspera, likewise served as chaplain of the presidio of Santa Cruz, eleven leagues distant from the mission, and visited the ranches of Comaguita, Costraca, and San Lorenzo. San Ignacio and Magdalena, with an estimated population of 1,300 *vecinos* and 36 O'odham, supported two friars in charge of five non-Indian pueblos and three ranches, covering seventeen leagues in all. Caborca, comprising three pueblos and a total resident population of 733 persons almost evenly divided between *vecinos* and Indians, also supported two missionaries.

The figures for grain production and livestock included in the table, in addition to Fray Pérez's comments, point to considerable stress in the mission economy of the Pimería Alta. Although each of the eight missions reported a wheat harvest, probably from irrigated fields, the failure of the maize crop, except in Bac and Cocóspera, appears particularly noteworthy. Maize supported native subsistence and, in mission cultivation, largely depended on rainfall during the growing season. Pérez, writing in December 1818, and Fray González, who reported the following year, explained in dramatic terms the effects of a four-year drought on agriculture, stock raising, and the hospitality traditionally given to the nomadic *gentiles* who visited the pueblos. The numbers noted for different classes of livestock were mere estimates for, due to the drought and the depressed economy, mission herds had wandered out of the grazing lands (*sitios*) belonging to the pueblos in search of pasture, much of it not branded and wild (*bronco y alzado*). Furthermore, the costs of rounding up the livestock proved prohibitive, more than what their products—tanned hides and dried and salted meat—would bring on the market. The environmental crisis was compounded by the disruptions in supply trains and curtailment of the missionaries' stipends occasioned by the wars for independence, reducing the circulation of minted coin and, if that were not all, burdening the missions with taxes, such as the *alcabala* on the sale and movement of goods, from which they had traditionally been exempt. All of these hardships had impoverished the Indians, threatening the viability of communities that for over a century had reproduced culturally and materially, dependent in part on mission resources.

> Payment is in cloth or goods (for there is no money, not even for the soldiers nor for the stipend with which our beloved Sovereign, may God bless him, supported us). The clothing . . . is not good enough even for the poor, nor can it cover so much nakedness and poverty as suffer these poor Indians in these calamitous times of insurrection, burdened with alcabalas, which they had never paid, and which they cannot understand. Cultivation, weaving, and other crafts are woefully behind for lack of the means to pay laborers and supply them with tools. The charity that we were accustomed to give the *gentiles* has ceased for the same reason.[26]

Faustino González, writing a year later, particularly lamented the missions' inability to supply the *gentiles* with food. These nomads had come to the pueblos of the Pimería Alta in ever greater numbers because of the prolonged drought that parched the desert, seeking help from the missionaries and their networks of O'odham kinfolk.

Drought, disease, insurrection, and the perennial hostilities occasioned by Apache raids on missions, ranches, and villas defined the independence struggle (1810–21), the short-lived imperial regime of Agustín de Iturbide (1821–23), and the constitution of the Mexican Republic (1824) as a time of crisis and hardship on the Sonoran frontier. The second half of the 1820s marked a period of political upheaval and institutional change that proved critical for the precarious economic administration of the missions. Sonora, Ostimuri, and Sinaloa were joined together in the State of Occidente (1825), reflecting the contours of the colonial intendancy of Arizpe, but separated into two states, Sonora and Sinaloa, in 1831. The state governors and fledgling legislatures under Mexico's first federalist constitution represented new political entities to which the missionaries turned, to little avail, for military protection and financial support.

TABLE 14 Spiritual and Temporal State of the Missions of Pimería Alta, Province of Sonora, 1818

Missions	*Indians*					*Spaniards and Castas*			
	Baptisms		*Marriages*	*Burials*	*Souls*	*Baptisms*	*Marriages*	*Burials*	*Souls*
	In pueblo	*Gentiles*							
Oquitoa	7	8	4	12	126	13	5	3	163
Tubutama	2	0	2	4	35	6	3	7	130
Sáric	0	3	0	1	25	4	2	3	55
Tumacacori	2	5	3	5	105	4	3	4	35
San Xavier del Bac	23	6	15	51	287	1	0	2	37
Cocóspera	9	3	5	4	87	2	2	0	36
San Ygnacio	2	1	1	3	36	78	25	78	1,300
Caborca, Visanig, and Pitic, *visitas*	20	31	11	26	393	11	4	6	175
Totals	65	57	41	106	1,094	119	44	103	1,931

Source: Archivo General de la Nación, Misiones, vol. 3, 1818–21.

Note:
The source for tables 14 and 15 named the dependent villages (*visitas*) for each mission, but reported the data for each mission as a whole. The *visitas* are Oquitoa and Atti; Tubutama and Santa Teresa; San Xavier del Bac and Tucson; San Ignacio and Magdalena; and Caborca, Visanig, and Pitic.

TABLE 15 Pimería Alta Mission Crops and Livestock, 1818

	Grains in fanegas					*Livestock*				
Missions	*Wheat*	*Maize*	*Beans*	*Chickpeas*	*Lentils*	*Cattle*	*Sheep/goats*	*Horses*	*Mules*	*Burros*
Oquitoa	241	0	6	2	3	3,050	408	406	68	20
Tubutama	300	8	15	0	4	3,000	700	500	16	15
Sáric	200	0	5	0	0	300	800	60	12	10
Tumacacori	150	0	40	2	0	5,000	2,500	600	89	15
San Xavier del Bac	340	53	34	8	16	7,000	1,186	243	43	10
Cocóspera	120	70	24	2	4	2,095	790	208	11	5
San Ygnacio	170	0	10	1	1	2,000	990	110	5	4
Caborca	340	7	6	0	0	700	500	454	11	11
Totals	1,861	138	140	15	28	23,145	7,874	2,581	254	90

Source: Archivo General de la Nación, Misiones, vol. 3, 1818–21.

Spain's refusal to recognize Mexican sovereignty fueled mistrust among Creole patriots and led to the national decree of December 1827, ordering the expulsion of all Spaniards, followed by similar laws enacted in the states.[27] During the spring of 1828, the military commander of the State of Occidente ordered the expulsion of all Spanish clerics, effectively leaving the missions without resident friars. In 1818, ten Franciscans served in the eight missions of the Pimería Alta; after 1828, the number of missionaries oscillated between one and four. From that time until December 1830, state authorities placed mission properties under the administration of lay commissioners, returning them to the care of the Franciscans of the Colegio de Querétaro after persistent lobbying by the designated president of the missions of the Pimería Alta, Mexican-born José María Pérez Llera, bolstered by the Indians' complaints and the recommendation of Manuel Escalante y Arvizu, political chief of the department of Arizpe.[28]

Pérez Llera corresponded assiduously with state officials, as well as his superiors in Querétaro, expressing in impassioned terms the critical state of his missions' economy and spiritual health. According to his appraisal, the commissioners had stripped the missions of much of their livestock and stored grains and supplies, due to corruption and mismanagement, and they had allowed *vecinos* to make new encroachments on communal lands. In only two of the missions, native governors served as commissioners on an interim basis: Ramón Pamplona at Tumacácori and Nicolás Martínez in Cocóspera, who were appointed by the friars before their forced departure. Pamplona, in particular, was praised for his caretaking of the herds and stored harvests of Tumacácori.[29] Pérez Llera's report of the mission inventories he received in 1830 indeed showed significant losses when compared with the figures recorded by his predecessors in 1818–19. Sáric, located at the headwaters of the Altar River upstream from Tubutama, had been abandoned following an especially severe Apache raid; similarly, Tumacácori had suffered heavy losses due to Apache raids on the mission's herds at Calabasas. Livestock in beef cattle, sheep and goats, and horses, mules, and burros numbered only in the tens and hundreds at all the missions. Cultivated lands, described as irrigated fields (*labores*) and orchards, were diminished, owing to lack of water in Atí and Cocóspera, and to the absence of ministers to oversee planting and harvesting in San Xavier del Bac. Tubutama and Santa Teresa had good orchards, but the fields were left unplanted.[30]

The abandonment of arable land hitherto considered mission property may be misleading, however, since a number of the missions had leased some of their fields to *vecinos*. Rental contracts, paid in a combination of crops and money, became more frequent during the administration of the civil commissioners. Cowboys, shepherds, and field hands, comprising Indians and mestizos, were paid in food rations, in conditions similar to hacienda peonage. Finally, the diminished livestock herds should not be attributed solely to drought and the daring incursions of Apache raiders since the commissioners had sold substantial portions of the herds to pay their own salaries and the wages of contractual laborers, as well as miscellaneous debts claimed against the missions.[31] It is telling that the detailed accounting sheets for the Chiquitos missions analyzed above, indicating the revenues they generated for the Santa Cruz prefecture, are not replicated for Sonora during the early republican period. Their absence indicates that the corporate economy of the missions figured scarcely, if at all, in the public finances of the postcolonial administration of Sonora. Positive evidence of the reduced importance of mission production and trade, parallel with the expansion of private landholding and commercial networks, comes from a series of published reports authored by Mexican officials and foreign travelers. Their common concerns focused on locating exploitable mineral and agricultural resources and on developing a profitable political economy that would generate capital assets and sustain commercial production.

Sonoran statesmen, entrepreneurs, military officers, missionaries, and parish priests, as well as outside observers, offered different visions of the region's economic resources and potential for development. During the early republican period, the district capitals of the States of Occidente and Sonora reverberated with complaints of hardship in the countryside. They cited losses of life, livestock, and crops due to raiding bands of Apaches from the northeastern sierras, Seris from the desert coast, and armed uprisings of Yaquis in the central mining and agricultural zones of Ostimuri and Sinaloa. Since the wars for independence and the unraveling of the colonial administration, the military garrisons that had maintained the region's presidios were severely underfunded. Soldiers went unpaid and deserted their posts; the Opata companies of Bavispe and Bacoachi and the Pima company of San Rafael de Buenavista, stationed at Tubac in 1787, were practically disbanded, their past military prowess unheeded. The peace encampments formed at the presidios of Janos (Chihuahua), Fronteras, Bacoachi, and Tucson (Sonora) during the 1790s—

where different bands of Apaches settled under the promise of food, clothing, tools, and planting seed to raise small plots of maize—had collapsed by the mid 1820s as the supplies on which they depended dwindled for lack of fiscal resources (and political will) to maintain them.[32]

The litanies of ranches and mining camps abandoned, their desertion attributed to hostile raiding, filled official and private correspondence from this period, similar to the Franciscans' reports for the Pimería Alta of 1818 and 1830. It seems probable that the rural settlements of the Sonoran piedmont and cordillera contracted from the 1820s to the 1840s, reducing as well the agrarian economy they had generated. José Francisco Velasco, writing at mid-century, chronicled the sorry state of the presidios, with their buildings deteriorated and their garrisons reduced to half or a third of their allotted strength. *Vecinos* who had the resources to move had migrated to larger and more secure towns and cities, as exemplified by this description of Bacoachi in 1850.

> The presidial company is formed by Opatas, who should reach the number established in the official regulations for presidios, but today they are severely reduced because of impoverishment, motives for abandonment, and lack of pay. . . . It is indeed a shame to see such good soldiers reduced almost to nothing. The civilian population ten years ago was considerable, reaching 2,000 persons in addition to the garrison, but today it is reduced to a fourth of that number or less, because all the wealthier families emigrated to the interior of the state: some have established themselves in the Sonora valley, others in Ures, and many in Hermosillo [colonial Pitic], for they could no longer suffer the continual hostilities of the barbarians.[33]

Sonoran Creoles insisted in referring to nomadic raiders as "barbarians," a term that concealed the cultural and material bases for the armed resistance mounted by different indigenous groups to the enclosed ranching properties that curtailed their movements through the *monte*. Furthermore, internecine political struggles among contending factions of the governing elite and the disarray of public finances had jeopardized the peace encampments and trade that had led to a cautious modus vivendi among Spaniards, settled Indian communities, and itinerant *rancherías* during the Bourbon administration, as was illustrated by the negotiations between Apache Captain Valdés and Captain Vildósola of the presidio of Fronteras, in 1762.[34] Ignacio de Zúñiga, an early republican statesman with distinguished military service, observed candidly in 1835 that the desperate state of the presidios should not be attributed to relentless Apache raids, but, rather, that increased raiding resulted from the collapse of the presidial system of defense and negotiation.[35]

These descriptions of desolation were countered, however, by reports of the growth of urban centers serving as nodes in the expanding trade circuits that, in turn, were linked to regional networks in northwestern and western Mexico. The port of San Fernando de Guaymas on the Sonoran coast of the Gulf of California developed as an effective entrepôt during the decades following Mexican independence. Commercial shipping frequented by American and European vessels connected Guaymas with Loreto, Baja California, and Mazatlán, San Blas, and Manzanillo, on the Pacific coast. Within the state, Guaymas supported two commercial axes, one leading directly eastward to Hermosillo and the inland valleys of San Miguel, Sonora, and Oposura, the other pointing southeast through the Yaqui and Mayo valleys to Alamos, a colonial mining center that retained its preeminence and urban status to the end of the nineteenth century. British navy lieutenant and mining prospector Robert Hardy, who traveled through the northwestern mainland and the Gulf in 1825–28, pronounced Guaymas to be the best harbor in Mexico, "surrounded by land on all sides and protected from the winds by high hills." Mexican statesman and judge José Agustín de Escudero concurred, reporting confidently that the Guaymas harbor was deep and spacious enough to accommodate up to one hundred seagoing vessels. Hardy was singularly unimpressed with the town of Guaymas, which he dismissed as a cluster of flat-roofed adobe houses; in 1828, its population did not reach one thousand souls, but by 1850, it had surpassed three thousand.[36]

The villa of Pitic, renamed Hermosillo and elevated to the status of a city in 1828, grew in population and importance due, in large measure, to its proximity to Guaymas. Hardy identified Pitic as "a place of considerable commerce and the chief residence of the most opulent merchants of Upper Sonora," estimating its population at 5,000. By 1830, that estimate had grown to 9,000, and a decade later, Velasco reported a census for Hermosillo of 13,655 persons, including as many as 2,000 Yaqui Indians who had not been counted. Despite the modesty of Pitic's architecture and dusty streets, its economy thrived on the local production of corn, wheat, sugarcane, fruit, vegetables, and wine and brandy. A network of irrigation canals watered the town and its gardens; livestock, grazing outside the villa, provided work animals (horses, mules, and burros), meat, hides, milk, butter, and cheese.[37] Overseas commerce brought luxuries to the Pitiqueño elite and stimulated overland trade to the interior by mule train. Hardy wrote, "Tea, coffee, chocolate, and white sugar, as well as china cups and saucers are all imported from the East Indies, Lima, and the United States. This place serves, too, as a depôt for effects of every description imported at Guaymas, and designed for the markets of Upper Sonora and New Mexico, receiving in return gold, silver, copper, serapes and wheat, which are respectively shipped at the port of Guaymas for Loreto, Mazatlán, San Blas, Acapulco, and China."[38]

Small farm plots (*labores de los vecinos*) and large haciendas in the San Miguel and Sonora river valleys supported towns of various sizes and categories (villas and pueblos), as well as a number of private estates with tenant families and sharecroppers: Alamito, Zacatón, Codorachi, Topahue, Agua Salada, Concepción, and De la Huerta. Hardy praised the well-run Hacienda de la Labor, owned and managed by Joaquín Astiazarán, where he observed orchards with grafted trees, well-ordered fields, and a library stocked with agricultural treatises.[39] Farther upstream, the villas of San Miguel de Horcasitas and Ures hovered between two and three thousand inhabitants, in which a network of notable families—including the Aguilar, Contreras, Pesqueira, Iñigo Ruiz, Bustamante, Moreno, and Escobosa clans—prospered, intermarried, and supplied most of the parish curates of Sonora.

In the Sonora river valley midway between Ures and Arizpe, the former Jesuit mission of Baviácora, a *visita* of San Pedro de Aconchi whose ledger books provided the basis for the analysis of mission economy presented in chapter 2, comprised a parish principally of Opata peasant cultivators, estimated at six hundred souls in the mid-1820s. Their priest, who came from the Escobosa family and was related to the powerful Aguilar lineage of San Miguel de Horcasitas, served as a deputy to the legislature of the State of Occidente.[40] The alluvial valley in which the Indians and *vecinos* of Baviácora and Aconchi irrigated their fields marked a northerly route to the pueblos of Huépac and Banámichi and the city of Arizpe, and a trail leading eastward through the Sierra de las Palomas to the valley of Oposura. Both of these routes passed near the vestiges of colonial mines of earlier fame, such as Motepore, San Juan Bautista, and Nacozari, and new encampments where prospectors hoped to recover old veins and discover new bonanzas. In the environs of Baviácora, findings of gold and silver at Dolores and San Antonio showed promise; in the hills surrounding the abandoned *real* of San Juan Bautista, La Descubridora, San Pedro de Vigilla, and San José indicated but three of many sites that had supported the traditional export wealth of Sonora. Their abandonment was attributed to disputed ownership, flooding, or the depletion of high-grade minerals.[41]

The pueblos, haciendas, and *rancherías* of Oposura and Tepache were represented in some detail in the parish censuses of 1803, analyzed in chapter 4. Both parishes showed a mixed population of Indians, mestizos, and *castas*, with an implicit social and economic range of elite proprietors, independent *rancheros*, and rural laborers. Their demographic profiles reflected the combined origins of these settlements in the missions, mines, and ranches of Sonora. Two decades later, Robert Hardy passed through Jamaica, the property of priest Julián Moreno of Oposura, and the villages of Jécori, Pivipa, and Térapa, where he was hosted by José Therán, before reaching the town of Tepache, "situated in a beautiful ravine, having groves of orange-trees in every part of it, and several waterfalls, which serve for turning flour-mills. Tepache has long been celebrated for the beauty of its females and for the valour of its Opata Indians, who have considerable possessions of cultivated lands near it." Hardy estimated its population at 800; in 1803, the entire parish comprised 1,636 souls, but the persons enumerated for the pueblo of Santa Ana de Tepache totaled 292.[42] In 1803, Oposura presented a total of 1,468 souls among Spaniards, Indians, and mestizos, distributed across two villages, three haciendas, and three *rancherías*. In 1822, Oposura registered a population of 2,534; at midcentury, now renamed Moctezuma, the villa had an estimated 2,000 to 3,000 inhabitants. José Francisco Velasco cautioned that Moctezuma had suffered numerous incursions of Apaches, to account for the migration of some of its citizenry to more secure places, while Agustín de Escudero praised its hardworking *vecinos* and noted its mixed economy of agriculture, woolen textiles, trade in tanned hides, and stock raising.[43]

Farther south, the formerly productive mines of Soyopa, Río Chico, and San Antonio de la Huerta, interspersed with Pima and Eudeve missions along the middle drainages of the Yaqui river, were reduced to pueblos by the mid-1820s. San Antonio de la Huerta had formed the hub of the eighteenth-century commercial network in which Jesuit missions exchanged their produce for trade goods, but its profitability as a silver mine diminished due to the costs of draining water out of its tunnels. Nevertheless, prospectors opened new gold placer mines within its environs.[44] The hills east of the Yaqui and Mayo rivers harbored two important colonial mining centers that continued to produce silver during the nineteenth century: Baroyeca and Alamos. In 1849, the villa of

Baroyeca, with an estimated 1,400 inhabitants, upheld its status as the parish seat and the administrative center of its district. Its mines produced over two thousand *marcos* of silver each year, wealth offset only by the aridity of its environment, which had made it necessary to dig deep wells to supply the town with water, leaving the outskirts devoid of vegetation.[45]

Two principal mines, Promontorio and La Aduana, supported the markets and urban stability of Alamos. Hardy reported that together they produced the equivalent of eighty thousand dollars worth of precious metals and required a large quantity of mercury imported through Guaymas. Although the estimated population of Alamos fell from 6,000 in the mid-1820s to 4,300 at mid-century, its mines produced twenty-five thousand *marcos* of silver per year and employed an average of three hundred day laborers, many of them Yaqui and Mayo Indians. Local commerce traded as much as four hundred thousand pesos annually in merchandise, in addition to the consumption of foodstuffs—grains, beef, and brandy. Alamos lacked running streams sufficient for irrigation, but the city was surrounded with citrus orchards and vineyards, as well as maize plantings dependent on rainfall.[46]

It was in the northwestern region of the Pimería Alta, however, that new mineral strikes led to significant bonanzas in the decades following Mexican independence. From colonial *reales* at San Ildefonso de la Cieneguilla, Santa Ana, and the legendary surface deposits of virgin silver at Arizona, prospectors developed new mining placers along the Altar River and in the vicinity of the O'odham village and ceremonial site of Quitobac. Mining entrepreneurs, in nearly all cases, followed indigenous *gambusinos*—usually identified as Yaquis—who made the initial discoveries. The profitability and duration of these placer mines depended to a large degree on the commercial networks of merchants and itinerant peddlers (*rescatadores*) who supplied the workers with food, water, tools, and dry goods and recovered the grains and nuggets of gold. Often it was the case that mines ceased to produce not because their wealth had entirely played out, but because a new bonanza lured laborers, *rescatadores*, and merchant backers away from established sites. The lucrative mines opened in the Altar Desert near Quitobac, from 1834 to 1844—at San Antonio, Sonoita, El Zoñe, La Basura, San Perfecto, Las Palomas, El Alamo, El Muerto, and Vado Seco—caused the abandonment of San Francisco, a center of gold placers and mines near Cieneguilla.[47]

José Francisco Velasco, whose detailed statistical report has been cited throughout this section, wrote knowledgeably and candidly about the fortunes quickly amassed, but often dissipated, and the layers of dependency that defined the structure of the mining industry. Velasco in 1807 served as the second administrative officer at the *real* of San Francisco, where he kept monthly registers of the gold collected by the merchants established there in order to remit these lists to the officers of the intendancy. Velasco estimated that during the height of production, from 1803 to 1810, San Francisco yielded 4 to 5 million pesos in gold each year. The sequence of exploitation there typified the pattern of mineral extraction in the desert washes and hills of northwestern Sonora. After the initial discovery of rich surface deposits, these placers would give out after a few years, requiring the excavation of shaft wells (*pozos*) and tunnels to follow the veins of precious metal until they gave out (*en borrasca*).[48]

Equally striking is the labor system employed in these mines, in practice a series of unequal exchanges in which the *rescatadores* acted as intermediaries between the workers and the merchant-proprietors who claimed ownership of the sites and supplied the tools and installations for their operation. Even in formal mining *reales* like Cieneguilla and San Francisco, where the ore was processed in large benefiting patios (*haciendas de beneficio*) and, as we have seen, production was controlled and taxed to a degree, there were parallel operations through the *rescatadores* who purchased the nuggets that workers could take out on their own on Saturdays. Velasco observed in Cienguilla that from these informal exchanges "more than 50 small patios were kept busy on a daily basis, called *maquila*, whose speculation made their operators wealthy." When the mines of Cieneguilla had passed their peak of high-grade production, the gold placers once again became the province of indigenous *gambusinos* who scratched out of the earth a daily yield of six to eight reales—perhaps three to four times the monetary value of rations and wages paid to rural day laborers.[49]

Velasco's history of the mines near Quitobac places the uneven quality of these exchanges in bold relief. Gold placers at San Antonio came into production in 1835, when a family of O'odham Indians, after consulting their missionary Faustino González, revealed the site to Dionisio González, a *vecino* of the villa of Guadalupe outside the presidio of Altar. The news spread quickly, "in an instant whites and Indians together began scouring the crevices and hillsides that seemed most promising."

> Indians from the immediate rancherías and vecinos from Altar and Caborca joined the first prospectors who had rushed to the site. During the early days of

> the discovery, for lack of weights and measures, those speculators who took money and copper coins, taking advantage of the innocence of the Indians, exchanged these for gold in bulk: for example, what could be held in the palm of the hand for 10, 15, or 20 pesos or, again, the gold dust that filled a small reed.

In this way, these *rescatadores* enriched themselves, while the women who sold the Indian workers food charged exorbitant prices, taking away a *marco* of gold in profits. At this time, Velasco observed dryly, Dionisio González made his fortune.[50]

This overview of the commercial economy of Sonora during the first half of the nineteenth century highlights the presence of different indigenous groups in the production of foodstuffs, extractive industries, and regional markets. Opatas, Yaquis, Mayos, and Pimas appear repeatedly in contemporary documents, but, for the most part, outside the institutional context of the missions. The communal agrarian holdings of Indian pueblos shrank under the pressure of legal measures to privatize the land, and native workers—whether peasant cultivators like the Opatas of Baviácora and Tepache, mine workers, muleteers, or hacienda laborers—sought their livelihood as individuals and families. At mid-century, according to Agustín de Escudero, small-scale farmers worked subsistence plots that at one time belonged to the towns where they lived and over which they began to acquire property or use-rights. Peonage defined the labor systems used in private estates.

> Haciendas and ranchos are served by peons who generally are Indians—Mayos, Yaquis, Opatas—or mulattoes, although the latter are less constant and faithful than the former. The peons' wages are five or six pesos a month and a weekly ration of two almudes of maize. It is customary to advance the peons their wages for five or six months, resulting in a form of bondage, but their flight restores a kind of balance, because it is very difficult to find them in the vast woodlands of the Yaqui River or wherever they have gone.

"This mutual deception between servant and master," concluded Escudero, "is a great obstacle for the advancement of industry."[51]

Native Sonorans did not merely acquiesce to the new landscapes created by the haciendas, nor did they easily abandon their communities. Julián Moreno, the entrepreneurial priest of Oposura who accompanied Robert Hardy to visit the mines in his parish, became embroiled ten years later in a bitter contest with the Opata governors "and all the *común*" of Oposura and Cumpas over the last remaining irrigated piece of communal land claimed by the pueblo of

Cumpas.[52] Through legal petitions, judicial litigation, and armed revolt, native communities claimed their place in the new economic and political orders of the Mexican Republic and challenged the discriminatory definition of citizenship that Sonoran political notables had written into the constitution and enforced through their militias and governing power.

Society, Citizenship, and Political Culture

The rush to find gold in the sand and garner the profits from mining bonanzas provided dramatic instances of material struggles over territory and resources among different ethnic and social contenders. Conflicts over space, livelihood, and profit shaped, in fundamental ways, the emerging political cultures of both Sonora and Chiquitos. Native governors, Creole politicians, landholders, and ecclesiastics, as well as peasants, artisans, and laborers, contested the meanings of citizenship, community, and identity in the changing institutional context of the early postcolonial states of Mexico and Bolivia. They fought wars of words on paper and in direct confrontations, erupting into armed conflict over land, water, livestock, precious minerals, and the rules governing their exchange in the marketplace. These opposing claims to the power of surveillance and taxation, on the one hand, and to autonomy and communal patrimony, on the other, employed languages of loyalty, allegiance, and reciprocity to different ends.

The philosophical and juridical foundations for these debates in both regions stemmed from the Spanish Constitution of Cádiz (1812) and its apparently radical departure from the dual structure of *república de indios* and *república de españoles* to profess the equality of all subjects within the imperial realm. After the violent defeat of the colonial order and the troubled beginnings of national republics in Mexico and Bolivia, the legacy of Cádiz led to comparable, but distinct, outcomes in their respective frontier provinces. The formative political culture of Sonora proclaimed universal citizenship, but it conferred its protections unevenly in a racially divided society. In Chiquitos, on the periphery of the Bolivian Republic, the constitutional definitions of citizenship created a civic and racial divide between Indians and non-Indians that persisted until the mid-twentieth century. These differences in the definition of citizenship in each region had a significant impact on postcolonial adjustments by native Sonorans and Chiquitanos to refashion their communities and forge linkages to the institutions of the nation-state. A key question facing indigenous leaders concerned the role of the *cabildo* in the new political climate that restructured elections and established new criteria for citizenship.

Sonora

Nearly a quarter century of historical research into the indigenous polities of nineteenth-century Mexico provides a reference point from which to approach the processes of conflict and shifting alliances in Sonora. Well-documented studies focused mainly on central, southern, and southeastern Mesoamerica—including the modern states of Puebla, Guerrero, Yucatán, the Huasteca region of Veracruz, and the Quiché peoples of highland Guatemala—have demonstrated a number of recurring themes to guide further research on the political cultures and social divisions of northwestern Mexico. Important postcolonial institutions that strengthened internal cohesion and provided leverage with municipal, state, and federal authorities evolved from religious, political, and military realms of public life.

The *fiesta-cargo* system of officeholding established a hierarchy of interlocking posts between the religious confraternities (*cofradías*) and indigenous village governance; each rank represented specific duties (*cargos*) and levels of prestige, usually associated with the organization and financing of religious festivals in honor of the titular saint of the pueblo. Clearly, the social and symbolic significance of religious fiestas to venerate local saints, following the Catholic liturgical calendar, had its roots in the colonial past of Christian evangelization. The formal structures of ranked offices, distinguished by age, gender, and ethnicity, came into place, however, during the postcolonial period of challenges to the political autonomy and communal land base of Indian pueblos. Comparable in some ways to the rituals of "folk Catholicism" described for a number of indigenous groups of northern Mexico in chapter 6, the *fiesta-cargo* system bestowed solemnity to the political rites of election and officeholding; established reserved privileges and prestige for the *pasados*, a village elite of older men and former officials; and confirmed the honored language of petition and address, in the name of the *común*, before Mexican authorities.[53] Nevertheless, it is telling that with the important exception of the Yaqui and Mayo pueblos, the structured institutions of ritualized officeholding known as the *fiesta-cargo* system were not replicated among the indigenous villagers of Sonora, whose political and religious cultures were nurtured in the missions.

National guard service comprised a second important institution that molded power relations within communities and involved them in the struggles between liberal and conservative factions of the political elite. Through the national guard, Indians participated directly in the defense of national sovereignty against the North American (1846–48) and French invasions (1865–67) of Mexican territory. Indigenous militias' claims to distinguished service provided important leverage for their communities before national and state governments often beholden to these irregular units but disparaging of their political integrity in the pursuit of policies to divide communal lands and recruit peasant cultivators as dependent laborers on private estates.[54] In significant contrast to the history of the national guards in central Mexico, Opata and Pima auxiliary troops in Sonora could not parlay the prestige they had garnered under the colonial General Commandancy of the Interior Provinces into an acknowledged political role in the early republic. Notwithstanding the need for militias to defend Sonoran pueblos, haciendas, and ranchos from the incursions of Apache raiders, the demise of the presidial system undercut the merits that would have substantiated indigenous soldiers' claims to citizenship.

The histories of Indian communities and the Mexican state in different regions have developed a number of interpretive principles that refer to the adaptation and reformulation of ethnic and cultural allegiances and to the contested nature of political hegemony.[55] The postcolonial developments of northwestern Mexico intersect with this literature to underscore the following points. First, these are not histories of a simple division between "white" elites and "Indian" plebeians and peasants, nor is it easy to define the indigenous community as a cohesive unit. Second, the defense of local customs (*usos y costumbres*) and communal lands, while genuinely motivated by compelling circumstances, as often as not gave expression to threads of disunity and complex entanglements of conflict and distrust within communities and the polities they claimed to represent.

New cleavages opened between native commoners and elites, the latter variously identified as caciques, village officers, landholders, or mercantile entrepreneurs. In reference to generational splits, service in the national guard provided an alternative ladder to prestige for young men parallel to the *fiesta-cargo* system of rankings dominated by their elders. Flight from the constraints of community politics coupled with the search for a livelihood led individuals and families into tenancy and sharecropping contracts on private estates, without necessarily renouncing their cultural allegiances to village or parish. These journeys through local spaces that crossed property boundaries immersed their travelers into the social and ethnic borderlands of *mestizaje*, as occurred with the *indios laboríos* (Indians who worked under contract) and *castas* listed in the 1803 censuses for the pueblos and ranches of Tepache and Oposura, and changed the political configurations of their home communities.[56]

Struggles over the symbols of power and political practices relating to elections, taxation, adjudication of disputes, and allocation of posts reached into the heart of local communities and linked them, in turn, with regional and national conflicts over church-state relations, administrative and territorial divisions, and opposing federalist and centralist formulas for constitutional government. Intraethnic and intracommunity divisions frequently led to interclass and cross-ethnic alliances, involving different and even opposing groups of elite politicians and indigenous leaders, militias, and commoners. What stands out in the histories of political struggles in nineteenth-century Mexico is the importance of locality in the articulation of spatial and socio-political differences. Their stories center on the resistance of peasant and indigenous political actors to the modernizing projects of commercial production and private aggrandizement that threatened their subsistence and cultural lifeways.[57] Without denying the salience of these structural issues, it is important to recognize that popular movements did not merely react to the innovations of modernization; rather, they coalesced around agendas of their own making, grounded in local concerns. Indigenous and peasant leaders articulated their versions of republican ideologies of citizenship and equality, fashioning their particular images of the nation.[58]

The state did not figure at the center of these narratives, especially during the formative decades of the Mexican Republic; communities had deeper histories, more textured and complex in their ethnic diversity and internal conflicts than either the institutional or social formations of the state.[59] Historical and theoretical challenges to the conventional dichotomies of traditional versus modern, or popular versus elite, emerge in bold relief when our attention turns to the nomadic frontiers of northern provinces like Sonora, Chihuahua, Coahuila, and Texas, so distant geographically and politically from the urban centers of power. Among the seminomadic Tohono O'odham, Cunca'ac, and Apache bands of the Sonoran deserts and sierras, as well as the Pima, Opata, and Cahita riverine cultivators, territorial conflicts concerned the disputed occupation of discrete spaces *and* the very definition of boundaries. Forced sedentarism by means of military sweeps through the *rancherías* of herders and foragers accused of raiding constrained their space and freedom of movement in a reenactment of colonial policies of *reducción*. Conversely, village agriculturalists impoverished through the appropriation of communal lands by private interests and drawn into armed hostilities with state and federal forces turned episodically to raiding and guerrilla warfare, adopting nomadic modes of survival. The Yaqui and Opata rebellions of 1824–33 and the Tohono O'odham uprising of 1840–43 constituted political protests focused on territory, local governance, and the integrity of indigenous military and civilian elected offices.

Opata warriors turned against the commandancy general that they had served so well for generations, enflamed by the issues of elections and appointments of their captains general and remuneration for their service. Opatas who had fought for the cause of independence in the final stages of the insurgency, without receiving wages, returned to their villages to find their families destitute and their fields abandoned. In 1820, Juan Dórame led the Opata company of Bavispe in rebellion, drawing into the movement additional warriors from the pueblos of Arivechi, Pónida, Sahuaripa, and Tónichi. Rebel forces brought their prowess in the use of firearms and their knowledge of the arts of warfare to their movement: they held their own in several skirmishes, took several high-ranking officers prisoner, and judged them in formal councils of war. In two dramatic confrontations in Arivechi and Tónichi, rebel troops held out for several days against superior forces, but they surrendered when their ammunition was completely exhausted and their numbers depleted. Their leaders were summarily executed.[60]

Opata rebels failed to gather widespread support for their movement among the separate presidial companies of indigenous warriors, but their anger was not silenced. In 1824, when aspiring politicians from the leading families of Sonora and Sinaloa debated whether to join the unified State of Occidente, Opatas rebelled against the federal military commander of Sonora, Mariano Urrea, who had arbitrarily deposed their captain general. This affront to their dignity capped long-standing grievances, and their demands were echoed in successive revolts during the following decades: the right to elect their own officials, the exclusion of non-Indians from their territory (with the significant exception of priests), and the restitution of communal lands that the Opatas claimed had been unjustly taken from their pueblos.[61]

Within a year, in 1825, the Yaquis rose up against federal forces. Although the rebels accepted a negotiated amnesty with the governor of Occidente, their movement continued to simmer and produced an extraordinary leader who galvanized indigenous resistance and left his mark on Sonoran history: Juan Ignacio Jusacamea, widely known as Juan Banderas. Banderas chose a banner of the Virgin of Guadalupe for his emblem, perhaps emulating Father Miguel Hidalgo, and cultivated a cross-ethnic alliance of Indians in Sonora and Sinaloa, from the Fuerte River to the Pimería Alta, to repossess their lands and recover their autonomy. Banderas did not achieve a lasting pan-Indian movement, and he faced opposition within the Yaqui pueblos, but he developed effective tactics for guerrilla warfare that repeatedly put beleaguered state and federal troops on the defensive, especially since they were divided between the Apache and the Yaqui campaigns. Through several phases of armed skirmishes, raids, and amnesties, Banderas upheld his stature as captain general of the Yaqui nation.

During the truce of 1827–28, the state legislature of Occidente promulgated three decrees that undermined the autonomy of self-governing indigenous pueblos and native militias, by abolishing the rank of captain general, and accelerated the privatization of communal lands.[62] Their import ran directly counter to the political demands articulated by Opata and Yaqui rebel leaders. In 1831–32, when Occidente dissolved into the two states of Sonora and Sinaloa and conflicts brewed in the Yaqui and Mayo valleys, Banderas and his principal rival, Juan María Jusacamea, communicated separately with Mexican authorities, negotiating through priests and *vecinos* who served as intermediaries. In May 1831, Yaqui leaders presented a proposal for self-government modeled on the colonial *cabildo*. They demanded to elect one judge, an *alcalde*, with authority over the whole valley; governors, or *regidores*, in each pueblo; and two military officers, a captain general and a lieutenant general, effectively reinstating their own militias. The Sonoran legislature appeared to accept the Yaquis' proposal, but instituted two non-Indian directors for the Yaqui and Mayo valleys, with powers to oversee the elections and actions of the indigenous *alcaldes* and *regidores*. A year later, Banderas led his forces in open rebellion, joined this time by significant contingents of Opatas and Mayos recruited largely among the Indians who worked in mines and haciendas outside the pueblos. The uprising flared throughout the state and seriously compromised the political and military credibility of the fledgling Sonoran government. In December 1832, however, Banderas, his Opata ally Dolores Gutiérrez, and twenty rebel leaders were captured; in January 1833, Banderas and Gutiérrez were condemned to death. Their executions brought closure to this phase of rebellion, but it did not resolve the conflicts that underlay rural revolt.[63]

The following three decades marked a tumultuous period of civil war and foreign invasions at both the national and state levels, during which a pattern of tactical alliances evolved in Sonora among different indigenous and Creole factions. Native leaders sought to defend their territory and freedom of movement, while emerging caudillos—patriarchs of notable landholding families—gathered a military following and pursued power through combat and control of public office. Sonoran political divisions followed the national schisms between Federalists and Centralists, then Liberals and Conservatives, but their internecine conflicts were personalized in the leading figures of propertied families that coalesced around the northern presidios, the mining district of Alamos, the Hermosillo-Ures axis, and the port of Guaymas. All the competing Creole groups shared the ambition to expand their holdings and retain Indian commoners as dependent laborers on their haciendas and ranches; they differed, however, in their strategies for containing assertive native polities among the Opatas, Yaquis, and Mayos. The Centralists and Conservatives, led by Manuel María Gándara, opted for a policy of partial accommodation and actively recruited indigenous allies into their militias. Federalists and Liberals, represented by the Elías González, Urrea, and Pesqueira patrilineages, among others, vigorously pursued the privatization of village lands and confronted indigenous militancy with armed repression.[64]

Native peoples, for their part, were divided among themselves, with different poles of leadership represented by the *alcaldes* and village councils, military captains, and bands of rebels. They entered into alliances with contending political factions in order to defend their lands and local autonomy. The liberal reforms crystallized in the constitution of 1857 removed the last threads of juridical protection for corporately held land in Mexico and pointedly legislated the division and sale of church properties and village holdings. During the ensuing civil war and the French intervention of 1864–67, Yaquis, Mayos, and Opatas entered the Conservative, pro-imperialist forces in support of the administration of the Archduke Maximilian, who had decreed protective guarantees for Indian communities to gain title to and claim restoration of their communal lands.[65]

The Tohono O'odham revolts of the 1840s in the Altar Desert, occurring beyond the radius of tactical alliances with Creole caudillos, expressed some of these same issues, but in different terms. Bands of O'odham hunter-gatherers and cultivators maintained a relationship to the land no less profound, but distinct, from that of the riverine pueblos of central Sonora and the alluvial deltas of the lower Yaqui and Mayo valleys. They practiced a kind of alternating nomadism in the arid environment of northwestern Sonora with seasonal migrations to the fields (*oidag*) they cleared at the mouths of washes during the summer rains and to the wells (*wahia*) in the hills for hunting and gathering during the winter months. The Tohono O'odham did not perceive the land as property, but rather as a living, nurturing environment. Theirs was an ecologically varied territory in which extended families established their campsites for procuring different resources in the desert and along the shallow arroyos according to the seasons. Their survival depended on the freedom to move through this territory and to gain access to the harvests of the riverine villages in times of drought and scarcity. As we saw in chapter 6, the O'odham made the desert their homeland and peopled it with spirits associated with mountains, wildlife, and livestock, and honored through the rituals of the devil way.

The extreme aridity of the Altar Desert provided a natural barrier to the expansion of commercial estates into the heart of the Tohono O'odham world. Nevertheless, the growth of private ranches and haciendas in the greater Pimería Alta, as the histories of disputed land titles in Ocuca and Santa Ana (in chapter 3) illustrated, and the loss of cultivated lands in the missions of the Magdalena and Concepción valleys significantly diminished the resources that sustained the subsistence patterns of O'odham villagers and nomads. We may recall Faustino Gonzalez's lament that during the severe drought of 1816–19, the missions of Pimería Alta could no longer offer refuge to the migratory Papagos (O'odham) who had come to the pueblos in ever greater numbers in search of food.

Mexican ranchers saw the desert as an unfenced range in which their herds could roam, impinging even further on the vegetation and scarce catchments of water in the area. Domesticated livestock occupied an important place in O'odham material culture, but in tension with wild game, as revealed in their cosmology and healing traditions from the mid-nineteenth century to the present.[66] The ecological and cultural impacts of ranching were compounded by the feverish rush to discover and exploit placer mines near Quitobac, bringing several thousand prospectors and itinerant merchants to the region. O'odham *gambusinos*, who together with Yaqui laborers were often credited with finding the mineral deposits, confronted a bewildering mixture of ethnic and social groups who moved from one bonanza to another in a landscape rapidly and radically transformed.[67]

Conflicts mounted during the 1830s over mutual accusations of cattle rustling between O'odham and *vecinos*. The difficulties of claiming ownership of livestock in the open range were complicated further by roaming bands of Yumas, Cocomaricopas, and Apaches, among other groups of *gentiles*, through the region. O'odham leaders complained to Sonoran authorities that Apaches from the peace encampment of Tucson moved westward into their territory, "from one direction to another, with no other objective than to explore the land and see for themselves the goods [livestock] that inhabit it and the number of men and weapons there."[68]

The presidio of Altar and its contiguous settlement, the villa of Guadalupe, constituted the administrative and military hub of this western portion of the Pimería Alta. Presidial officers attempted to extend their authority over the *vecinos* and Indians who lived in the villages of the Altar and Concepción rivers, from Tubutama to Caborca, and over the shifting unruly mining camps of the Altar Desert. To that end, consonant with Sonoran state policies toward the Opatas and Yaquis, commander Antonio Urrea appointed governors of different *rancherías*, as well as captains and lieutenants over the entire "Papago nation," in an effort to establish a hierarchy of accountability among the seminomadic bands of O'odham.[69] The ephemeral connection between these *rancherías* and particular places, however, added to the frequent turnover of native officers. In this nomadic frontier Mexican officers found it difficult to distinguish between *gentiles* and the Tohono O'odham who lived seasonally in the pueblos, mines, and ranches of the zone, or to enforce the appointment of ranked indigenous authorities to carry out orders emanating from the presidio.

Armed conflict began in May of 1840, the "painful month, when the saguaro seeds turn black" (*Kai chukalig mashad*),[70] before the onset of summer rains, when Mexican forces carried out an unauthorized attack on the O'odham *ranchería* of Tecolote. Convinced that these Indians had stolen over three hundred head of cattle and horses over the preceding two years, the subprefect of the villa of Guadalupe, Dionisio González (who had reaped his fortune at the placer of San Antonio de Quitobac) and the second commanding officer of Altar, Rafael Moraga, led 360 armed troops into Tecolote and opened fire, killing 11 O'odham civilians, 1 indigenous auxiliary, and 6 militia soldiers. Governor Manuel María Gándara reprimanded González and removed him from his post, noting pointedly that it would be "malicious and imprudent" to open another hostile front with a "numerous tribe that had kept good faith with the government" at a time when Federalists and Centralists contended for control of the state government and regular troops were engaged in sporadic skirmishes against bands of Yaquis and Opatas and in continuous confrontations with Apaches.[71]

The massacre at Tecolote had repercussions throughout the western Pimería. Many of the burgeoning placer mines were temporarily abandoned as *vecinos* and Indians fled to the pueblos of Caborca and Pitiquito. Yaqui workers left the mines, fearful that they would be implicated in the violent events of Tecolote; entire families "suffered hunger and thirst" without paid work and deliberated whether to return to the Yaqui valley or move eastward to the central portions of the state. The Tohono O'odham took refuge in a number of *rancherías* and migrated as far as the Pima villages in the Gila river valley. Rumors continued to excite the *vecinos* of Guadalupe, Cieneguilla, and the riverine pueblos of Altar-Concepción over stolen livestock and the threats of imminent rebellion by O'odham, Gila Pimas, and Yaquis.[72] Isolated reports of skirmishes and raids continued throughout the decade, associated with cattle theft and followed by punitive expeditions, but without achieving a lasting resolution.

The concentration of *vecinos* in the presidio and the pueblos contrasted with the dispersal of Indians in diverse *rancherías*, under the leadership of multiple chiefs and governors, where they defied orders on several occasions to turn over suspected rustlers to presidial officers. It appears that in these encampments of extended families the O'odham found strength in the familiarity of their desert environment. Sonoran authorities opened negotiations in the region after the events of Tecolote and authorized native governors to travel to the state capital to present their demands and complaints. The language of civilian prefects and military commanders, however, required of the Indians their submission to the "superior government and the other authorities established by law."[73] State officials undertook to resettle the O'odham in permanent villages, following the model of mission *reducciones*, with the intent of obligating them to obtain passports in order to go on hunting expeditions or visit distant *rancherías*. In one telling episode, Tohono O'odham sent messengers to the presidial commander of Altar to request a license to harvest the fruit of the saguaro cactus "to make their syrup and sweets from the seeds, as was their custom when they were at peace." The O'odham offered to reveal the location of three new gold placers they had discovered, thus appealing to what they believed most interested the Mexican authorities.[74]

What do these episodes of armed conflict, negotiation, and shifting alliances show us about the political cultures that developed in Sonora during the early formation of the Mexican Republic? The foregoing stories express in words and actions the contested meanings of citizenship and their practical applications in the uneven distribution of rights and obligations through the web of relations that developed between the state and different segments of regional society. Indian communities were undergoing profound changes in their demographic composition, internal governance, and built landscapes. The heterogeneous population of *laboríos* who lived and worked in haciendas and mines were numerically and culturally as significant as the *hijos de pueblo* (residents of indigenous communities) in the villages. They comprised discrete categories in official censuses and correspondence, but the same people—individuals and extended families—occupied more than one of these categories in their lifetimes and in the course of this troubled period.

The environment is woven into the narrative through climatic events signaled by repeated droughts and crop failures that punctuated the accounts of armed confrontations and through the cycles of planting, harvesting, hunting, and gathering that marked the continuities of subsistence. Environmental perceptions underlay the conceptual crossovers between geographical space, expressed through the physical contours of the land, and the political spaces so bitterly disputed for the symbols of rank and prestige and for the exercise of power. What was at stake in the nineteenth-century rural movements of Sonora brought together the major issues of land and the struggle for control over resources; ethnic allegiances and national citizenship, particularly in reference to local governance, elections, and indigenous militias. Reported variously as revolts, rebellions, or uprisings, they were *not* caste wars between unified blocks of Indians and whites, notwithstanding the racialized language that filled the documents.[75] Rather, the seemingly contradictory succession of armed encounters and strategic alliances across ethnic boundaries strongly suggests a dynamic of internal discussions and disputed leadership within each of these contending groups.

Creole notables who split into rival factions in their bid to control state offices sought to aggrandize their landholdings and convert indigenous peasants into dependent laborers. Yet, time and again, they turned to Indian auxiliaries to defend the frontier or to augment their partisan militias. State authorities and private citizens recruited indigenous laborers, but attempted to control their mobility by imposing systems of passports and licenses to travel. Native polities, represented by the *común* of Opata and Yaqui pueblos, as well as by the dispersed *rancherías* of the O'odham, tenaciously defended their political autonomy and the territories in which they dwelt and from which they drew their material and spiritual livelihoods. Their stance for autonomy asserted through armed rebellion and articulated in their verbal and written presentations did not mean isolation. Each of these indigenous groups, to a greater or lesser degree, lived *in* Sonoran society, and they sought recognition in the political institutions of the Mexican Republic.

Chiquitos

Republican institutions and distinct political cultures developed slowly in the eastern Bolivian borderlands, shaped by their internal and external frontiers and by the colonial legacy of the missions, which, in large measure, supported the economy of Chiquitos province and of the entire prefecture of Santa Cruz. The social and spatial frontiers that guided this process hardened the separation between the Creole elite based in Santa Cruz (Cruceños) and the indigenous population of the pueblos—who constituted, by far, the majority in the region. Internal frontiers followed the rivers and overland trails that mapped the administrative divisions between departments and provinces within the lowlands and between them and the Andean highlands. External frontiers, marking Bolivia's shifting international boundaries with Peru, Brazil, Paraguay, Argentina, and Chile, played a decisive role in the formation of the nation-state and, in particular, of the territorial and political configuration of Chiquitos.[76]

Cruceño merchant and landholding families, much like their counterparts in the highlands, elaborated a political discourse rooted in the colonial distinctions between *vecinos* and *naturales*, but informed by liberal doctrines of universal participatory citizenship and popular sovereignty. The rural elite of Santa Cruz sustained a polarity between citizen (*ciudadano*) and barbarian (*bárbaro*, *salvaje*) intended to legitimate the Cruceños' control of public office in the new political order. Their sense of entitlement was challenged, however, by the established communities of Chiquitos and Moxos, representing an alternative vision of citizenship within the constitutional model of a republic. The complex fabric of mission culture woven from the strands of Catholic ritual, labor discipline, and internal governance centered in the *cabildo* had evolved in the pueblos for over a century and developed a language for addressing both ecclesiastical and secular authorities.[77] These competing discourses of deference and reciprocity, on the one hand, and of proprietorship and service, on the other, emerged from the colonial foundations of mission and *encomienda*, in which both the Chiquitano and Cruceño cultures were steeped.

The internal contradictions of inclusion and exclusion inherent in the Creole dialectic of citizen and barbarian permeated the Bolivian constitutions of the nineteenth century and gave rise to different juridical and social definitions of citizenship and the nation. From the constitutions of 1826 and 1834, as well as the civil and penal codes promulgated by Andrés de Santa Cruz in 1830–32, emerged the guiding precepts for exercising political rights and obligations in Bolivia. Based on the laws and codes emanating from the Spanish Cortes of 1812 and 1822, these foundational documents distinguished explicitly the faculties ascribed to citizens and Bolivians in the construction of the republic. The category of Bolivians comprehended all those born within the territory of Bolivia and was extended to the children of Bolivian parents born outside the country, but who stated legally their desire to live in Bolivia; foreigners who obtained a letter of naturalization or who had established three years of residency (*vecindad*) in Bolivia; veterans of insurgent troops in the battles of Junín and Ayacucho; and, finally, slaves who, by the constitution of 1826, were declared free. *Citizens* designated those Bolivians who were married heads of household or persons over twenty-one years of age (legal adults), knew how to read and write, and had employment or independent means of livelihood without being in the service or pay of another. By implication, if not stated specifically, the status of citizenship was reserved for men. All Bolivians were enjoined to live under the constitution and the laws of the land, respect and obey the legitimate authorities, contribute to public expenses, keep watch over the conservation of public liberties, and sacrifice—when necessary—their means and their lives for the health of the republic. Only citizens in full exercise of their rights (*ciudadanos en ejercicio*) could vote or be elected to public office.[78]

Following the principles of the Cortes of Cádiz, the Bolivian constitutions and basic codes of governance did not specifically differentiate Indians within the categories of Bolivian and citizen. By implication, however, very few indigenous commoners could fulfill the qualifications of citizenship: to be literate in Spanish, own property or have independent status as an artisan, professional, or merchant. Furthermore, the absence of *indio* as a discrete category in these foundational documents removed the juridical protections enshrined in colonial laws and court practices, by which Indians occupied a tutorial status in relation to the crown and the institutions of justice established in the colonies. The colonial regime had subjected Indians of the Andean highlands to forced labor (*mita*) and tribute payment—and, in the mission pueblos of the eastern lowlands, to simulated tributary and labor systems—but had granted Indians legal standing in the corporate ownership of land and afforded them access to colonial courts (at reduced cost) through the protector of Indians. These statutory conditions ceased to operate in the formative institutions of the Bolivian Republic, although specific laws and decrees addressed the social and legal status of Indians, primarily in terms of tribute payment and of their treatment as *pobres de solemnidad* (worthy poor) for reduced payment of legal fees.[79]

In contrast to the colonial order, in which Indians held a special, if subordinate, status, the projected neutrality of the republican *civitas* as an undifferentiated body politic of electors undermined the ancestral customs and cultural integrity of indigenous communities and threatened to dissolve their corporate patrimony. Significantly, the reinstitution of tribute in 1831 and the legal recognition of corporate lands until 1874 postponed the implementation of a liberal vision of political homogeneity and the privatization of wealth in the highlands until the closing years of the nineteenth century.[80] In the tropical forests and savannas of the Chiquitanía, however, the economic legacy of mission livestock and the rangeland that supported it passed into the hands of Cruceño *estancieros* through government sanctioned *denuncias* of land and officially auctioned leases of cattle, beginning in the 1850s.[81]

The contested meanings of postcolonial political practices in the nineteenth-century Andean republics have been discussed widely in the literature, principally around the themes of tribute, corvée labor, and forms of peonage. Through the material and ceremonial dimensions of tribute payment, public works, and the *fiesta-cargo* system of religious and political officeholding at the village level, historians have explored ways in which indigenous communities renegotiated their relationship to the state. Conversely, a number of historians have underscored the internal contradictions in the Creole discourses of nationality and nationalism in response to Benedict Anderson's assertion that Latin American political elites articulated the intellectual foundations of nation-states, meant to encompass heterogeneous ethnic communities, even prior to their counterparts in Europe. Clashes between the ideas and practices of modernity in the Latin American republics surfaced in the discordance between a republic of citizens constituted in terms of juridical equality and hierarchical societies in which ethnicity, family lineage, and personal relations of clientelism and servitude persisted as indelible markers of prestige and governance. Qualifications for citizenship centered on property, formal education, and European culture narrowed effective political participation to a class of Hispanicized notables to the exclusion of the majority of the population, thus belying the national sovereignty that they claimed to lead.[82] These dichotomies underlay political conflicts in the Spanish American republics, but they did not constitute a simple binary division between the governing elite and a disenfranchised mass of plebeians. The oft-mentioned Creole elites were frequently divided among themselves and locked in factional disputes for power, while the popular classes of indigenous *comuneros* (commoners) and mixed-race workers and peasants fought to exercise the rights of citizenship on local issues and joined shifting political alliances with different caudillos and political parties.[83]

Thematic analyses of changing political and social structures within communities and the wider Creole society, their relations of mutual dependency, and their different points of reference to the institutions of the state, all familiar topics among Andeanists, have been applied only sparsely to the lowlands, where these issues were less visible. The emerging militancy of Amazonian tribal peoples in the late twentieth century, however, recast the terms of political debate in Ecuador, Peru, and Bolivia and opened new perspectives on the conflicted history of nation building in their tropical frontiers. Archival material conserved from the prefecture of Santa Cruz brings into focus the ambivalent nature of citizenship among Creoles and Indians in the pueblos and *estancias* of Chiquitos during the period of early national formation in Bolivia.

Two dramatic episodes of 1815 and 1825 illustrate the contentious processes of politicization among native peoples, Cruceños, and colonial officials in the Chiquitos borderlands. In the course of active fighting during the wars for independence, Indians were divided between royalists and patriots. In 1815, when the insurgent forces of Ignacio Warnes were pitted against the royalists led by Juan Bautista de Altolaguirre and Marcelino de la Peña, the royalists fell into an ambush in Santa Barbara, a thickly wooded valley within three leagues (twelve kilometers) of the pueblo of San Rafael. Attacked from the rear, the Indians dispersed and fled to the surrounding savannas; only thirty troops and four officers survived the battle, among them de la Peña, who escaped into the *monte*. In his account, as recorded by Alcides d'Orbigny, "the carnage was fearful among the dead and wounded who covered the plain. Weary of the killing, Warnes decided to dispose of the corpses by setting ablaze the dry pasture, thus burning alive the unfortunate [victims] who were still breathing . . . more than a thousand Indians died that day." This painful memory, retold in a number of versions, cast a bitter shadow over the independence process in lowland Bolivia.[84]

Ten years later, Chiquitano *cabildo* officers were drawn into a confrontation of sovereignty and allegiance across the international boundary with Brazil. As the Ejercito Libertador Colombiano under the command of Mariscal José Antonio de Sucre entered the territory controlled by the *audiencia* of Charcas in February 1825, a remnant of the royalist forces commanded by general Pedro Antonio de Olañeta remained in Potosí. Simón Bolívar and Sucre feared a counterrevolution through Brazil, spearheaded by the European Holy Alliance (formed in Vienna in 1815), to advance the territorial aims of the Brazilian empire and restore monarchical rule in South America. The royalist governor of Chiquitos, the Spanish colonel Sebastián Ramos, requested protection from the imperial authorities of Mato Grosso in March 1825, when the insurrectionary forces in Cochabamba and Charcas broke his lines of communication to Olañeta in Potosí. Ramos served under the authority of José Videla, intendant governor of Santa Cruz, who had joined the insurgent cause. Perhaps fearful of reprisals after patriot forces took Santa Cruz, or desirous of keeping his post in Chiquitos, Ramos took two contradictory steps: he sent a letter of adherence to the cause of independence to Sucre from Santa Ana on 13 March, and on 28 March, he signed a treaty of capitulation in Mato Grosso, ceding the province of Chiquitos to Brazilian authorities. In his attempt to annex Chiquitos to Mato Grosso, Ramos apparently took as many as three hundred Chiquitos Indians across the Brazilian frontier.[85]

Brazilian troops under the command of Manuel José de Araujo e Silva invaded Chiquitos, reaching as far as the pueblo of Santa Ana, and threatened Governor Videla, should he impede their advance, that he would "strip the land and execute the garrison." Sucre, when notified, prepared to counterattack in order to recover Chiquitos and warned Araujo e Silva that he would invade Brazil in retaliation. Bolívar supported Sucre's determination to repel the Brazilian forces from Chiquitos, but admonished him not to invade Mato Grosso. Bolívar doubted that the imperial court of Pedro I in Rio de Janeiro or the European Holy Alliance supported Araujo e Silva's invasion of Chiquitos, and he wanted to avoid a military confrontation on this border parallel with the territorial dispute between Buenos Aires and Rio de Janeiro over the Banda Oriental of the Río de la Plata, later to become the Republic of Uruguay. The emperor ordered Brazilian forces to withdraw, as Bolívar had predicted, and annulled the capitulation of 28 March 1825, removing Araujo e Silva from his post. In their retreat, Araujo e Silva's forces plundered the marshes and plains of eastern Chiquitos through which they passed, taking livestock and Indian captives with them.[86]

This bizarre incident of frontier geopolitics bears on the discussion of citizenship in the conflicted transition from colony to republic in two important ways. First, had the capitulation engineered by Ramos and Araujo e Silva been enforced, it would have contemplated nothing less than the transference of all the productive resources of the mission pueblos, including their livestock and even the sacred vessels of their churches, to the imperial authorities of Brazil. Interestingly, however, it stipulated the conservation of the "political and ecclesiastical institutes" by which the *naturales*, administrators, and priests were governed. Second, Ramos attempted to legitimate the capitulation signed in Villa Bella on 28 March by convening an assembly of prominent *vecinos* and Indian officers in the pueblo of Santa Ana on 24 April, ostensibly to ratify the document and recognize the sovereignty of Pedro I over the province of Chiquitos. In the company of the priest and several militia officers, the presence of the indigenous *cabildo* lent solemnity to the act, co-signed by *alférez* Fabio Parabás, *corregidor* Manuel Rocha, *teniente* Pablo Matos, *comandante* Manuel Ticoi, and *alcalde* José Suquiriqui.[87] Ramos's pretense to unite the provinces of Chiquitos and Mato Grosso in allegiance to the Catholic monarchies of Spain and Brazil, replete with the celebration of Mass, Te Deum, and ringing of bells, may have persuaded some of the Indian leaders to join him, even as they had supported the royalist cause during the early stages of the wars for independence. The active role of the *cabildo* marked the institutional continuity between the colonial era and the nascent republic.

In accordance with the constitutional provisions for public administration, the prefecture of Santa Cruz became a department, Chiquitos remained a province, and the ten pueblos were designated cantons. Reports submitted by the governor of Chiquitos to the prefect of Santa Cruz on local governance during the 1830s listed two principal officers for each pueblo: the *corregidor*, who came from the Indian *cabildo*, and the *ecónomo juez*, signifying a new title for the administrators of mission economy: during a five-year period, 1835–39, the names presented varied only slightly. Priests were included in the roster of

TABLE 16 Local Governance in the Cantons of Chiquitos, 1835 and 1839

Canton (pueblo)	*Corregidor (from* cabildo*)*	*Ecónomo (juez de paz)*	*Curate (priest)*
1835			
San Xavier	Juan Chovirus	C. José Ignacio Ortiz	José Manuel Rodríguez
Concepción	Mateo Coasas	C. Telesforo Baca	José Jacinto Paz*
San Miguel	Felipe Cahuis	C. Mariano Rocabado	José Vicente Dorado
San Ignacio	José Chubes	C. Pedro José Baca	Pedro Nolasco Ortiz
Santa Ana	Ignacio Arayuru	C. Fran Xavier García	José Lorenzo Zespedes
San Rafael	José Poicees	C. José Manuel Bazán	Pedro Pablo Velasco
San José	Miguel Taborga	C. Juan José Montera	José Ignacio Mendez**
San Juan	Juan Bautista Chubes	C. Andrés Marquez	Urbano Vicente Baca
Santiago	Tomás Macanas	C. Toribio Segura	José Manuel Egues
Corazón	Baltazar Yobios	C. José Manuel Herrera	José Merino
1839			
San Javier	Juan Bautista Quiaras	C. José Ignacio Ortiz	José Manuel Rodríguez
Concepción	Mateo Coasas	C. Telesforo Baca	José Manuel de la Roca
San Miguel	Carlos Soriocos	C. Ramón Suárez	Miguel Antonio Urtubes
San Ignacio	José Chubes	C. Mariano Rocabado	Pedro Nolasco Ortiz
Santa Ana	Cristobal Sola	C. Mariano Bazques	José Lorenzo Zéspedes
San Rafael	José Manl Hurtado	C. José Manuel Baca	Pedro Pablo [?]
San José	Francisco Tomichas	C. Antonino Paz	José Ignacio Méndez
San Juan	Nicolas Chores	C. José Ma de Velasco	Urbano Vicente Baca
Santiago	Miguel Méndez	C. Toribio Segura	Miguel Jeronimo Duran
Corazón	Juan de Dios Motores	C. José Manuel Herrera	José Merino

*vicar for the northern pueblos
**vicar for the southern pueblos

Source: Universidad Gabriel René Moreno, Museo de Historia Fondo Prefectural, vol. 1, exp. 15. Libro Diario Año 1834, vol. 1, exp. 17, 1835.

public employees; their salaries figured as part of the provincial expenses for the province of Chiquitos and were paid out of the revenues generated from the sale of mission products (see table 16). It is noteworthy that only the Creole administrators were accorded the rank of citizens, indicated by the "C." that preceded their names, and further, that they held the dual titles of *ecónomo* and *juez de paz* (justice of the peace), the latter suggestive of the judicial functions of magistrates associated with the *alcaldes* of Hispanic colonial institutions.

If indeed the Constitution of Cádiz and its application in the colonies marked a turning point in the configuration of municipal corporations (*ayuntamientos*) as the primary sites of local governance and electoral assemblies, there were no Creole municipalities parallel to the mission *cabildos* in the Chiquitanía prior to independence. The intersection between the indigenous *cabildos* and the representatives from each of the cantons who, in turn, chose the electors for the province of Chiquitos surfaces in the official documents that recorded these elections, an indirect mode of suffrage in which Indians and Creoles participated during the first decade of republican rule.

From the *actas* (official records) for the electoral assemblies drawn up by the presiding governor of Chiquitos, conserved for five different occasions between the years of 1826 and 1837, it is clear that the electors from the cantons included Chiquitano men, but the electors chosen for the province invariably were Creoles, most often the priests who served in the pueblos. Each assembly required all participants to present their credentials and the appointment of a president, secretary, and examiners (*escrutadores*) to certify the electoral process, as illustrated by the following passages:

> San José, 19 May 1831. Having convened the canton electors Juan José Tomichás for the pueblo of San Miguel, Ignacio Arayuru for Santa Ana, the administrator C. Juan de Dios Baca for San Rafael, Benito Xores for San Juan, and for this pueblo [San José] Matías Taseos, and after having shown their legitimate powers of representation, they appointed two examiners and a secretario, who were Ignacio Arayuru, Matías Taseos, and C. Juan de Dios Baca, they proceeded to cast their votes for the electors of the province and by a plurality of votes, they chose as electors: the First Curate of San Rafael Presbítero José Ygnacio Biana, the Second Curate of San Miguel José Vicente Dorado, and C. Lucas José de Mercado, thus concluding the present Act, presided by me, the Governor of the Province, Teniente Coronel Marcelino de la Peña, cosigned by Matías Taseos, Juan de Dios Baca, and Ignacio Arayuru.

> Santa Ana, 7 May 1837. Canton electors convened by Governor Peña, except those from San Ignacio, Santiago, and Santo Corazón who could not attend the assembly due to the flooded roadways, proceeded according to Article 20 of the Electoral Law to install the Provincial Junta with the electors who were present: for San Xavier, C. Francisco Tacoons, for Concepción, C. Mateo Coasase, and for San Miguel the ecónomo and juez de paz C. Ramón Mariano Suárez, for Santa Ana, C. Julián Arayuru, for San Rafael, the parish priest Pedro Pablo Velasco, for San José C. Matías Taseos, and for San Juan, C. Alejandro Pesoa. And after the Governor presented the credentials of these electors, which were proven to be legitimate, they appointed the president, examiners, and secretary of the assembly, who were C. Ramón Mariano Suárez, President; C. Julián Arayuru and

> C. Francisco Tacoons, examiners; and Presbítero Pedro Pablo Velasco, secretary. In virtue of this, they proceeded to cast their votes for the electors of this province publicly, one for one, and by a plurality of votes, they chose as electors the curate of the pueblo of San Miguel, Presbítero José Vicente Dorado, C. Lucas José de Mercado, and the *ecónomo* for the pueblo of San Xavier, C. José Ignacio Ortiz as alternate, thus concluding the present Act.[88]

The indigenous electors who participated in the provincial assembly, most probably officers of the respective *cabildos* in their pueblos, were designated as citizens only in the last of these acts, corresponding to 1837. Their literacy, extending at least to the ability to sign their names to the document, and their effective political participation in these local elections may have entitled them to the rank of citizen—a prerogative limited, however, to the municipal level of the canton.[89]

Indigenous voices from the cantons are no longer heard in the acts for the election of deputies and electors to represent the province of Chiquitos in the departmental junta of Santa Cruz de la Sierra for 1839. The institutional presence of the *cabildo* and the participation of indigenous *jueces* receded at the provincial and departmental levels of suffrage, overtaken by the preferential civic role of Creole citizens. During the month of April, elections took place in the city of Santa Cruz, in the pueblos of Valle Grande and Cordillera, and in San Ignacio de Chiquitos. In Santa Cruz, the popular assembly convened on 17 April 1839 selected a board of five electors (*mesa electoral*) among two parishes; the proceedings lasted an entire day in which 1,103 votes were cast verbally, "one by one, spoken out loud."[90] One week later, these five chose three of their number to constitute the provincial electoral board, despite a disagreement over the credentials of one of the electors.

In San Ignacio, the provincial junta began formal proceedings at ten o'clock in the morning on 24 April 1839, presided over by the governor of Chiquitos. The six electors in attendance represented only four of the ten pueblos, including governor Pedro José Urtubey; Miguel Santos Rivero, the teacher appointed for the province (*juez de letras*); the priest of San Rafael, Pedro Pablo Velasco; the *ecónomo* of San Miguel, C. Ramón Mariano Suarez; the *ecónomo* of San Ignacio, C. Mariano Rocabado; and the priest of Concepción, José Manuel de la Roca, who did not attend the junta, but sent his sealed vote. Following the presentation of credentials and the appointment of a president, a secretary, and two examiners, the votes were cast unanimously in favor of Dr. Rivero for first elector and, by a majority of votes, for the priests Pedro Pablo Velasco and José Vicente Dorado (although not present) as second and third electors.[91] Rivero represented Chiquitos in the departmental elections of 19 May 1839 in Santa Cruz.

The political practices of nineteenth-century Chiquitanos were deeply anchored in the *cabildo* and traditions of local governance. The postcolonial political culture of Chiquitos aspired to the prerogatives of citizenship in the new republican order, but found its expression in the readaptation of the colonial institutions founded in the mission polity. Despite the entrenched ethnic boundaries between Indians and Creoles and among different indigenous groups, and in response to the suggestion by some scholars of a brief period of greater autonomy for the Chiquitos pueblos during the early nineteenth century, the analysis presented here of the economic and political processes that defined the early republican period suggests that Indian communities did not retreat into separatism nor that they were left in rural isolation by the prefectural and national authorities of Bolivia.[92] Through indirect and direct taxation of the Chiquitanos' craft manufactures, livestock, and extraction of salt deposits, their political economy continued to support the commercial circuits and public finances centered in Santa Cruz. The indigenous *cabildo* maintained a structural presence in the civic culture of the Bolivian lowlands during the early republic, as evidenced in the cantonal elections recorded in Chiquitos.

Ethnicity, gender, and class defined social differences in these new political settings that replaced the imperial boundaries between colonizer and colonized. Spanish *vecinos* in the pueblos and private properties of Chiquitos, although relatively few in number, opened important conduits between the province and the Hispanicized Creole society that developed in Santa Cruz. These pathways of economic interdependency and cultural exchange became more heavily traveled after Bolivian independence, and the social networks they fostered formed a visible part of postcolonial transitions in this lowland frontier. As we have seen in chapters 2 and 3, the priests and lay administrators appointed to oversee the religious and temporal life of the pueblos came from the established Creole families of Santa Cruz. Their trade in the commodities produced by the Indians and their management of mission assets, especially livestock, enhanced the commercial and landowning interests of their families and those of the closely interrelated Cruceño elite. Furthermore, these networks rested on the strong institutional presence of the Catholic Church in Chiquitos.

The bonds (*fianzas*) that curates and administrators were required to present to the prefect of Santa Cruz in order to secure their appointments provide clear evidence of the kinship networks that hovered around the Chiquitos pueblos. A few examples, from many instances recorded in the prefectural archives of Santa Cruz, illustrate these ties. Priest José Gregorio Bazán, stationed in San Rafael, had given power of attorney to his brother José Manuel Bazán who, in turn, named two guarantors for his administration of the pueblo, Miguel Añes and José Escalante. Curate José Ignacio Méndes relied on his own brothers José Gregorio and Juan Bernardo to post bond for his appointment to the mission of San José. (Curiously, in this case, the fiscal officer who approved the guarantors accepted Juan Bernardo Méndes solely on condition that his wife cosign the document, because it was well known that the assets with which he could cover his brother's administration of the mission were her property.) Finally, the *cura rector* (priest in charge) of the Santa Cruz cathedral, José Vicente Durán, nominated on behalf of his brother, priest Miguel Jerónimo Durán, two guarantors well known in Cruceño society, Manuel Suárez and Juan José Baca Molina.[93]

In a few instances, these webs of mutual interest translated into settlement and property ownership in the environs of the Chiquitos pueblos. José Lorenzo Suárez, a priest "domiciled in the bishopric of Santa Cruz and living in the province of Chiquitos," where he had served in different pueblos for thirteen years, had developed his own cattle ranch in the area of San José. In August 1840, Suárez declared his intension to expand his property by adding to it a *potrero* (pasture for breeding cattle) called Yquita that bordered on his ranch, located between another *potrero* called San Estévan and the *estancia* that belonged to the Indians of the pueblo of San José. Arguing that during the rainy season, when grazing lands were flooded, mission cattle became mixed with his as the herds migrated to Yquita, Suárez asked for entitlement to an extension that measured five square leagues, as certified by the administrator of San José, Juan José Montero, and three witnesses: Miguel José Zéspedes, Cayetano Egües, and Juan Pérez. The property was adjudicated to Suárez, although none of the Chiquitanos who kept the mission herds was called to testify in the proceedings. The following year, two *vecinos*, Miguel Santos Rivero and Ramón Antonio Paz, received title to a piece of land called Concepción, carved out of the communal lands of San José.[94]

These fleeting references to Creole landholding in the province of Chiquitos do not, in themselves, substantiate processes of racial mixture or cultural

TABLE 17 Population of the Department of Santa Cruz, 1835–39

Place	*Population*				
	1835	*1836*	*1837*	*1838*	*1839*
City of Santa Cruz	4,596	5,066	5,530	5,634	5,316
Cantones Cercado	15,010	14,195	15,220	15,155	15,485
Valle Grande	16,313	16,047	16,610	16,610	16,610
Chiquitos	16,896	17,332	17,452	17,452	18,219
Moxos	—	—	—	—	—
Total (without Moxos)	54,381	54,457	56,730	57,036	57,984
Arithmetic totals	52,815	52,640	54,812	54,851	55,630

Source: Universidad Gabriel René Moreno, Museo de Historia, Fondo Prefectural, vol. 1, exp. 15.

hybridization. Global population counts for the department of Santa Cruz during the 1830s provide a general profile of the demographic distribution and growth for the prefecture and its provinces (table 17). The data were reported for the "Guide for Foreigners," along with a description of the prefecture and a list of the public officials in each year. Inexplicably, no figures were reported for the province of Moxos, and the totals given in the source are approximately two thousand persons more than the arithmetic totals of the provinces listed. The duplicate numbers for Valle Grande and Chiquitos in 1837–38 suggest that new censuses were not taken in these regions during 1838. In the absence of data from Moxos, Chiquitos appeared as the most populous of the provinces in the prefecture.

Evidence of *mestizaje* surfaces in the mid-nineteenth century in Santa Cruz de la Sierra, in conjunction with labor demands and the movements of Indians between the pueblos and the provincial capital. The history of *encomienda* presented in chapter 4 during the colonial period points to the convergence of a heterogeneous population of different native groups and their commingling with Hispanic settlers and African slaves and freed persons in the environs of Santa Cruz, under the conflictive conditions of contractual and bonded labor. The construction of the Cathedral of Santa Cruz beginning in the 1840s, which required fired bricks, limestone for mortar, and a steady supply of manual labor, gave rise to the systematic employment of workers paid with public money. Diverse ethnic origins arise in the ledgers of weekly labor drafts that brought an average of sixty workers to the construction site and the limestone quarries, all of them paid in a combination of wages and food rations.

The sample ledger page, reproduced below, shows characteristic indigenous

TABLE 18 Limestone Quarry Workers Paid in Santa Cruz, 1840
Week of October 19–30, 1840

Worker	*Ethnicity*	*Wages*		*Rations*		*Combined Wages and Rations*[a]	
		ps.	*rs.*	*ps.*	*rs.*	*Pesos*	*Reales*
Mariano Osinaga, bricklayer	Hisp/Mest	4	4		5.25	5	1.25
Simón Barba	Hisp/Mest	2	1	1	.25	3	1.25
José Manuel Mansilla	Hisp/Mest	1	5	7	.5	2	4.5
Gregorio Torriso	Hisp/Mest	2	1	1	.25	3	1.25
Narcico Torres	Hisp/Mest	2	1	1	.25	3	1.25
Guillermo Durán	Hisp/Mest	4			2.25	6.25	
Catalino Flores	Hisp/Mest	4			2.25	6.25	
José Ma Yobios	Chiq/Guar[b]	1	6	1	.25	2	6.25
Ylario Surubis	Chiq/Guar	1	5.5		7⅞	2	5⅜
Santiago Simies	Chiq/Guar	1	6	1	.25	2	6.25
Antonio Surubis	Chiq/Guar	1	5.5		7⅞	2	5⅜
Andrés Surubis	Chiq/Guar	1	6	1	.25	2	6.25
José Quiaras	Chiq/Guar	1	6	1	.25	2	6.25
Simón Socores	Chiq/Guar	1	5.5		7⅞	2	5⅜
Pablo Chobirus	Chiq/Guar	1	5.5		7⅞	2	5⅜
Pascual Chenees	Chiq/Guar	1	6	1	.25	2	6.25
Asemio Mocheres	Chiq/Guar	1	6	1	.25	2	6.25
Juan Paraiti	Chiq/Guar		6		3.75	1	1.75
José Manl Maniqui	Chiq/Guar		6		3.75	1	1.75
José Mariano Pimental	Hisp/Mest		2		1.5		3.5
Andres Serrama	Hisp/Mest		2		1.5		3.5
Lorenzo Otondo	Hisp/Mest	1			4.5	1	4.5
Francisca Cortes	Hisp/Mest		4.5	1	4.75	2	1.25
María Lucía Morales	Hisp/Mest	4		1	3.25	1	7.25
Gregoria Mendoza	Hisp/Mest	4		1	3.25	1	7.25
Calculated Totals		35	2.5	27	7.0	62	9.5

Source: Universidad Gabriel René Moreno Museo de Historia, Fondo Prefectural, vol. 30, 1841.

Notes:

[a] Workers were paid in a combination of wages and rations, calculated in pesos and reales (8 reales = 1 peso). The values appear here as recorded in the document, and the arithmetic totals are estimated based on the proportion of reales to pesos.

[b] Chiq = chiquitano; Guar = guaraní.

surnames that are markers of Chiquitano, Chiriguano, Guaraní, and Guarayo ethnic identities mixed together with Hispanic surnames, in a revolving labor force that may have been recruited by barrios of Santa Cruz and from different pueblos. The *corregidor* of Porongo, for example, was ordered to send successive drafts of fifty Indian laborers to the quarry, to be relieved each month. In the sample given here, the bricklayer was paid five reales a day; the manual laborers one peso a week; and the women (the last three entries) one peso a month (one peso equals eight reales).[95] The mixture of ethnic origins in a shifting labor force brought together men and women from different localities who may have spoken different languages. These networks of labor recruitment overlapped with the lineages of propertied Cruceño families that reached westward into the Andean foothills of Porongo and Buenavista and eastward to the Chiquitano pueblos. Yet, at mid-nineteenth century, local officials acknowledged the presence of Cruceños in the pueblos, as well as the movement of Indians from mission villages and ranches to the provincial capital, in terms of social and racial distance, rather than as *mestizaje*.

Conclusions

This chapter has explored the meanings of postcolonial landscapes in terms of changing environments and shifting sociopolitical boundaries marked by ethnicity, class, and gender. It has examined rival claims to territory, armed conflict, and contested practices of citizenship and community in the geographical and cultural contexts of Sonora and Chiquitos. The processes narrated here revisit some of the central themes developed in the histories of nineteenth-century Latin American nation-states, from the spatial and cultural perspectives of two borderlands that constituted regions in formation. Their common points of departure underscore the local nature of indigenous polities, reenacted in the *cabildo*, the lines of interdependency between different segments of Creoles and Indians in each region and, conversely, the divisions within and between contending sets of political actors. The histories of northwestern Mexico and eastern Bolivia diverged during the early republican period in the specific content of territorial disputes and the limits of political participation in each area. The Yaquis and Opatas of Sonora defended particular tracts of land, linking their claims to citizenship explicitly to ethnic boundaries, while the Tohono O'odham lived and moved through different environments defined, in part, by their vegetation and seasonal characteristics. Somewhat similarly to the O'odham cultural patterns, different ethnic *parcialidades* in Chiquitos occupied distinct ecological zones of the forest and savanna, moving their *chacos* as the soil required and following the rhythms of hunting and gathering. Some tribal groups may have returned to the forest, but the *monte* no longer served as a permanent option for the native peoples of Chiquitos, in view of their cultural roots in the pueblos and the important roles that commodity exchange and livestock played in their internal economy and in their traffic with the Cruceños.[96]

In both northwestern Mexico and eastern Bolivia the combined effects of stockbreeding, agriculture, mining, and road building changed the physical environments and altered the range of cultural alternatives for the indigenous communities. The differences between the territorial and social configurations of these two regions emerge from both their colonial and postcolonial formations, especially in reference to population, economy, and social ecology. Nineteenth-century Sonora represented a larger population of Hispanicized *vecinos* and a more diversified economy than did Chiquitos during the same period; in the former province, land tenure disputes were more acute and erupted into violence more frequently than they did in the latter. In Chiquitos, by way of contrast, the corporate economy of the mission pueblos assumed fiscal and social importance for Indians and Creoles until the last third of the nineteenth century, clearly absorbing the attention of provincial governors and defining the terms of debate among *vecinos*, pueblo administrators, and *cabildo* officers.

The postcolonial histories of both regions open new avenues of discussion concerning citizenship and the formation of political cultures in different nation-states of the Americas. Through the material reviewed here for Sonora and Chiquitos on elections and public office, as well as through the content and form of political conflicts, we can affirm that the textual analysis of early republican charters, with roots in the European revolutions of the eighteenth and nineteenth centuries *and* in the cultural soil of the Ibero-American colonies, is important but not sufficient to explain the exercise of civic rights and the differential claims of individuals and social groups to the civil status of citizens/*vecinos*. Political cultures develop in spatial dimensions with historical linkages to changes in the land. Status and rank, with political, ethnic, and social overtones, mattered locally and regionally; at mid-nineteenth century, however, the spatial configuration of citizenship did not yet comprehend the nation-state for either Creole or indigenous political actors. Local geographies and regional frontiers constituted the boundaries of their territorial base and the limits of their social networks and communities.

External frontiers intersected with their world, however, such that *territory* comprehended mapping, foreign exploratory expeditions, and colonization projects that reconfigured ethnic spaces and reallocated resources away from indigenous communities and into private commercial ventures and national fiscal revenues. Territorial concerns became inseparably linked to the drawing and redrawing of national boundaries, international conflicts, and military defense of borders. Paraguay and Brazil figured ever more prominently in the postcolonial landscapes of Chiquitos, while the invasion of Mexico by the United States, and its aftermath in mining speculation and the binational boundary commission, proved pivotal events for the development of Sonora. These themes constitute the concluding chapter of this study, focused on the contested formation of borderlands on the frontiers of Paraguay-Brazil-Bolivia and Mexico–United States.

CHAPTER 8

Contested Landscapes in Continental Borderlands

Cultural landscapes are grounded in particular places that give them meaning for the daily business of living, for the political construction of communities, and for molding historical memories and oral traditions of storytelling. The descriptive qualities of local landscapes endow places with the moral import of stories, thus strengthening social identities based on language and kinship in different settlements and among ethnic peoples. Natural settings that conditioned the cultures and technologies of Sonoran and Chiquitano peoples and were, in turn, altered by them, defined the symbolic construction of cosmic space. They figured importantly in the performance of rituals for communing with the powers of nature, healing disease, and enforcing ethical conduct. The O'odham saint way and the devil way, rooted in the "staying earth," Yaqui and Rarámuri concepts of wilderness, and Chiquitano beliefs in the *jichis* of plants, animals, and watercourses all expressed cultural landscapes that resembled their territorial environments and reflected historical processes of change and conflict.[1]

This final chapter connects the ecological and cultural meanings of specific landscapes to the spatial dimensions of regional borderlands. It shifts our attention to the external boundaries and international conflicts that molded the nation-states of Mexico and Bolivia, particularly in reference to the formation of regions on their peripheries. Sonora and Chiquitos, the focal points of this study, developed as colonial provinces with distinct geographic and cultural features over three centuries through complex historical processes of warfare and conquest, enslavement and missionization, trade and commercial exploitation. Yet it was during the nineteenth century that the perception of regionality took hold in both of these frontiers, when the interior provinces of Iberian imperial cartography became borders in the geopolitical contests for territory and power among the emerging nation-states of North and South America.[2] Internal boundaries framed the social limits of citizenship by ethnicity, rank, and gender and merged with the external borders of national frontiers in the gradual definition of regions whose content was both geographic and political.

Regional representations of Sonora and Chiquitos evolved through mapping and commercial circuits that established their spatial limits and drew connections beyond them to neighboring provinces and distant trading partners. Colonial institutions provided the framework for the administrative hierarchies and land routes along which people, goods, and knowledge moved, creating networks that linked different parts of the Americas to one another and to their European metropoles. The economies created by the Jesuit missionaries in Sonora and Chiquitos exemplified these imperial webs through the production of commodities and trade routes, connecting the former to the principal mining centers and markets of northwestern New Spain and the latter to the Villa Rica de Potosí and the *audiencia* of Charcas. During the late colonial and early republican periods, as we have seen, formal boundaries and land titles defined the spatial dimensions of communities and established municipal and provincial authorities. National states asserted their power to govern these frontier regions through military colonies and missions, as well as through private concessions for exploration, commercial transport, overland routes, and riverine navigation. Their intent was to subjugate or displace so-called savage tribes—like the Apaches of northern Mexico and the Yuracarés, Guarayos, Chiriguanos, and other nomadic groups of eastern Bolivia.

Indigenous concepts of territory brought communities of peasants, pastoralists, and foragers into conflict with national and provincial elites who sought to turn land into property and draw regional boundaries according to their competing claims to political power. At stake were juridical concepts of communal and private ownership of real estate, usufruct of forests and watercourses, and the possession of movable assets such as livestock, harvested crops, and commodities like cloth, leather goods, and wax. Native peoples of Sonora and Chiquitos articulated clear ideas of their place in the nation-state and their role in the market economy that reshaped provincial territories as the quasi-protective boundaries of mission compounds and mercantilist barriers gave way to modified regimes of free trade. For the most part, Sonoran and Chiquitano *rancherías* defended their freedom of movement to different resources and locations even as they asserted the integrity of their communities. In political movements, armed uprisings, and the ongoing search for livelihood, indigenous cultivators, laborers, and hunter-gatherers modified their physical environments and created cultural landscapes from traditional pathways of seasonal migrations and cropping patterns and from new industries like the gold fields of Quitobac, in Sonora, and the salt flats of southern Chiquitos.

Of the many episodes of governmental and private projects for commercial exploitation and administrative control documented for both frontier regions, the following sections will consider the attempted colonization of southeastern Chiquitos by Manuel Luis de Olíden (1832–42), the military colony of Tremedal established on the border with Brazil by Sebastián Ramos (1837–47), and the commercial projects undertaken by the Sociedad Progresista de Bolivia (1864–74); for Sonora, the Treaty of Guadalupe-Hidalgo, the Gadsden Purchase, and the binational Boundary Commission following the war between Mexico and the United States (1846–54), as well as the so-called filibustering expeditions that invaded the state at mid-century in pursuit of mineral wealth and territories to rule. These stories were chosen because of their impact on the landscapes and communities of both regions, in reference to their spatial and cultural implications for the changing contours of international borderlands.

Postcolonial Landscapes and the Creation of Borderlands

At the beginning of the rainy season in October of 1835, José Manuel Herrera, the administrator of the pueblo of Santo Corazón, wrote an eloquent description of the landscape on the eastern frontier of Chiquitos in compliance with governor Marcelino de la Peña's assignment to assess the potential value of new salt beds discovered in the flatlands of the Jaurú river.

> In all my experience traveling in the Province [of Chiquitos] and beyond it, I have not seen a countryside to equal this one, nor do I believe that it could exist, for words fail me to describe such beauty. Hunting game is so abundant that it could never be exhausted: deer of different kinds as well as livestock wander in large herds. The abundance of ducks and other water fowl along the banks of the lagoon [*curichi*] form unending flocks, something I have never seen before. Grasslands for grazing all kinds of cattle are abundant and of superior quality, such that the best lands in all Bolivia have been vacant and unknown up to now. I have become so enthralled with the beauty of this place that, if you will permit me, I shall settle and populate these fields. To that end, I have sent for over one hundred head of cattle from Villa María [Brazil], and if you and the other

> officials in the province were to see this place, you would do the same, for hunting alone will suffice to support [human settlement] without butchering a single steer.[3]

While Herrera extolled the beauty of the wildlife in these wetlands and promoted their economic potential for breeding livestock, Peña centered his attention on the quality of salt—nearly pure sodium—obtained in the first samples taken from the beds found in the vicinity of Santo Corazón. The salt quarries of San José and Santiago were managed as a state resource and a potentially lucrative industry; thus Peña alerted the prefect of Santa Cruz to the need to build roads that would connect the salt beds of Santo Corazón to the central pueblos of Chiquitos—San Miguel, San Ignacio, and Santa Ana—and from there to Concepción and San Xavier, in order to forestall their exploitation by Brazilians and secure their revenues for Bolivia.[4]

The Province of Otúquis in the Wetlands of Chiquitos

Readers familiar with travel literature and landscape descriptions will recognize the hyperbole in Herrera's prose: forest game and flocks of waterfowl so abundant they could not be counted or depleted and, significantly, his assumption that these were empty lands, unused and awaiting settlement for breeding livestock. In Herrera's view, this was a landscape ripe for commercial exploitation; its beauty resided in the potential commodification of its resources. Herrera had served for most of the decade at Santo Corazón, bearing the titles of *ecónomo* and *juez de paz*, and, presumably, he knew the territory well.[5] His post required him to administer the marketable production of this, the easternmost of the Chiquitos pueblos, where the corporate economy depended substantially on quarrying and transporting salt, as well as on breeding and grazing livestock.

Herrera's portrayal of these grasslands as vacant and unknown seems striking since indigenous cultivators, foragers, and herders from the *parcialidades* known as Chiquitos, Zamucos, Curavés, and Otukés lived in the pueblo and provided the labor to gather salt and herd cattle. They were well-acquainted with the wildlife of the forests and streams since hunting and fishing provided a large part of their subsistence. These peoples had settled in the Jesuit mission of Santo Corazón over a century earlier, in 1717, marking a cultural and ecological frontier between the eastern marshlands of Chiquitos and the drier savannas of the Chaco Boreal. Intermittent warfare between different Chiquitos and Guaycurúan bands prompted the courageous odyssey of Father Joseph Sánchez Labrador in 1767, traveling northward along the Paraguay River from the Mbayá mission in the Chaco to the pueblo of Santo Corazón. In 1779, native officers of Santo Corazón successfully petitioned colonial authorities to move the pueblo from a swampy, unhealthy location northward to a new site along the Boturiquis River. Only three years prior to Herrera's communication, in 1832, Alcides d'Orbigny described the pueblo favorably, located in a woodland watered by streams, where he found a population of over eight hundred villagers; in the census of 1842, Santo Corazón registered over one thousand souls.[6]

The beauty and potential wealth of the marshy grasslands, streams, and salt mines of eastern and southern Chiquitos that so attracted local administrators and provincial officers caught the attention of the Bolivian national government and foreign entrepreneurs as well. President José Ballivián (1841–47) designed an integral plan for the Bolivian *orientes*—eastern lowlands extending from the rain forests of Amazonia to the arid grasslands and scrub forests of the Chaco—and took positive steps to bring the vast department of Santa Cruz, including the provinces of Moxos, Chiquitos, Vallegrande, and Cordillera, under national sovereignty. The military prowess of General Ballivián, gained in his successful defense of Bolivian territory against the annexationist ambitions of Peruvian forces in 1841, added luster to his political image and to his projects for colonization and road building. Ballivián's victory at the Battle of Ingavi was duly commemorated in Chiquitos by assemblies of appointed officials, *vecinos*, and *naturales* in the different cantons. The surviving acts for the pueblos of San Ignacio and San Miguel give testimony to the presence of native *corregidores* and *síndicos* in ceremonies that served to enhance the authority of the president and link this frontier province to the highland capital of the republic. Chiquitanos and Creoles recorded their names in the official documents, thus participating—if symbolically—in the imagined community of the Bolivian nation.[7]

President Ballivián linked the economic progress of Bolivia to the development of the eastern lowlands with a view to opening riverine navigation to the Atlantic. In 1842, Ballivián established the department of Beni by presidential decree, effectively separating the province of Moxos from the department of Santa Cruz. Significantly, Ballivián extended full political rights of citizenship to the Indian communities of Moxos, a measure intended to enhance the settlement and colonization of this new administrative unit. He sponsored numerous exploratory expeditions and commissioned the first authoritative maps of Bolivian territory. Ballivián's vision of incorporating the *oriente* into the nation called for establishing military colonies and religious missions to secure the frontier, control nomadic tribes of *salvajes*, and promote the commercial development of cacao plantations and the lucrative gathering of quinine bark.[8]

Subsequent administrations reiterated Ballivián's substantive measures—notwithstanding renewed episodes of civil strife at the national level and boundary disputes among the departments of Santa Cruz, Beni, La Paz, and Cochabamba—intended to enhance trade, extend communications, and integrate the tropical frontiers of northern and eastern Bolivia into the political nation. Following the first national census of 1855 and the reconfigurement of the administrative territory of Beni, the tribute (*temporalidades*) owed collectively by the Indian communities of Moxos was suspended in 1856, a measure most likely designed to hasten their incorporation into the commercial economy of the region as free laborers. In the mid-1860s, president José María de Achá authorized a concession to build an overland wagon road from Santa Cruz to the right bank of the Paraguay River, and president Mariano Melgarejo awarded a concession to José Domingo Vargas, who led an expedition from the pueblo of Santiago in Chiquitos to build a road and found a port on the Bolivian side of the upper Paraguay. In the following decade, entrepreneur Miguel Suárez Arana, who registered the Empresa Nacional de Bolivia in Paraguay, and the Sociedad Progresista de Bolivia obtained authorizations to cut roads through the forest and establish ports on the Paraguay River.[9]

Among these many projects and concessions, the colony initiated by Manuel Luis de Olíden, creating the province of Otúquis in the *pantanal* of southern Chiquitos, calls our attention in view of its ecological and social impacts for the pueblos of San Juan, Santiago, and Santo Corazón. A native of Buenos Aires, a merchant, and a militia officer, Olíden established his early career in the lucrative trade routes that linked the highland colonial centers of La Plata, Potosí, and La Paz. An early adherent to the cause of independence launched in Chuquisaca in 1809, Olíden obtained a concession from the Bolivian government to twenty-five square leagues of land and authorization to establish a port at the headwaters of the Paraguay River in 1832, in recompense for the losses he had suffered due to the royalist occupation of Potosí during the wars for independence. Olíden's grant included exclusive political jurisdiction over the territory he would govern, communicating directly with the federal Ministry of the Interior, and reduced import and export duties on the products moving in and out of his colony—a measure aimed to promote commerce from this frontier outpost overland to the interior of Bolivia and by river transport to foreign markets.[10]

Olíden arrived in the province in 1833 and began to lay out the town site for the port that would carry his name at the confluence of the Tucavaca and Rafael streams, whose combined flow formed the Otúquis affluent of the Paraguayan headwaters. Olíden received assurances of assistance and cooperation from the governor of Chiquitos, Marcelino de la Peña, for this new venture and, in 1836, Peña gave Olíden formal possession of the province of Otúquis, with the authority to colonize, distribute land allotments, and govern the pueblos of Santiago and Santo Corazón, which were located within his territorial concession. In accord with his official status, Olíden issued two decrees offering potential colonists land grants for house lots in the projected town of Olíden, as well as for farm sites and cattle ranches in the marshlands and savannas of Otúquis. At the same time he assured the Indians and *vecinos* of the pueblos of Santiago and Santo Corazón that they would retain use-rights to their houses and surrounding communal croplands within two square leagues. Olíden anchored his political status to the canton of Santiago, where he assumed the presidency of the municipal council, and assured his livelihood with the establishment of two haciendas: Sutós and Rinconada. During the following eight years, Olíden devoted his efforts to building roads and promoting the commercial development of agriculture and stock breeding.[11]

The broad ambition to develop Otúquis as a commercial nexus between Bolivia and the Atlantic trade routes depended on the successful navigation of the Paraguay and Plate rivers. José León de Olíden, Manuel's son, undertook a six-month exploration of numerous streams and tributaries, traveling northward into Brazil and then turning south in search of the union of the Otúquis and Paraguay rivers. His journey proved arduous and revealed two unyielding obstacles to the project, the first embedded in the natural terrain and the second arising from international political economy. The Otúquis stream did not flow into the Paraguay River, but emptied into unnavigable swamps one hundred miles west of the main channel. The younger Olíden traveled downstream from the Brazilian port of Coimbra, hoping to find the mouth of the Otúquis, but when he arrived within sight of Fort Olimpo, cannon shots forced him to halt his journey. Paraguayan foreign policy asserted exclusive claims to navigation on the river and would not permit the commercial transport of products to or from Bolivia along the waterway.[12] Olíden's grand design foundered, then, in the swampy wetlands where three hundred years earlier Alvar Núñez Cabeza de Vaca had first attempted to push Spanish imperial claims northwestward into the Chaco Boreal and lost his governorship to intertribal warfare among different bands of Guaraní, Guaycurú, and Chiquitos peoples, and to the internecine politics of the *vecinos* of Asunción.[13]

What we know of the province of Otúquis comes from the reports and pamphlets published by Olíden's secretary, Moritz Bach, and the articles placed in numerous periodicals of the period.[14] We have no direct evidence of the number of colonists who settled in Olíden's concession, but the calls for development of plantations and *estancias* directed to foreigners and Bolivian nationals alike, and to local officials in the pueblos of Chiquitos, promoted ever increasing herds of livestock and the expansion of tropical agriculture.[15] Prior to his association with Olíden, Bach had attempted his own plantation in the recently missionized province of Guarayos between the provinces of Moxos and Chiquitos, where he laid plans to contract Indian labor and cultivate cacao, grapevines, cotton, tobacco, coffee, and local food crops. Two disastrous fires consumed his investment and terminated the enterprise, foiling his dreams of independent prosperity.[16]

As occurred with private ranching throughout the region, colonization projects like Olíden's Otúquis set in motion ecological processes counter to the seasonal rhythms of *chaco* cultivation, fallow, and forest recovery. Extensive clearing of spaces measured in square leagues, road building, and intensive planting of cotton, sugar, cacao, and coffee accelerated the alteration of humanly crafted landscapes in the *pantanal* of southern Chiquitos: the tension of opposing processes emanating from subsistence patterns of cultivation, hunting, and gathering; the corporately managed economies of horticulture, handicrafts, and gathering in the pueblos; livestock grazing and tropical agriculture—all wrought physical changes in soils, vegetation, and the morphology of lagoons and streams. Just as palpable as these changes in the land, if less immediately quantifiable, were the cultural transformations of community. The gendered and generational composition of pueblos and *rancherías* shifted and work took on new meanings as indigenous laborers were recruited for private estates and public projects to build roads and riverine ports.

The border between Chiquitos and Mato Grosso was porous and poorly defined—traversed by Indians and Creoles alike, seeking everyday supplies like

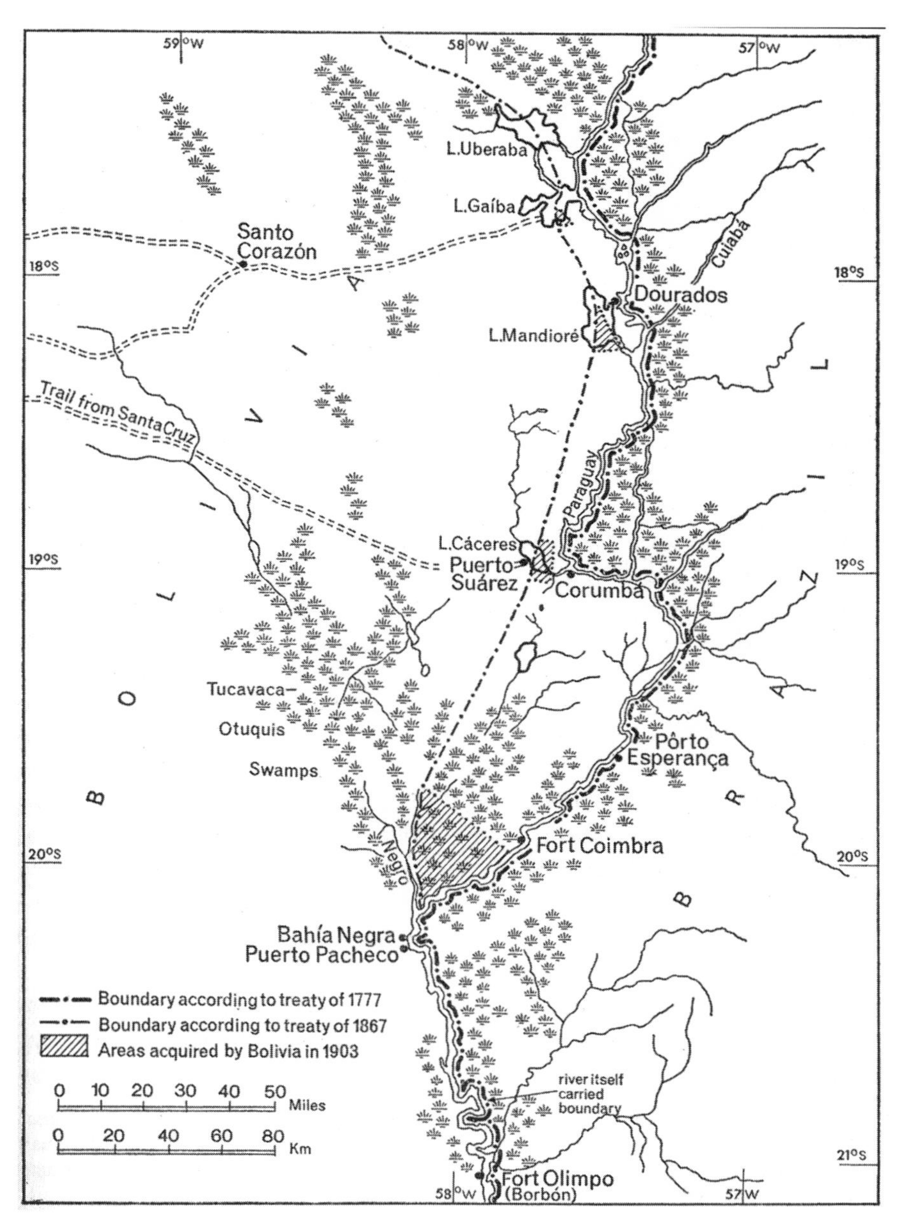

Map of the upper Paraguay River, 1777–1903. From Fifer, *Bolivia: Land, Location, and Politics*, 193. Courtesy of Cambridge University Press.

iron implements, lantern oil, and even soap—yet it constituted the binational boundary between Bolivia and the Brazilian empire.[17] From east to west, muleteers, cattle drivers, merchants, fugitive slaves, and others "of the miserable class" passed from Brazil to the department of Santa Cruz in the hopes of selling their wares or taking refuge in the pueblos and *estancias* of Chiquitos.[18] This loosely defined boundary, itself a route of human transit, blurred into overlapping cultural and political borderlands that encompassed a natural terrain of salt beds and grassy marshes in the eastern perimeter of the province. Within this meandering corridor—traversed by Brazilians, Bolivians, blacks, village Indians, and nomadic bands of different ethnic lineages—the pueblo of Santo Corazón became the point of reference for Tremedal, a military colony that followed the earlier experiment of Olíden at Otúquis, established primarily to defend Bolivia's contested fluvial border with Brazil.

The attempt to establish a settler colony of stockbreeders and farmers at the confluence of the Jaurú and Paraguay rivers focused on the Tremedal stream and a deep lagoon (*curichi*) that flowed into the oxbow lake of Uberaba and served to divide Bolivian and Brazilian territory. Located initially on the east bank of the Tremedal *curichi* facing a low range of hills and a Brazilian military outpost, the tiny Villa del Tremedal was intended to bring together Indians (*naturales*) of San Juan and Santo Corazón—including nomadic Bororós—together with mestizo colonists.[19] The colony arose from a combination of government sponsorship and private initiative, in a mixed economy of salt gathering, barter, and stock raising that brought together Brazilian and Bolivian officers and aspiring entrepreneurs, cowboys, soldiers, and prospectors. Correspondence generated around Tremedal recorded border confrontations, illicit expeditions to recover fugitive slaves, and individual acts of violence, at the same time as it attested to informal business arrangements among Brazilians and Bolivians.[20]

The brief history of the colony of Tremedal restored the colorful and controversial figure of Sebastián Ramos to the eastern frontier of Chiquitos. We may recall from chapter 7 that during the final phase of the wars for independence, Governor Ramos, then a royalist officer, attempted to transfer the province (and his command) from the viceroyalty of Río de la Plata to the Brazilian empire. His actions led to a military standoff and his own exile in Brazil; the capitulation of 28 March 1825 that would have annexed Chiquitos to Mato Grosso was annulled, however, and Chiquitos remained formally a part

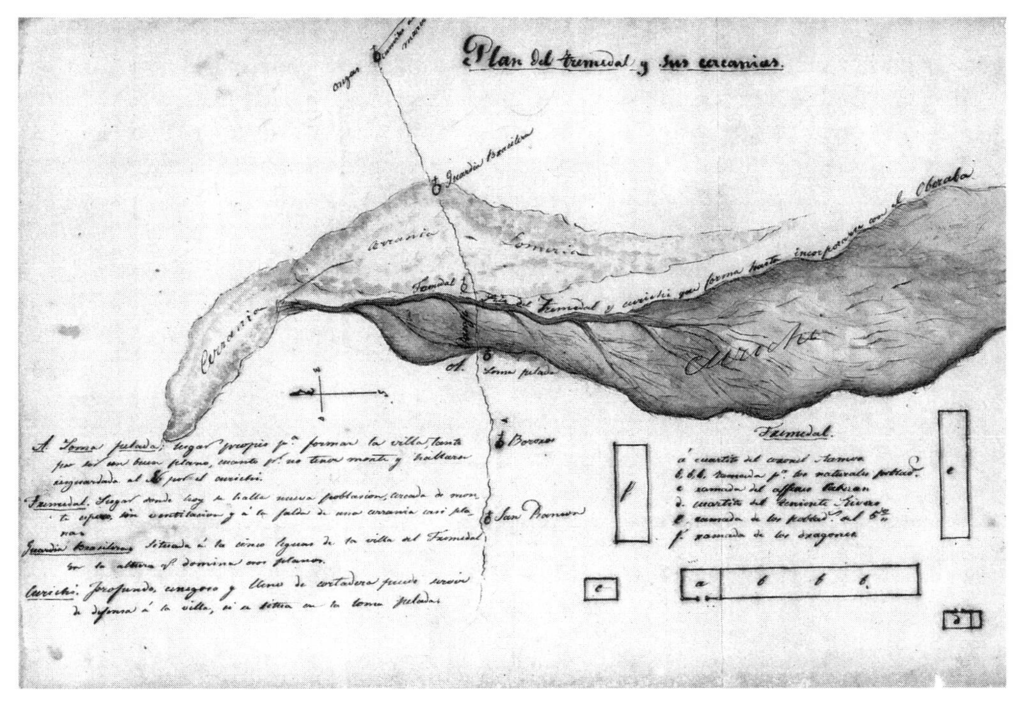

Plan del Tremedal y sus cercanías. Courtesy of Universidad Gabriel René Moreno, Museo de Historia, Fondo Prefectural 1/36.

of the Republic of Bolivia. On at least one occasion, in 1832, Ramos had attempted to return to Bolivia, but his petition for a pardon was denied. Five years later, however, Ramos recovered from his treasonous act and accepted the post of *juez territorial* in Marco de Salinas, within the canton of Santo Corazón. In view of the federal government's keen interest in developing the frontier regions of Moxos and Chiquitos for purposes of defense and commerce, it is likely that Ramos convinced Bolivian authorities to overlook his past errors on the strength of his knowledge of the local environments and peoples of the borderlands shared and disputed with Paraguay and Brazil. Never given to modesty or conformity with superior officers, Ramos indignantly protested the summons he had received to pay the personal tax on his properties and retainers in the Marco de Salinas, offering a robust defense of his services to the nation.

> Government employees in easy posts should look on me with respect, because since 15 January 1837, I have served without pay as juez territorial of these frontiers, my person and my property always at the service of the nation. In 1831, I discovered the Marco de Salinas and from then to the present [1842] I have expanded the area in a semicircle, lately cutting open a road from San Juan to Marco, mindful that the King of Spain spent thousands of pesos in engineers (matemáticos) and troops for all of these discoveries, without attaining them. I am unflagging in my efforts to congregate and settle the barbarian bororós, helping them in what I can to keep them under control so that they will not attack the villagers of Santo Corazón, as they used to do in years past.[21]

Five years later, in 1847, Ramos brandished the impressive titles of lieutenant colonel and superior chief of the eastern frontier. In this capacity he advised the prefect of Santa Cruz to transfer Tremedal from its location east of the *curichi* to a hill located on the west side of the swamp, which would afford the fledgling settlement greater security from a hypothetical attack by Brazilian forces and relief from the insects that plagued the colonists "in a dense forest without any ventilation." In this same letter and the accompanying map, Ramos pointedly referred to the precarious construction of housing in the colony, consisting of unwalled ramadas, and promised to lay out a more substantial village in the new location while supervising the clearing and planting of new *chacos* to raise food for the colony.[22]

Prefect Manuel Rodríguez Magasinos authorized Ramos's proposal to move the colony. That same year, however, reports filtered back to the prefecture of the colonists' abandonment of Tremedal in the face of protracted disagreements between Ramos and the governor of Chiquitos, Ramón García Lemoine. Rodríguez Magasinos issued rigorous instructions to return the wayward colonists to Tremedal and recruit new settlers, supplying them with livestock from the herds in each of the ten pueblos—the *ganado de secciones*—which, in theory, belonged to the Indians. While denying that it was his intention to divert the Chiquitanos' collective and individual property to Tremedal, the prefect insisted that this was the best way to assure the survival of the frontier colony and fulfill the priority set by the national government.[23] Notwithstanding the prefect's orders and Ramos's concerted efforts, it appears that Tremedal did not endure beyond the decade, thus shattering the ambitions of the Bolivian state to establish a defensive colony on the Jaurú river.

Ramos's map reveals particular geographical features of this frontier outpost, focusing on the actual and proposed locations of Tremedal and two separate *rancherías* for the Bororós and the secondary colony of San Ramón in relation to the swamp and the hilly country east of the stream. The Brazilian military post (Guardia Brasilera) is prominently labeled midway between Tremedal and the road to Marcos, suggesting that the salt beds Ramos had discovered and begun to exploit were located in Brazilian territory. His sketch of the settlement revealed the spatial and racial segregation of the villa, distinguishing the room he had assigned for himself (*a*), the ramadas assigned to the Indian settlers (*b*), two additional rooms reserved for his subordinate officers (*c* and *d*), the ramadas allocated to non-Indian settlers (*e*), and, finally, the ramada for the soldiers (*f*). Had the colony prospered, Tremedal would have reproduced in microcosm the social and ethnic divisions that defined the contours of Chiquitos province as these had developed over two centuries of colonial rule and shaped the political foundations of the Bolivian republic.

The Bolivian national government continued its efforts to secure its borders with Brazil, Argentina, and Paraguay through diplomatic initiatives and private concessions during the third quarter of the nineteenth century, with important repercussions for the regional configuration of Chiquitos. At issue were international rivalries for exclusive territorial and navigational claims to different segments of the Paraguay river in anticipation of new technologies in steamship transport and railways. In 1853, Bolivia declared the freedom of commerce on all the rivers that traversed or bounded its national territory, in an attempt to protect Bolivian access to shipping on the Amazonian and Paraguayan river systems.[24] International pressures became critical during the War of the Triple Alliance (1865–70) that pitted Argentina, Brazil, and Uruguay against Paraguay. The war concluded in a series of treaties submitted to external arbitration that maintained Paraguay as an independent nation—despite its devastating military defeat—and traced new boundaries through the Chaco Boreal, dividing this territory among Argentina, Bolivia, and Paraguay.[25] The peace treaty of 1876 and the Hayes Award of 1878 established the Pilcomayo river as the boundary between the two nations in the Chaco. Over the next half century, Bolivia and Paraguay continued to dispute their binational border, giving rise to repeated confrontations that erupted into the Chaco War of 1932–35, a crisis of dramatic environmental and political consequences for both countries and the peoples living in these contested lowlands.

Concurrent with these international disputes, private concessions for colonization, road building, and riverine ports complicated even further the conflicting territorial claims to land and commercial navigation. In 1864, the

Sociedad Progresista de Bolivia successfully concluded treaty negotiations with the government (on the strength of the 1853 decree that declared free navigation on Bolivian waterways), which conceded ample territorial and commercial powers to the company in return for opening overland roads for wheeled vehicles and mule trains that would communicate Santa Cruz with the Brazilian port of Corumbá, passing through Santiago and Santo Corazón. The Sociedad Progresista promised to build a port and establish a military outpost on the western margins of the Paraguay River and, further, to operate a steamship line with regular trips to carry freight, passengers, and mail to Asunción (Paraguay) and Corrientes, Entrerríos, Rosario, and Buenos Aires (Argentina). By 1866–67, the Santo Corazón–Corumbá route had been cut through the swamps by Chiquitano Indian workers, and a documentary trail among Bolivian, Brazilian, and Paraguayan officials attested to the passage of correspondence and goods along these combined terrestrial and fluvial roadways. The commercial advances made by the Sociedad Progresista de Bolivia occurred under the shadow of the War of the Triple Alliance and a web of bilateral and triangular diplomacy among Bolivia, Brazil, Paraguay, and Argentina.[26]

In terms more ambitious than those awarded to Olíden to govern the province of Otúquis three decades earlier, the Sociedad Progresista de Bolivia assumed the exclusive trading concession for the export of wax and governing powers over the Chiquitos pueblos included in its territorial grant. Five key articles of the contract between the company and the Bolivian government effectively transferred administration of the *naturales* and their labor from the institutions rooted in the pueblos to the company. The government would provide a priest "of known moral standing" to fulfill the presumable religious needs of the Indians and colonists brought to the region under the company's auspices, and twenty-five armed soldiers to guard the roads and port. All other religious orders were prohibited from entering the company's concession "on account of the strong divisions among the *naturales* of that country." The Sociedad Progresista was authorized to recruit forty to fifty indigenous men from the region, along with twelve musicians and their instruments, for construction labor, enjoined only to pay them the same daily wage assigned to local inhabitants until they returned to their pueblos. Finally, "all Indian tribes that were reduced" to settled life would remain under the orders of the company for the duration of the contract. These same Indians should be remunerated in their personal service and would enjoy the civil and political rights guaranteed to all Bolivians by the constitution, but without the additional prerogatives ascribed to citizens.[27]

The presence of the Sociedad Progresista in the borderlands between Bolivia, Paraguay, and Brazil, with delegated authority to discipline Chiquitanos and dispose of their labor and territory, directly challenged the corporate structures of the Chiquitos pueblos. Successive development projects that opened roads, promoted colonization, and gained territorial concessions at the expense of the pueblos fueled the uneven advance of Creole society in this region marked by ethnic mixture and geographic ambiguity. The Indians' loss of control over the livestock that had constituted their communal patrimony—the *ganado de secciones*—began with government-auctioned leases in the 1850s and was compounded by private concessions for grazing land and the commercialization of salt and wax gathered by the *naturales* as a form of tribute.

Economic and political processes merged in the 1871 *Reglamento* (regulation) for establishing a new mission regime under state auspices to Christianize and civilize the "savage tribes" of the tropical lowlands, and the 1874 presidential decree that ordered the privatization and division of corporate lands.[28] While the republican sponsorship of Franciscan missions proved significant for the Guarayos and Chiriguano provinces, its application in the Chiquitanía was negligible. The direct impact of the 1874 legislation fell on the *altiplano*, where the expansion of private haciendas (latifundia) gained legal force over the indigenous communities. Nevertheless, the colonization of the Chiquitano frontier, in the name of nationalizing and developing territories and peoples considered to be poorly integrated into the Bolivian nation-state, disbanded the very communities that had sustained a regional political culture in their internal governance and corporate economy.

These combined measures emanating from the state and private initiative shifted the fulcrum of economic and political power and undermined the autonomy of the indigenous *parcialidades* who had maintained their towns as heterogeneous communities within the cultural bonds of Catholicism for over two centuries. Regional historiography and anthropology point to the rubber boom of the late nineteenth century as the violent disruption of indigenous communities in the Amazonian lowlands.[29] Notwithstanding the global significance of rubber, the processes analyzed here for Chiquitos, fueled by Bolivian and foreign projects for colonization and commercialization of the forests and savannas in the middle third of the century, opened irreparable fissures in

Mexico and Guatemala, 1834. From *Tanner's Universal Atlas*. Shows topographical features and basic territorial divisions prior to the United States–Mexico war and the Gadsden Purchase (Treaty of Mesilla). Courtesy of Map and Geography Information Center, Centennial Library, University of New Mexico.

Mexico and Guatemala, 1834, Sonora insert from *Tanner's Universal Atlas*. Notice references to different bands of Apaches and colonial presidios identified as "forts." Courtesy of Map and Geography Information Center, Centennial Library, University of New Mexico.

corporate native polities that rendered them vulnerable to peonage on the cattle *estancias* of the region and to the international demand for rubber and its dehumanizing forces.

Sonora: Mapping Territorial Boundaries in the Desert

During this same middle period of the nineteenth century, Mexico faced a series of crises that threatened to disband the nation-state, arising from foreign incursions and internal divisions that led to repeated episodes of civil war. Sonora remained on the fringes of the political revolts and internecine struggles that shook the Mexican capital, but the Sonoran ruling classes mobilized to confront intermittent warfare with bands of nomadic tribes, rebellions by indigenous and peasant communities over land and political autonomy, factional conflicts among the elites, and invading forces under different guises pursuing mineral wealth, commerce, and territorial annexation.

The ecological contours of the Sonoran Desert, distinguishing the arid northwestern plains of the Pimería Alta from the eastern cordilleras, shaped the bordered lands of Mexico and the United States as these emerged from nearly a decade of warfare in 1846–54. The binational boundary survey established a mapped line reinforced with stone landmarks in the desert, a theoretical demarcation that crossed numerous streambeds and arroyos that served as seasonal corridors for the movements of people, cattle, and goods through the region. For over three decades, until the completion of the railway line between Nogales and Guaymas, the boundary constituted not so much a clearly defined border as a zone of contact characterized by a network of footpaths, mule trails, and wagon roads extending from the Magdalena and Altar-Concepción valleys eastward to the Santa Cruz and San Pedro drainages and northward to the presidial outposts of Tubac and Tucson.

The final years of attenuated mission administration in the Pimería Alta, analyzed in chapter 7, presented a bleak picture of abandoned ranches and rural settlements due to the fear of Apache raiding reinforced by Ignacio de Zúñiga's report on the decline of the presidios. These alarming views of desolation contrasted with new mining bonanzas, the commercial development of the port of Guaymas, and the growth of inland towns that sustained the expanding economic and political power of the principal landholding families. Sonoran political culture was forged in the material struggles over land and community autonomy that pitted Cahitan (Yaqui and Mayo), Piman, and Opata village leaders against the Creole elites who, in turn, vied among their ranks for control of the legislature and governorship. Colonial administrative boundaries, based largely on the ethnic provinces defined by the mission districts—such as the Opatería, the Pimería Alta and Baja, and Ostimuri—were redrawn under the municipal and fiscal institutions of the states of Occidente and Sonora.

The central argument of this narrative of postcoloniality in the Sonoran frontier is that Indians sought corporate economic and political spaces in the new republic, while Creoles maneuvered to restrict their participation to contractual obligations of peonage and wage labor. These internal struggles for territory, political representation, and recognition of the indigenous *común* and village *cabildo* were compounded further by national conflicts over the federalist and centralist constitutional formulas for governance—each championed by rival political leaders (caudillos) and their followers in Sonora—foreign invasions and, after 1854, the ineluctable military and economic presence of the United States in the newly established Territory of Arizona.

In this context, the U.S. and Mexican boundary commission, carried out over four years (1850–53), produced an important source of firsthand infor-

Mapa de los Estados Unidos de Méjico. New York, 1847. Courtesy of Map and Geography Library, University of Illinois, Urbana-Champaign. Boundaries show the U.S. annexation of Texas, but represent the territories of Alta California, Sonora, and New Mexico as they were prior to the U.S. invasion of Mexico and the renegotiation of borders in 1848 and 1853.

mation and impressions of northwestern Mexico, including its natural and cultural landscapes, and marked significant changes in the geopolitical configuration of the Sonoran Desert. The boundary survey established a political border that intersected diverse ecological and ethnic frontiers, altering the social and cultural meaning of the borderlands derived from the colonial internal provinces through migrations, land disputes, and changing identities of citizenship and mixed allegiances. The regional connotation of the U.S.-Mexican frontier grew out of its division between two nation-states, shaped by parallel ranching and mining economies in Sonora and Arizona, but distinguished by the institutional frameworks that bound them to their respective national centers. By the close of the nineteenth century, major technological innovations in hydraulic irrigation, mining extraction, and ore processing, together with urbanization and the railroad, brought transforming rhythms of modernization to the entire frontier region, but with contrasting social and political impacts in Arizona and Sonora.[30]

Table 19 provides a schematic overview of major events that marked the histories of the Mexican nation and the Sonoran region from independence to the consolidation of the Porfirian regime. A series of private expeditions seeking adventure, wealth, and territories to govern occurred between the U.S. and French invasions of Mexico. Known as *filibusteros* in Sonoran history, their appearance in the region was linked in part to the California gold rush, from which disappointed fortune seekers turned southward into Mexico, and in part to the sectional cleavages between northern and southern interests in the United States.

Three of these expeditions have received special attention because of their size and armed confrontation with Sonoran troops: that of Gastón Raousset de Boulbon, the French prospector who first contracted with the banking firm of Jecker Torre to restore the silver mines of Arizona and then mounted a military invasion of Guaymas (1852–54); William Walker's invasion of Baja California (1853–54); and Henry Alexander Crabb's overland incursion into the Altar valley, defeated in Caborca (1856–57). The memory of these and numerous smaller expeditions, occurring in the midst of political turmoil in Mexico, reinforced a historical trope of defensiveness vis-à-vis foreign ambitions to extract the mineral wealth of Sonora and establish minirepublics and personal fiefdoms in Mexican territory. Notwithstanding the egregious ambitions of these and other adventurers, it is noteworthy that Boulbon had sought diplomatic and business credentials in Mexico and that Crabb had cultivated per-

TABLE 19 Significant Turning Points in Nineteenth-Century Sonora and Mexico, 1824–83

Mexico	*Date*	*Sonora*	*Date*
Federalist constitution	1824	State of Occidente	1825
Expulsion of Spaniards	1827	Expulsion of Spanish friars	1828
First cholera Epidemic in Mexico	1833	Yaqui and Opata rebellions	1824–33
Centralist constitution	1836	State of Sonora	1831
Texas War of Secession	1835–36	Department of Sonora	1836
U.S. annexation of Texas and boundary dispute	1845	Tohono O'odham uprising	1840–43
U.S. invasion of Mexico	1846–48	U.S. occupation of Guaymas	1847–48
Treaty of Guadalupe-Hidalgo	1848	Gold prospecting in California draws Sonoran migrants	1849–51
Second cholera epidemic in Mexico	1851	First cholera epidemic documented in Sonora	1851
Gadsden Purchase under U.S. threat of renewed war	1853	Binational boundary commission	1853–53
Revolution of Ayutla and Liberal constitution	1855–57	Governor Ignacio Pesqueira	1857
War of the reform	1857–61	Yaqui and Opata rebellions against colonization projects	1858–61
French intervention	1864–67	French occupy Guaymas and ally with Yaqui and Mayo leaders	1865–67
Porfirio Díaz takes Presidency Plan of Tuxtepec	1876	Porfirian general Luis E. Torres elected governor of Sonora	1879
Law of Vacant Lands (survey and subdivide communal lands)	1883	Railroad line Tucson-Guaymas (Southern Pacific Railroad)	1882

Sources: James Officer, *Hispanic Arizona, 1536–1856* (Tucson: University of Arizona Press, 1987), 204–61; Evelyn Hu-Dehart, *Yaqui Resistance and Survival: The Struggle for Land and Autonomy, 1821–1910* (Madison: University of Wisconsin Press, 1984), 56–93; Thomas E. Sheridan, *Los Tucsonenses: The Mexican Community in Tucson, 1854–1941* (Tucson: University of Arizona Press, 1986), 111–30.

sonal ties to some of the leading families of Sonora.[31] Furthermore, even as Sonoran politicians and military officers mobilized to repulse the *filibusteros*, they pursued policies of colonization that promoted foreign immigration and disavowed the traditions of land use and village governance sustained by indigenous communities.

The two principal reports published from the boundary survey reveal the commissioners' intentions for their assignment, as well as their preconceptions concerning the territories and communities through which they passed. Their descriptive texts indicate how the North American surveyors read the

Sonoran landscape and open significant questions for us about the direction of change in the environment. John Russell Bartlett, U.S. commissioner during 1850–53, acknowledged the depredations committed by Americans traveling through Sonora even without the *filibusteros*' pretensions of taking over Mexican territory.

> The quiet people here have been so much annoyed by the conduct of California emigrants who have passed through the country, as to make them shy of all Americans. These reckless adventurers often set at defiance all law and propriety, and we had many accounts of their shameful and brutal conduct. The fields in this country are seldom fenced, and it is no uncommon thing for a party of these men to encamp and turn their animals into a field of corn, on which the helpless ranchero and his family are probably depending for their chief support. They will enter a house, pistol in hand, demanding whatever it affords; frequently they help themselves, without the ceremony of paying for what they take; and commit other outrages which make one who had any national pride blush to hear recited.[32]

Bartlett's itinerary shows that his commission went beyond mapping the U.S.-Mexican border. He traveled through many ranches and towns, moving inland from the partially abandoned presidios of San Bernardino and Fronteras in the northeast through the Sonora river valley to Hermosillo and Guaymas. Bartlett repeated entertaining stories from rancheros of lassoing feral cattle as he gathered statistics systematically and persistently on crop production, prices, and herds of livestock. While commenting obliquely on the poverty of Sonoran rural people, as evidenced in the simplicity of their diet, the spartan frugality of their homes and furnishings, and the illnesses they bore, Bartlett's descriptions of cultivated landscapes bore witness to the horticultural legacy of the colonial missions in grain production and orchards of peach, pear, apple, mulberry, quince, and pomegranate trees. In Bacoachi, once a mission *visita* and Opata presidial garrison, Bartlett saw "a beautiful valley a mile in width . . . with luxuriant fields of wheat, corn, and [chick]peas. It was intersected by a broad acequia, the course of which was marked for a mile or more by a line of cotton woods and willows."[33]

Fronteras, the oldest presidio of Sonora, had been abandoned and reestablished as a military colony where "a new population, including many of its former inhabitants, have taken possession. . . . Acequias have also been opened, large fields of wheat and corn cover the beautiful valley, numerous cattle graze on the meadow, and the importance which the place once enjoyed seems about to return." Until its harvests matured, however, the Fronteras colony subsisted on dried meat, roasted agave roots, and *pinole*.[34] Arizpe, a colonial mission and villa elevated to the capital of the State of Sonora in 1832–38, had fallen into decline, which Bartlett attributed to internal civil strife, Apache raiding, and the flow of Mexican migrants to the gold fields of California.[35] Bartlett's detailed description of Tepahui, the private estate of Manuel María Gándara, governor and caudillo of Sonora (1838–56), contrasted the productivity and wealth of this expansive hacienda with the poverty of peasant cultivation in many of the villages he had visited. Bartlett was escorted through the buildings and lands of Tepahui by Frederick Augustus Ronstadt, Gándara's manager, who affably showed his American visitor the account books for the entire estate, including *fanegas* of harvested grain, herds of livestock, wagon loads of sugarcane, woolen and cotton textiles, and a silver mine. Bartlett noted approvingly that "the books of this gentleman were admirably kept, exhibiting a correctness of system which would be creditable in the counting-room of one of our New York merchants."[36]

In general, Bartlett and William H. Emory, who succeeded him as boundary commissioner, condescended to the Sonorans, attributing their material wants to a lack of industry and poor hygiene, but reserved special comments for Yaquis, Opatas, Papagos (O'odham), and Seris. Both authors constructed *Indians* as a social and racial category, but developed distinguishing vocabulary for different indigenous communities they observed, employing words like *savage*, *hostile*, or *docile*. Bartlett took pains to write down words from the Yaqui, Opata, and Seri languages; in addition, he interviewed and sketched Opata leader Tánori and his wife, "who always accompanied him."[37] Bartlett's description of Yaqui laborers bespeaks his prejudices as well as his assumptions about the course of cultural assimilation.

> The laborers of Ures and of other towns in the central and lower parts of Sonora are the Yaqui Indians. They fill the same place and perform the same duties as the lower class of Irish do in the United States. I was told that they are invariably honest, faithful, and industrious, traits of character which cannot be said to belong to the lower order of Mexicans. . . . These Indians were in the early history of the country extremely warlike; but on being converted to Christianity, their savage nature was completely subdued, and they became the most docile and tractable of people. In the civil wars of the State, some thirty years since, they took part with one of the factions [in the service of Gándara]; and when this

> strife had passed away, it was not easy to subdue again the dormant propensities for war which had thus been aroused. They are now very populous in the southern part of Sonora.[38]

Bartlett deemed it both natural and logical that native peoples like the Yaquis and Opatas should accept their place in the Mexican nation-state. Furthermore, he assumed a "propensity for war" that was racially inherent in the Yaquis. Emory wrote in similar terms of the O'odham in the western deserts of the Pimería Alta, mixing history with legend.

> The town of Sonoyta is the door of the State of Sonora from the California side. It is a resort for smugglers and a den for a number of low, abandoned Americans who have been compelled to fly from justice. Some few Mexican *rancheros* had their cattle in the valley near by. It is a miserable poverty-stricken place, and contrasts strangely with the comparative comfort of an Indian village of Papagos within sight. The Papagos wander over the country from San Javier as far west as the Tinajas Altas. They were at one time a formidable tribe, and waged unceasing war against the Mexicans. Having sustained repeated losses, they at length sought their God, who is said to dwell upon the high peak of Babuquivari, to ask his aid and countenance in their last grand fight with their enemies. They assembled their families and herds of horses and cattle within an amphitheatre enclosed by the mountain ridges, and battled it manfully for many days at its entrance; but their God could not turn the fate of war, and they suffered an overpowering defeat; since that time they have been quiet and peaceable. . . . This tribe is comparatively well off in worldly goods; they plant and grow corn and wheat, and possess cattle, and many fine horses.[39]

Emory observed further that "the women, who do all the labor," carried water to their *rancherías* in clay vessels, walking long distances. "The women are better dressed than most Indian women; they all wear skirts of manta or calico, covering the body from the hips down. They appear to be a good, quiet, and inoffensive tribe. A subtribe of the Papagos, called Areneños [Hiach-ed], live on the salt lakes near the Gulf of California, and principally subsist upon fish."[40]

The story that Emory summarized of Papago-Mexican combat in a mountain pass, where the O'odham had brought together their families and their herds, recalls our discussion in chapter 6 of the importance of cattle and horses for their moral economy and cosmology, symbolized in the songs and poetry of shamanic healing in the devil way.[41] In the views of both commissioners, "good" Indians worked hard, cultivated the land, and—especially if they were women—wore clothes. Bartlett's striking analogy comparing the Yaquis of Sonora with the Irish immigrant workers in the United States wove together codes of difference based on both class and ethnicity. The men's renditions of historical encounters between Yaquis, Papagos, and Mexicans fashioned a historical progression in which previously fierce and warlike tribes had succumbed to domination or peaceful coexistence. In the process they lost the aura of bravery, but they gained the merits of Christianity and joined the march of civilized and tractable peoples toward order and modernity. It is well to note that Bartlett and Emory had entered Mexico not as neutral travelers, but as representatives of the U.S. government, the authority of their commission underscored by recent military victories of U.S. forces over Mexican troops. Their narratives assumed a tone of intellectual and political superiority, implying that Indians and Mexicans alike should integrate themselves into the industrial economy that the United States had consolidated in North America.

On these very points, Bartlett and Emory found the Mexican rural landscapes sadly wanting in stewardship and commercial development. Their reading of the Sonoran countryside pointed to signs of decline from former stages of prosperity, evidenced in the deterioration of adobe buildings, the large herds of feral livestock, and the neglect of orchards and gardens. Both commissioners passed through San Bernardino, once a presidial garrison and a cattle hacienda, and Emory's description of it seems especially noteworthy for the implied direction of change, reverting from a cultivated landscape to a wild, unkempt environment.

> It is very possible the whole of the new territory, except the . . . desert . . . , may be brought under the influence of artesian wells and made productive; but until that is the case, agriculture must be confined to the beds of the river, where the land is below the water level. . . . Throughout the whole course of the San Pedro [river] there are beautiful valleys susceptible of irrigation and capable of producing large crops of wheat, corn, cotton, and grapes; and there are on this river the remains of large settlements which have been destroyed by the hostile Indians, the most conspicuous of which are the mining town of San Pedro and the town of Santa Cruz Viejo. There are also to be found here, in the remains of spacious corrals and in the numerous wild cattle and horses which still are seen in this country, the evidences of its immense capacity as a grazing country. Removed from the river beds at the base of the mountains where perpetual springs are found, are also to be seen the remains of large grazing establishments; the most famous of

> which is the ranch of San Bernardino, which falls half in the United States and half in Mexico. I have been informed that this establishment was owned in Mexico, and when in its most flourishing condition boasted as many as one hundred thousand head of cattle and horses. They have been killed or run off by the Indians, and the spacious buildings of adobe which accommodated the employees of this vast grazing farm are now washed nearly level with the earth.[42]

This long passage is particularly expressive of the cultural assumptions with which Bartlett and Emory observed the landscapes of the Sonoran desert and uplands and imagined their condition in the past. The "new territory" to which Emory referred comprehended all the land taken from Mexico by the United States, in his words, awaiting North American technology to become productive. Riverine floodplain agriculture, sustained by "miles of acequias" and living fence rows, yielded "highly cultivated fields of grain and sugar-cane" and "gardens loaded with richly flavored fruit of the tropics." In their judgment, however, these agrarian landscapes of the "tierra caliente" of Sonora had not reached their full potential.[43]

These impressions were based in large measure on what the surveyors were told as they traveled through the region in stories that emphasized the Sonorans' losses of livestock to the Apaches and the retreat of ranching families to safer locations in the riverine towns. They resonated with the warnings issued by Mexican military officers, statesmen, and observers like Ignacio de Zúñiga, José Francisco Velasco, and Agustín de Escudero, whose impressions guided our discussion in chapter 7 on the uneven development of the region during the first half of the nineteenth century. Bartlett and Emory, however, attributed the abandoned ruins of haciendas and ranches uncritically to destructive raids and counterattacks by Mexicans and Apaches, and to the abandonment of rural hamlets by gold seekers—like the *gambusinos* of Quitobac—who had migrated to California. What they overlooked were the short-term crises provoked by the U.S. invasion itself, and the hastily mobilized military response in Mexico, demanding requisitioned livestock, cloth, and crops, in addition to climatic cycles of drought and the fearful cholera epidemic of 1851. Furthermore, their assessment of the Sonoran countryside as half-empty and underexploited implied its measurement against a standard of temperate-climate farming in close proximity to urban markets. In fact, landowners in the Sonora and San Miguel valleys had expanded their production of surplus grains and other crops to supply the new gold fields and settlements of California, as well as their established markets in Sinaloa and the Sonoran cities of Guaymas and Hermosillo.[44] Less visible to the inquisitive gaze of the commissioners was the long-term transition from communal to private landholding and the spread of peonage that diverted land and labor to the commercial production of surpluses on haciendas, as exemplified by Bartlett's description of Tepahui, but reduced many indigenous and mestizo peasants to dependent subsistence and impoverished the rural communities of Sonora. In ways that echoed the descriptions of early Spanish explorers and missionaries, Bartlett and Emory and their contemporary Mexican counterparts misread the Sonoran landscape and imposed different historical scripts on its ecological and cultural features. Seventeenth-century Spaniards like Francisco Eusebio Kino and Juan Mateo Manje saw abundant grasslands waiting to be filled with livestock and river valleys whose crop cycles were ripe for expansion with European fruit trees and grains.[45] Nineteenth-century narratives of foreign travelers and local observers, by way of contrast, portrayed a trajectory of arrested development in postcolonial agrarian landscapes in which levels of commercial flowering had been followed by partial abandonment.

Landscapes and *Paisajes*, Borderlands and *Fronteras*

We find the Spanish equivalents of *landscape* in the terms *paisaje* and *contorno*, defined variously as the artistic (visual) aspects of a given terrain or as the territory that surrounds a particular place. *Landscape*, understood here as historical process and as cultural space created and observed through human agency, is rendered by the Spanish meanings of *paisaje* and their connotations of a spatial environment that characterizes a locality or region. It is well to remember, however, that these concepts are not exactly translatable; nor are their terms of reference identical in the Iberian and Anglophone traditions. Their differences point to the need for caution when moving from one culture to another; at the same time, pausing to reflect on their implications enriches our discussion of landscapes, borders, and regions.

To conclude this comparative history of changing environments and cultural identities in the imperial borderlands of two American continents, it is useful to contrast the overlapping and nuanced meanings of *borderlands* and *frontiers* in the extensive English-language literature with Ibero-American understandings of these terms and the historical experiences that have informed them. North American historiography on frontiers and borderlands has been overshadowed by the prolific pens of Frederick Jackson Turner and Herbert Eugene Bolton, whose work defined the early paradigms of twentieth-century scholarship and inspired several generations of revisionist historians. Recent published work in history, anthropology, and geography has challenged Turner's narrative of Anglo-American continental expansion and the imperialist profile of Bolton's "rim of Christendom," recasting the once familiar categories of frontier and borderlands with cultural significance and underscoring the ambiguities of conquest.[46] Significant theaters of imperial encounters and transculturation that created borderlands took place in the Great Lakes (the French *pays d'en haut*), the Great Plains—including distinct ecological and ethnographical regions between the Mississippi-Missouri drainages and the Rocky Mountains—and the contested imperial borders that marked the interior provinces of northern New Spain.[47] The wheel has turned again, however, with renewed emphasis on the dimensions of power in the political and territorial rivalries of European imperial states and Native American tribal confederations to balance the revisionist constructions of intercultural frontiers, hybridity, and social mobility.[48]

Shifting our vision to South America provides alternative models for imagining the evolution of borderlands and frontiers. Whether understood as territorial boundaries or as zones of cultural transfer, South American borderlands are anchored in the dramatic geographic features of mountain ranges and river systems, and imagined over time through the historical encounters of diverse imperial powers, native chiefdoms, and mixed populations arising largely from colonial labor systems, both slave and free. The Andean cordilleras and their extended transitional ranges scaling eastward into the tropical *selva* constitute a myriad of human ecologies that have been theorized as archipelagos at different altitudes, presenting diverse patterns of crop production, pastoralism, gathering, and trade.[49] Conversely, the Amazonian and La Plata river basins comprehend much more than the streams that carry their names. Comprised of complex networks of tributaries converging from the headwaters of the Andes, flowing eastward and southward, they water vast interior lowlands of forests, savannas, swamps, and floodplains as diverse ethnically as ecologically.[50] Multiple South American borderlands with political, environmental, and cultural connotations help us think about frontier societies in new ways and view the North American model in a wider framework. Specifically, these cross-regional perspectives in Ibero-America lead us to question the analytical usefulness of the distinction between imperial borderlands and cultural frontiers, as suggested recently for the North American continent.[51]

Borderlands may also be thought of as networks for exchanging symbolic and commercial goods, initiated by indigenous peoples, European colonists, and African American slaves and freed persons. These networks figure prominently in the frontier histories of both North and South America, in the theaters of imperial encounters referred to above, and beyond the formal boundaries of colonial dominion.[52] Indeed, the regional identities of Sonora and Chiquitos may well be more accurately and fully expressed in terms of their webs of relationships than of their tangible borders. Indigenous paths of migration—with exchanges of persons, foodstuffs, and trade goods—constituted central features of subsistence strategies, ethnic conquests, and historical change in the Sonoran Desert, linking the O'odham and Opata villagers to the Hohokam and Casas Grandes trading centers, and in the Bolivian lowlands, where the Moxos chiefdoms to the north and the Inca frontier to the west shaped the cultural configuration of Chiquitos prior to the invasion by European forces. The colonial economy of both provinces created circuits that connected Sonora to the mining centers of northern New Spain and Chiquitos to Charcas and Potosí. Postcolonial developments in the nineteenth century placed Sonora in the broad sweep of new political alignments and economic prospects ranging from California and the Gila and Colorado drainages to the Sierra Madre Occidental, and enmeshed Chiquitos in webs of commerce, road building, and navigation extending eastward across the Paraguay river into Brazil and southward into the Gran Chaco.

How do these multicentric webs of imperial rivalries, commercial exchange, and cultural encounters and adaptations translate into Iberian conceptualizations of *la frontera*? Do the words and their commonly understood uses signify borders of confinement or zones of liminality? The definition of *frontera* refers to the limits of a spatial region or a political state (*confín de un estado*), and its application to one of our colonial frontiers is illustrated by the name given to the first presidial garrison in Sonora: Santa Rosa de Corodéguachi de Fronteras, located in a mountain pass leading to the Sierra Madre Occidental. During the seventeenth and eighteenth centuries, Fronteras marked the division between Sonora and Nueva Vizcaya, guarding the passage for mule trains and wagon roads that carried loads of silver and trade goods to the missions and mines of both provinces. A similar meaning of *frontera* was evoked for nineteenth-century Chiquitos by Sebastián Ramos when he referred to himself as "territorial judge of these frontiers" (*juez territorial de estas fronteras*) in the military colony of Tremedal established on the banks of the Paraguay river.

Frontero/a en frente, meaning the juxtaposition of two points or entities facing one another, suggests the relational quality of frontiers as zones of commingling and exchange. Conversely, the hardening of *la frontera* into *la línea*, a corridor of surveillance fraught with danger that can be crossed only by a deliberate act, is largely the historical expression of twentieth-century developments along the United States–Mexico border.[53] Even so, the borderlands surrounding this corridor have generated mixed, but conflicting identities among Mexican American and Latino populations in the United States and northern Mexico–*poblaciones fronteras en frente*–who often share the same cross-border networks of kinship, migration, and political economy. *La frontera* connotes gendered borders as well, experienced in labor markets, migratory routes, and zones of violence where frontiers become the layered identities of society, family, and sexuality.[54]

Recent attempts to distinguish between *frontier* and *borderland* in an effort to separate the cultural and the political in our conceptualization of imperial borders are not entirely convincing. The terminology does not neatly bifurcate the historical processes of encounter, assimilation, and redefinition of differences. When we translate the discussion into the Spanish language and Iberian frames of reference, we find that frontiers and borderlands are, at times, coterminous and, at other times and places, obliquely juxtaposed. If, indeed, Spanish colonial geographies reified and mapped ethnic provinces such as Papaguería, Opatería, Apachería, and Chiquitanía, these were recognized as shifting territories whose boundaries changed as people moved through the desert and migrated in short cyclical patterns or over long distances. The intersection of frontiers and borderlands in eastern Bolivia is evidenced in the territorial mosaics of different *parcialidades* and personalized in the 1767 journey of Joseph Sánchez Labrador to mediate between Guaycurúan and Chiquitano warring ethnic polities. The postcolonial colonies of Otúquis and Tremedal, intended to define and defend Bolivia's territorial borders with Brazil and Paraguay, were themselves transitory settlements in cultural and political frontiers.

The transition from bordered lands to borderlands, suggested from a North American perspective, running parallel to the nineteenth-century emergence of nation-states from competing imperial spheres, is not, in my view, a linear process. The troubled U.S.-Mexican border has become, at the turn of the twenty-first century, an armed international boundary; at the same time, however, it supports a dynamic network of commercial exchange embedded in a zone of changing intercultural contact through different ethnic and gendered flows of migration. Comparing this process with the frontiers that arose historically in and around Chiquitos, it is instructive to pause in the Gran Chaco. This transitional zone of ecological, cultural, and economic importance disputed by Bolivia, Paraguay, and Argentina was the scene of numerous boundary commissions, erupting into the bitter Chaco War of 1932–35,[55] yet it has remained a space of cultural resilience among the Ayoreóde, Tobas, and other indigenous peoples.

Transnationalism has acquired importance in contemporary journalism and academic fields of study to express borders transcended and transgressed through literature, art, music, commerce, and the physical movement of people. For either *transnational* or *transcultural* to have meaning, however, it is necessary to recognize borders that represent barriers to be crossed or zones of contact where differences meet. The chronological extension of this comparative history crosses the formal lines of demarcation between the colonial administrations of New Spain and Upper Peru and the formation of nation-states in Mexico and Bolivia. For Sonora and Chiquitos, *borderlands* provides a comprehensive frame of reference for the changing human landscapes and contested webs of power fashioned and reworked in both of these frontiers. The emergence of regions in the nineteenth century with recognizable spatial and social coherence in Sonora and Chiquitos proved integral to the transition from imperial to transnational flows that created frontiers of difference and contention, but also of confluence and synthesis. Regionality became increasingly important for constructing identities that denied or superseded the histories of conflict and the multiple local identities that had given depth and meaning to both areas. The multiple narratives woven into this story have shown, however, that regions are not homogeneous spaces; rather, their formation is a dynamic process linked to changing landscapes and the historical, contested production of culture.

I began this project with a question: What difference does the environment make to our understanding of historical process? I have concluded that nature matters in very important ways to the telling of history, as the foundation of material culture and the source of human livelihood and as the template for the construction of meanings and identities. In the course of writing and comparing these two stories of Sonora and Chiquitos, however, I have turned the question around. Nature matters not so much because the environment conditions what people do, but because the landscapes through which we perceive nature are themselves the work of human agency.

The places I chose to study reflect the second foundational question that guided this study: Are peripheries and borderlands meaningful for our understanding of integrative historical themes? Approaching history through the environment and recasting historical narratives of resistance to empire, contested nation-states, and the resilience of local communities has allowed me to bring the periphery into the center of the story. Reconstructing the particular histories of two borderlands through broad comparative themes has helped me to reaffirm the spatial dimensions of history and to revisit the conventional temporal divisions used to frame the human comedy. Furthermore, it has confounded the distinction between core and periphery and deepened my conviction that borderlands are not marginal to the stories that matter in global projections of history. The Sonoran deserts and highlands contrasted with the Amazonian rain forests and savannas, revealing layered cultural landscapes that bespeak the depth of overlapping histories of conquest, transition, and renewal.

Notes

Preface

1. Radding, *Wandering Peoples*, represents my work on Sonora.

2. Representative works include Braudel, *Civilization and Capitalism*, esp. vol. 3, *The Perspective of the World*; and Wallerstein, *The Modern World-System*. Two succinct works focused on imperial systems in the Americas are Stein and Stein, *The Colonial Heritage of Latin America*; and Seed, *Ceremonies of Possession*.

3. Dobb, *Studies in the Development of Capitalism*; Aston and Philpin, *The Brenner Debate*; Hilton and Sweezy, *The Transition from Feudalism to Capitalism*; Hobsbawm, *The Age of Capital*; and Tracy, *The Rise of Merchant Empires*. See also Kula, *Problemas y métodos de la historia económica*; and Tenenti, *La formación del mundo moderno*. These same issues are viewed from an Americanist perspective in Andrien and Johnson, *The Political Economy of Spanish America in the Age of Revolution*.

4. Wolf, *Europe and the People without History*. Wolf's distinguished oeuvre on peasantries, rebels, and colonized peoples has had a formative influence on historians and anthropologists of several generations and many different fields. Representative of them is research on the political economy and culture of labor in colonial settings. See Lovejoy and Rogers, *Unfree Labour in the Development of the Atlantic World*. Recognition of the force of racialized and class distinctions for creating social inequalities in colonial and postcolonial settings has been sharpened and complicated by recent studies that employ gender as a central analytical category. See Dore and Molyneux, *Hidden Histories*.

5. Kupperman, *America in European Consciousness*; Cañizares-Esguerra, *How to Write the History of the New World*; Natalie Zemon Davis, "Metamorphoses: Maria Sibylla Merian," in Davis, *Women on the Margins*, 140–202, shows how science and gender crossed the European and American boundaries of the early modern Atlantic world.

6. Van Young, "Cities, Hinterlands, and Marches."

7. The notes in this volume have been edited for brevity to conform to space limitations. The full set of endnotes, including additional commentary and the original Spanish wording of direct quotations translated into English in the text, can be found in electronic format in the University of New Mexico institutional repository at http://hdl.handle/1928/467.

Introduction: Savannas and Deserts

All translations are by the author unless otherwise indicated.

1. Historically these Indians were—and are, today—known as Papagos. They prefer to call themselves *tohono o'odham*: "people of the shining earth" (Nabhan, Cultures of Habitat, 163). I have used these two terms, mostly the latter, throughout the book.

2. Fernández, *Relación historial*, 34–35. A number of the products that the Jesuit Fernández names in this passage are not native to the Americas: sugar, rice, and bananas were brought to Chiquitos and to other regions of Spanish America from the Mediterranean, Asia, and Africa.

3. MacCameron, "Environmental Change in Colonial New México"; Zimmerer and Bassett, "Approaching Political Ecology."

4. Crosby, *The Columbian Exchange*; Crosby, *Ecological Imperialism*; McNeil, *Plagues and Peoples*; Diamond, *Guns, Germs, and Steel.*

5. Grove, *Green Imperialism.*

6. Several outstanding studies for North and South America that employ environmental frameworks include Dean, *With Broadax and Firebrand*; Melville, *A Plague of Sheep*; Cronon, *Changes in the Land*; White, *The Roots of Dependency*; Whitehead, "Ethnic Transformation and Historical Discontinuity"; Saeger, *The Chaco Mission Frontier*; Adelman, *Frontier Development*; Bell, *Campanha Gaúcha*; Frank, "The Brazilian Far West"; Ouweneel, *Shadows over Anáhuac.*

7. Carl von Linné [Linnaeus], "Specimen academicum"; Humboldt and de Bonpland, *Personal Narrative of Travels*; and Worster, *Nature's Economy*, 31–55.

8. Worster, *Nature's Economy*, 388–433; Real and Brown, *Foundations of Ecology*; and Stauffer, "Ecology in the Long Manuscript Version."

9. For a critical review of the discourse of ecological management, see Escobar, "Constructing Nature." Zimmerer and Bassett, "Approaching Political Ecology," insist on the dual importance of biophysical processes and "socially mediated" understandings of nature (3).

10. Clifford Geertz's now classic formula of "thick description" remains relevant for the historical study of cultural encounters in colonial situations, but it has been reworked by, among others, Sewell, "The Concept(s) of Culture"; Sahlins, *Culture in Practice*; and James Clifford's emphasis on translation, movement, and becoming in *Routes*. Following James Clifford's metaphorical use of routes and translations, *culture* is rendered in its adjectival form, *cultural*, to comprehend distinctive modes of action and reflection among different peoples. See Geertz, *The Interpretation of Cultures.*

11. Sahlins, *Culture in Practice*; Sahlins, *Islands of History*, 34; Clifford, *Routes*, 11; Shaw, *Colonial Inscriptions*, 14–16.

12. Peet and Watts, "Liberation Ecology," 4–5. Recent exemplary studies that track the impact of social change on traditional cultures, stemming from colonialism, technological innovation, and the complex forces of modernization include Orlove, "Ecological Anthropology" and *Lines in the Water*. Applied anthropology has recognized local knowledge about the environment with the term *traditional ecological knowledge* (TEK). See Sluyter, *Colonialism and Landscape*, 7.

13. The term was first coined in foundational works by Steward, *Theory of Culture Change*; and Geertz, *Agricultural Involution*.

14. Basso, *Wisdom Sits in Places*, 66–70.

15. Cultural ecology and the supposed divide between a balanced nature untouched by human intervention and a disturbed environment have been discussed widely in the literature on the history of environmental change in colonial and postcolonial Africa. See, for example, Fairhead and Leach, *Misreading the African Landscape*, 9–14.

16. The term was adopted by Ramachandra Guha and earlier generations of historical sociologists as a kind of shorthand for environmentally oriented sociology. See Guha, *Social Ecology*, 1–18; Mukerjee, *Social Ecology*; Mukerjee, "Ecological Contributions to Sociology"; and Martínez-Alier, "Ecology and the Poor."

17. Radding, *Wandering Peoples*, 3. Amith, "The Möbius Strip," examines territoriality both conceptually and historically for central Guerrero, Mexico.

18. Gow, "Land, People, and Paper," 59.

19. Please see the glossary for further explanations of terms used throughout this book.

20. Daniels and Cosgrove, "Introduction," summarize the contextual interpretations of symbolic imagery elaborated by European art historians and philosophers such as Ernst Cassirer, Erwin Panofsky, and John Ruskin. In this same essay Daniels and Cosgrove point to the critical connections between art history and anthropology for the study of iconography, a theme developed further by Hirsch, "Landscape."

21. D'Orbigny, Bach, and Castelnau traveled through eastern Bolivia during the 1830s and 1840s: D'Orbigny, *Viaje a la América Meridional* (1835); Castelnau, *Expédition dans les parties centrales de l'Amérique du Sud* (1851); Bach, *Descripción de la nueva provincia de Otuquis en Bolivia* (1843). The reports of Hardy, *Travels in the Interior of Mexico* (1829); Zúñiga, *Rápida ojeada al estado de Sonora* (1835); and Bartlett, *Personal Narrative of Exploration and Incidents* (1854) provide different perspectives on Sonora. On the characterization of "insider" and "outsider" views of landscapes, see Cosgrove, *Social Formation and Symbolic Landscape*; and Williams, *The Country and the City*; both are quoted in Hirsch, "Landscape," 13–14.

22. Sauer, "The Morphology of Landscape." Sluyter, *Colonialism and Landscape*, 6–7, recognizes Sauer's clarity in emphasizing native modifications of precolonial landscapes.

23. Gow, "Land, People, and Paper," 47–53. Gow narrates these processes of associat-

ing family relationships with places and modified environments as "landscape implication." See also Ramos, *Sanumá Memories*, 19–177, on space-time and the construction of historical memory in the landscapes of Sanumá communities.

24. Fontana, *Of Earth and Little Rain*, xi.

25. The Bourbon colonial regime designated all of northern New Spain as the "Internal Provinces" under a military commandancy in 1779. The term was not used officially in the viceroyalties of Perú or Río de la Plata, but the forest-and-savanna lowlands extending eastward from the *audiencia* of Charcas, today comprising portions of Bolivia, Paraguay, and Brazil, became an internal frontier between the Spanish and Portuguese dominions of South America. See Santamaría, "Fronteras indígenas del oriente boliviano," on intermediate territories between the Spanish and Portuguese realms of South America.

26. On *longue durée* views of comparative imperial histories in Mesoamerica and the Andes, see Collier, Rosaldo, and Wirth, *The Inca and Aztec States*; and Harvey, ed., *Land and Politics in the Valley of Mexico*.

27. Scott, "The State and People Who Move Around"; Scott, *Seeing like a State*, 1–13; and Van Young, *Mexico's Regions*.

28. Stern, "Feudalism, Capitalism, and the World-System," argues persuasively that the core/periphery paradigm envisioned by Immanuel Wallerstein is tempered by the exercise of agency within the Ibero-American colonies; see Wallerstein, *The Modern World-System*. More recently, Anthony Ballantyne, "Introduction: Aryanism and the Webs of Empire," in Ballantyne, *Orientalism and Race*, uses the metaphor of webs to describe intercolonial relations in the British Empire. I find the metaphor instructive, but condition it to emphasize the tension between webs and hubs, representing parallel centripetal and centrifugal pressures within imperial systems.

29. Adelman, *Colonial Legacies*; and Voss, *Latin America in the Middle Period*.

30. Anderson, *Imagined Communities*.

31. Wolf, *Europe and the People without History*; Sahlins, "Cosmologies of Capitalism"; and Prakash, *After Colonialism*.

32. Said, *Orientalism*; Said, *Culture and Imperialism*; Clifford and Marcus, *Writing Culture*; Shaw, *Colonial Inscriptions*; Feierman, "Colonizers, Scholars, and the Creation of Invisible Histories"; O'Hanlon, "Recovering the Subject"; Burton, *At the Heart of the Empire*; Stoler, *Race and the Education of Desire*; and Afzal-Khan and Seshadri-Crooks, *The Pre-occupation of Postcolonial Studies*.

33. Natalie Zemon Davis's beautifully crafted *Women on the Margins* takes up the task of decentering European history for the early modern period; the second biography of that book, "New Worlds: Marie de l'Incarnation," 63–139, admirably captures colonial subjects; see also Davis, "Iroquois Women, European Women," 350–62. A different approach for the modern period and from the South Asian perspective appears in Chakrabarty, *Provincializing Europe*.

34. Seed, "Colonial and Postcolonial Discourse"; Mallon, "The Promise and Dilemma of Subaltern Studies: Perspectives from Latin American History"; Mallon, *Peasant and Nation*; Guardino, *Peasants, Politics, and the Formation of Mexico's National State*; Gould, *To Die in This Way*; Grandin *The Blood of Guatemala*; Chassen-Lopez, "Maderismo or Mixtec Empire?"

35. Todorov, *The Conquest of America*; Gruzinski, *La guerra de las imágines*; MacCormack, *Religion in the Andes*; Mills, *An Evil Lost to View*; Griffiths, *The Cross and the Serpent*; Schwartz, ed., *Implicit Understandings*; Mignolo, *The Darker Side of the Renaissance*; and Cañizares-Esguerra, *How to Write the History of the New World*.

36. León Portilla, *Visión de los vencidos*; Wachtel, *The Vision of the Vanquished*; Gibson, *Tlaxcala in the Sixteenth Century*; Gibson, *The Aztecs under Spanish Rule*; Farriss, *Maya Society under Colonial Rule*; Clendinnen, *Ambivalent Conquests*; Gruzinski, *The Conquest of Mexico*; Lockhart, *The Nahuas after the Conquest*; and Carmagnani, *El regreso de los dioses*.

37. Three recent collaborative volumes that illustrate the insights to be gained from comparative research are Langer and Jackson, *The New Latin American Mission History*; Guy and Sheridan, *Contested Ground*; and Griffiths and Cervantes, *Spiritual Encounters*.

38. See Hulme, "Including America"; and Shohat, "Notes on the Postcolonial."

39. The constitutions of Bolivia and Ecuador were modified during the 1990s to include specific language that defined these countries as "multiethnic and pluricultural" nations. The San Andrés Accords and the controversial Law of Indigenous Rights and Cultures, arising from the Chiapas peasant movement in Mexico, challenge the political definition of the Mexican nation set forth in the constitution of 1917. Recent scholarship that develops these themes for Mexico includes Warman, *Y venimos a contradecir*; Vanderwood, *The Power of God against the Guns of Government*; Van Young, *The Other Rebellion*. For the Andes, it includes Thurner, *From Two Republics to One Divided*; Walker, *Smoldering Ashes*; Guerrero, "El proceso de identificación"; and Abercrombie, *Pathways of Memory and Power*. On contemporary movements, see Collier with Quaratiello, *Basta!*; Whitten, *Sicuanga Runa*; Whitten, "Commentary"; Whitten, Whitten, and Chango, "Return of the Yumbo"; and Becker, "Comunas and Indigenous Protest."

40. The disciplinary implications of blending economic and cultural history are discussed in a special issue of the *Hispanic American Historical Review* 79.2 (1999), entitled "Mexico's New Cultural History: ¿*Una Lucha Libre?*" Contributing authors are Eric Van Young, William E. French, Mary Kay Vaughan, Stephen Haber, Florencia E. Mallon, Susan Migden Socolow, and Claudio Lomnitz. See also Beezley, Martin, and French, *Rituals of Rule, Rituals of Resistance*. On the limits of cultural history, see Roseberry, *Anthropologies and Histories*.

41. Escobar, "Constructing Nature," 46.

42. Cronon, *Changes in the Land*; Cronon, "A Place for Stories"; Cronon, ed., *Uncommon Ground*; White, *The Roots of Dependency*; and White, *The Middle Ground.*

43. My reading of White's use of *dependency* is consonant with the conceptual frameworks employed by Mallon, *The Defense of Community*, 3–7; and Larson, *Cochabamba*, 322–26.

44. Williams, *Linking Arms Together*, 40–61.

45. Jones, "Comparative Raiding Economies"; Hall, "The Río de la Plata and the Greater Southwest"; Hall, *Social Change in the Southwest*; Saeger, *The Chaco Mission Frontier*; and Hämäläinen, "The Rise and Fall of the Comanche Empire."

46. García Martínez, *Los pueblos de la sierra*; García Martínez and González Jácome, *Estudios sobre historia*; Melville, *A Plague of Sheep*; Ouweneel, *Shadows over Anáhuac*; Rojas Rabiela, *Agricultura indígena.*

47. Siemens, *A Favored Place*; Sluyter, *Colonialism and Landscape*; Amith, "The Möbius Strip."

48. Nabhan, *The Desert Smells like Rain*; Nabhan, *Gathering the Desert*; Sheridan, *Where the Dove Calls*; Sheridan, introduction to Sheridan, *Empire of Sand*; Bowen, *Unknown Island*; Villalpando, "Los que viven en las montañas"; Braniff, *La frontera protohistórica Pima-Ópata*; Hers et al., *Nómadas y sedentarios*; and Braniff, *La Gran Chichimeca.*

49. Murra, *Formaciones económicas y políticas del mundo andino*; Spalding, *Huarochirí*; Mayer, *The Articulated Peasant*, 239–78; Tristan Platt, "Fronteras imaginarias en el sur andino"; Gade, *Nature and Culture in the Andes*; Zimmerer, "Discourses on Soil Loss in Bolivia."

50. Dean, *With Broadax and Firebrand.*

51. Adelman, *Frontier Development*; Bell, *Campanha Gaúcha*; and Frank, "The Brazilian Far West."

52. Larson, *Cochabamba*; Mallon, *The Defense of Community*; Mallon, *Peasant and Nation*; Guardino, *Peasants, Politics, and the Formation of Mexico's National State*; Gould, *To Die in This Way*; Abercrombie, *Pathways of Memory and Power*; and Grandin, *The Blood of Guatemala.*

1. Ecological and Cultural Frontiers

1. Adorno and Pautz, eds., *Alvar Núñez Cabeza de Vaca*. Hereafter cited as *Relación*, with volume and page numbers. He is referenced in the bibliography as Núñez Cabeza de Vaca, but in the text his name is shortened to Cabeza de Vaca. Translations of passages from the *Relación* included in the text come from this edition by Adorno and Pautz. Outstanding literary and ethnohistorical studies of Cabeza de Vaca, usually centered on the *Naufragios* (*Relación*), include Rolena Adorno, "The Negotiations of

Fear"; Ahern, "The Cross and the Gourd"; Reff, "Text and Context"; and Rabasa, *Writing Violence on the Northern Frontier*, 31–83.

2. *Relación*, 1:211–13.

3. Ibid., 1:197–215.

4. Ibid., 1:217.

5. Oviedo y Valdés, *Historia general y natural de las Indias*.

6. *Relación*, 1:209–11.

7. Ibid., 237–43.

8. Ibid., 231.

9. Di Peso, *Casas Grandes*; Braniff, *La frontera protohistórica Pima-Ópata en Sonora, México*; Riley, *The Frontier People*, 39–96.

10. The relationship between Casas Grandes and eastern Sonoran peoples remains an issue of scholarly debate concerning the direction of influence and whether the connections between these two cultural areas were adversarial, dependent, or merely parallel. See Alvarez, "Sociedades agrícolas," 244–45; Riley, *The Frontier People*, 48; Newell and Gallaga, *Surveying the Archaeology of Northwest Mexico*; and Doolittle, *Pre-Hispanic Occupance*. Similar hypotheses of late precontact decline, based on combined archaeological and ethnohistorical documentary evidence, concern the chiefdoms of the mound peoples of the southeastern United States and northern Mexican Mesoamerican trading and ceremonial sites like La Quemada in Zacatecas. See Johnson, "From Chiefdom to Tribe"; Batres, "Visit to the Archaeological Remains."

11. Hammond and Rey, *Narratives of the Coronado Expedition*; and Obregón, *Obregón's History of 16th Century Explorations in Western America*; Bandelier, *Contributions to the History of the Southwestern Portion of the United States*, 106–78; Flint and Flint, eds., *Documents of the Coronado Expedition*.

12. *Chilchilticale* means "red big house" in Nahuatl. Carroll Riley's interpretation of the Nizza and various Coronado narratives, together with the archaeological work by Charles Di Peso, suggests that Chilchilticale was located on the Pima-Yavapai border, perhaps on the Salt River in eastern Arizona. Riley, *The Frontier People*, 49, 110–20; Flint, *Great Cruelties Have Been Reported*, 253–54, 262.

13. Ibid., 119–22, stresses the significance of shell trade routes that crisscrossed the *serrano* and desert regions.

14. Pennington, *The Pima Bajo*, 1:145–46; Pérez Bedolla, "Geografía de Sonora."

15. See the chronicles cited in note 11 above; Radding, "Domesticating Rural Space."

16. Pérez Bedolla, "Geografía de Sonora," 136–69.

17. Montané Martí, "Desde los orígenes," 203–7; Pennington, *The Pima Bajo*, 1:247–52; Nabhan, *Cultures of Habitat*, 166–71.

18. Pennington, *The Pima Bajo*, 1:143–51.

19. Nabhan, "Amaranth Cultivation," 130–31; Nabhan, *Gathering the Desert*, 101–3.

20. Radding, *Wandering Peoples*, 51; Sheridan, *Where the Dove Calls*, 54–56; Nabhan

and Sheridan, "Living with a River." See the accurate description of floodplain irrigation techniques, emphasizing the labor required to maintain them, in Nentvig's *Descripción geográfica*, 139–41.

21. Doolittle, *Pre-Hispanic Occupance*, 36–43. Doolittle argues convincingly that the location of village sites obeyed both environmental and political considerations. The former refer to proximity to water and arable land, or, especially in the early phase, to wild plant resources; the latter refer to surveillance from hilltops and control of segments of the floodplain.

22. The Hohokam canal irrigation and town centers of the Gila River valley flourished during the first millennium C. E. Like Casas Grandes (see note 10), Hohokam archaeology has yielded a complex history. Haury, *The Hohokam*; Gumerman, *Exploring the Hohokam*; Ackerly, Howard, and McGuire, *La Ciudad Canals*.

23. Doolittle, *Pre-Hispanic Occupance*, 29–35; Doolittle, *Cultivated Landscapes of Native North America*, 254–346; Doyel, "The Transition to History"; Bandelier, *Final Report of Investigations*, 449; Bandelier, *A History of the Southwest*, fig. 28.

24. Doolittle, *Pre-Hispanic Occupance*, 10–22; Pérez de Ribas, *History of the Triumphs*, 87–88.

25. Hayden, *The Sierra Pinacate*. See also Hayden's extensive bibliography (83–87); Ives, "The Pinacate Region"; and Kozak and López, *Devil Sickness and Devil Songs*, 85.

26. Bowen, *Unknown Island*, 16; Sheridan, ed., *Empire of Sand*, 10–12, 27 n. 20, 21, 108 n. 20.

27. Crosswhite, "Desert Plants, Habitat, and Agriculture"; Fontana, "Pima and Papago"; and Fontana, "History of the Papago."

28. *Relación*, 1:235.

29. Felger and Moser, *People of the Desert and Sea*, 20–28.

30. See description of Tohono O'odham seasonal migrations and *'ak-ciñ* (field prepared for planting) in desert washes in Radding, *Wandering Peoples*, 49–50, summarized from Castetter and Bell, *Pima and Papago Indian Agriculture*; Rodríguez and Silva, *Etnoarqueología de Quitovac, Sonora*; and Nabhan, "*Ak-ciñ* and the Environment of Papago Indian Fields."

31. The following geographic and ecological summary descriptions are based on Killeen, García E., and Beck, *Guía de árboles de Bolivia*; Birk, *Plantas útiles en bosques y pampas chiquitanas*; and Centurión and Kraljevic, *Las plantas útiles de Lomerío*. The term *subtropical* refers to the transitional nature of vegetation, between the Amazonian forests and the arid *matorral* (scrub forest) of the Chaco, although the latitudinal location of the Chiquitanía places it firmly within the South American tropics (Killeen, García E., and Beck, *Guía de árboles*, 6).

32. D'Orbigny, *Viaje a la América Meridional*, 3:1195.

33. Fischermann, "Contexto social y cultural"; Riester and Fischermann, *En busca de*

la loma santa, 150–54; Schwarz, *Yabaicürr, yabaitucürr, chiyabaiturrüp*, 59–60; Quiroga and Fischermann, "Pueblo viejo—pueblo nuevo." *Puquio* (as in Quechua) refers to a natural source of water; *paurú* to a water hole that is excavated or dug by people for domestic use. Information obtained from Helaine Silverman, University of Illinois, 2001.

34. Centurión and Kraljevic, *Las plantas útiles de Lomerío*, 29, 387; Killeen, García E., and Beck, *Guía de árboles de Bolivia*, 276; Birk, *Plantas útiles en bosques y pampas chiquitanas*, 3.

35. Birk, *Plantas útiles en bosques y pampas chiquitanas*, 182, 194.

36. Antonio and Isabel Tomichá, interview by the author, San Javierito, 9 August 2000.

37. Ramón Conrado Morón Tomichá, interview by Justa Morón Macoñó, San Ignacio de Velasco, November 2001. Rice (*Oryza sativa*) is an exotic to the Americas, converted into a basic foodstuff and cultivated under varying ecological conditions as a dryland, rain-fed crop or as wet rice in coastal lowlands, river deltas, and swamps. The dryland variety is cultivated in Chiquitos, and it is susceptible to drought. See Lugar, "Rice Industry;" Carney, *Black Rice*, 73–78.

38. Fischermann, "Contexto social y cultural," 29–30; Schwarz, *Yabaicürr, yabaitucürr, chiyabaiturrüp*, 106; Riester, "Los Chiquitanos," 128–33.

39. Lynch, "Earliest South American Lifeways," 241; Arrien, *Cacería y pesca*.

40. Ramón Conrado Morón Tomichá, interview, November 2001.

41. Fernández, *Relación historial*, 39–40.

42. Riester, "Los Chiquitanos," 148; Quiroga, "Matrimonio y parentesco entre los Chiquitanos."

43. This description is culled mainly from Toledo, "Inventario de una unidad doméstica," 429–31, from conversations with Rosa María Quiroga, and from the author's personal observations in a number of Chiquitano communities.

44. See, for example, Arrien, Salvatierra, and Pessoa, "El kúyuri." The Augustinian friar Yldefondo Vargas y Urbina reported on moving the pueblo of San Juan to a healthier location on 6 April 1776. Archivo y Biblioteca Nacionales de Bolivia (hereafter ABNB), Rück 64. *Paurú* is first explained in note 33 above.

45. This summary is based on Roosevelt, "The Maritime, Highland, Forest Dynamic and the Origins of Complex Culture."

46. Fernandez, *Relación historial*, 125–26. Freyer, *Los Chiquitanos*, 13, questions the inclusion of Manasicas linguistically and culturally as part of the Chiquitos nation.

47. Barragán Romano, "*¿Indios de arco y flecha?*"; Presta, *Espacio, etnías y frontera*. Zamaipata is part of a series of fortified ruins in the eastern Andean foothills between the Parapetí and Guapay (Grande) rivers, associated with cave paintings and stone structures at Incahuasi, Batanes, and Saipina. Erland and Olga Nordenskiöld visited these sites in 1913–14, reporting on them in *Forschungen und Abenteuer in Südamerika*.

48. *Adelantado* is, literally, "one who goes before." The dual title implied that Cabeza de Vaca would establish new colonies by means of conquest and exploration, and then would govern them.

49. Núñez Cabeza de Vaca, *Comentarios*, 99–100, 168–69. A ducat was a gold coin that circulated in Spain until the end of the sixteenth century, valued around 7 pesos. Translations of passages cited in the text from the *Comentarios* are by the author.

50. Adorno and Pautz, *Alvar Núñez Cabeza de Vaca*, 1:388–90. Their summary of this stage of Cabeza de Vaca's career and his legal defense in Spain is based largely on Oviedo y Valdés's *Historia general*, book 23, chap. 16. Oviedo interviewed Cabeza de Vaca on two occasions to ask him about his experiences in Florida and Río de la Plata.

51. References to *patos* are probably domesticated Muscovy ducks (*Cairina moschata*), while *gallinas* may be the indigenous *Gallus inauris* or European chickens previously introduced by Spaniards. See Garavaglia, "The Crises and Transformations of Invaded Societies," 2:4.

52. Saeger, *The Chaco Mission Frontier*, 5–6; Núñez Cabeza de Vaca, *Comentarios*, 130–40. The quotation comes from *Comentarios*, 137.

53. *Encomienda* holders were entitled to tribute and labor from the Indians assigned to them. Cabeza de Vaca's royal capitulation empowered him to grant *encomiendas* to colonists within his administration and, presumably, those who followed him on new conquests.

54. Núñez Cabeza de Vaca, *Comentarios*, 111.

55. Ibid., 111.

56. Ibid., 174–75; quotation, 174.

57. Ibid., 154–55, 176–81; René-Moreno, *Catálogo del Archivo de Mojos y Chiquitos*, 243–44; Adorno and Pautz, *Alvar Núñez Cabeza de Vaca*, 1:392. Fátima Costa, in *Historia de um país inexistente*, argues that the Lago de los Xarayes, repeated in so many colonial Spanish maps—see figure 13 in this book—never existed as such; rather, the Portuguese place name of Pantanal was geographically more accurate.

58. Núñez Cabeza de Vaca, *Comentarios*, 182–85.

59. Ibid., 67–73, 198–210. The illness, described as recurring fevers and associated with mosquitoes, may have been malaria.

60. See Sahlins, *Islands of History*, 5–26; Gutiérrez, *When Jesus Came, the Corn Mothers Went Away*, 19, 50–51; and Brooks, *Captives and Cousins*, 5–9, who argue this point similarly for different geographic and historical contexts.

61. Núñez Cabeza de Vaca, *Comentarios*, 209. The supposedly enlightened policies assumed by Cabeza de Vaca have been contested in various ways. Rabasa, in *Writing Violence on the Northern Frontier*, 31–33, 43–59, energetically dismissed Cabeza de Vaca's laurels as the protagonist of "peaceful conquest" concerned with fair treatment for the Indians. Catherine Julien, in a 2003 personal communication in Santa Cruz de la Sierra, pointed out Cabeza de Vaca's failure to understand the importance of trade

for the Guaraní and other Paraguayan tribes, arguing that his prohibitions against commerce among the other expeditionaries and the Indians and the taking of women were misguided. I interpret Cabeza de Vaca's political failures in this light, to suggest that he misread the cultural landscapes of the Guaraní-Guaycurúan frontier, allying himself only with certain chiefs, and failed to recognize the significance of trade beyond the exchange of goods and the procurement of food.

62. Núñez Cabeza de Vaca, *Comentarios*, 204–5.

63. Ibid., 181; Garavaglia, "Crises and Transformations of Invaded Societies," 2; Adorno and Pautz, *Alvar Núñez Cabeza de Vaca*, 1:386. The Solís expedition of 1516 was the first to explore the Río de la Plata and met defeat.

64. René-Moreno, *Catalogo del Archivo de Mojos y Chiquitos*, 190–213. San Lorenzo was close to the present-day location of Santa Cruz de la Sierra.

65. Denevan, "Stone versus Metal Axes"; Fontana and Matson, *Before Rebellion*, 17–18.

2. Political Economy

1. These social and ethnic processes of change are the subject of chapter 4. Opata and Eudeve peoples shared similar material cultures, but were distinguished by their languages and historical identities (Pennington, ed., *Arte y vocabulario de la lengua doema, heve o eudeve*).

2. I have chosen the terms *Sonoras* and *Chiquitanos* for convenience, based on the colonial conventions that referred to the provinces and to their native inhabitants by these names, when it is cumbersome to refer to the separate ethnic identities that peopled both regions. See Querejazu and Barbery, *Las misiones jesuíticas de Chiquitos*; Hoffmann, *Las misiones jesuíticas entre los Chiquitanos*; and Pfefferkorn, *Sonora*.

3. ABNB, Ramo Mojos y Chiquitos (hereafter MYCH), vol. 24, exp. 6, folios 114, 132, 154, 167, 185, 202, 219, 236, 251, and 266. Comprehensive censuses were recorded in the Chiquitos missions in 1767 and 1768, occasioned by the expulsion of the Jesuit missionaries.

4. Knogler, "Relato sobre el país y la nación," 161–65; Fernández, *Relación historial*, 135–57.

5. *Cacique*, a Taino Caribbean term, spread generally throughout Spanish America because it was used by Spanish ecclesiastical, civil, and military officials to distinguish native leaders of local or regional chiefdoms or to designate persons to act as interlocutors with colonial authorities. Alvar Núñez Cabeza de Vaca did not use the term in either his *Relaciones* or *Comentarios*, written in the mid-sixteenth century; rather, he referred to indigenous leaders as *principales*. By the mid-seventeenth century, *cacique* appears in Jesuit texts like Pérez de Ribas's *History of the Triumphs*.

6. Fischermann, "Viviendo con los paí"; and Santamaría, " 'Que reconozcan el triste estado de sus almas' "; Fernández, *Relación historial*; ABNB, MYCH, vol. 34, folios 241–48v.

7. Pérez de Ribas, *History of the Triumphs*, 111–24.

8. Gerhard, *The North Frontier of New Spain*, 245–48; Río, *La aplicación regional*, 83–115. The coastal provinces that became integrated into the Sonora-Sinaloa governorship were Rosario, Cosalá, Culiacán, Sinaloa, Ostimuri, and Sonora.

9. McCarty, *A Spanish Frontier in the Enlightened Age*, 5–51; Río, *La aplicación regional*, 143–45.

10. Ortega Noriega, "El sistema de misiones jesuíticas," 41–94; Polzer, *Rules and Precepts*, 13–37; Gerhard, *The North Frontier of New Spain*, 248.

11. Block, *Mission Culture on the Upper Amazon*, 11–54; Parejas Moreno and Suárez Salas, *Chiquitos*, 65–71.

12. Radding, "Voces chiquitanas"; ABNB, MyCh, vol. 23, exp. 7, 1751–53.

13. Father Pedro Méndez to Andrés Pérez de Ribas, quoted in Pérez de Ribas, *History of the Triumphs*, 412–13. Jesuit Méndez first visited the Sisibotari and Batuco peoples of central Sonora in 1621, beginning the *reducción* in 1628. The three pueblos to which he refers are Sahuaripa, Arivechi, and Bacanora.

14. Pérez de Ribas, *History of the Triumphs*, 209–13, quotations on 212.

15. Arches may have had autochthonous as well as Christian symbolic meanings. See Sheridan, *Empire of Sand*, 42 n. 44.

16. Pérez de Ribas, *History of the Triumphs*, 44, 168–69.

17. Knogler, "Relato sobre el país y la nación," 147–53.

18. *Relación* I: 163, 195–99; see chapter 1, notes 2 and 3. Ahern, "The Cross and the Gourd"; Radding, "Crosses, Caves, and *Matachinis*."

19. Pérez de Ribas, *History of the Triumphs*, 96–97, 116–17, 123.

20. Ibid., 297.

21. Ibid., 297.

22. Ibid., 298–99.

23. Reff, *Disease, Depopulation, and Culture Change*; Jackson, *Indian Population Decline*; Deeds, *Defiance and Deference*, records numerous epidemic crises for Nueva Vizcaya. The classic source on population decline for northwestern Mexico is Sauer, *Aboriginal Population of Northwestern Mexico*, summarized in Gerhard, *The North Frontier of New Spain*.

24. *Vecinos* refers to Hispanic or Hispanicized persons of mixed racial ancestry who distinguished themselves from the Indians. See Radding, *Wandering Peoples*, 5, 314, 362; Río, *La aplicación regional*, 122–30, on the meaning of *vecino* and *gente de razón* and on the changing demographic structure of Sonora.

25. García Recio, *Análisis de una sociedad de frontera*; Garavaglia, "The Crises and Transformations of Invaded Societies"; Whigham, "Paraguay's *Pueblos de Indios*."

26. Juan Bautista de Escalante, "Diary of Juan Bautista de Escalante, 1700," in Sheridan, *Empire of Sand*, 36–96. The counterpoint of violence and negotiation chronicled in Escalante's diary brings to mind the performance of French military officers and civilian governors as so-called alliance chiefs in the Great Lakes region of North America as portrayed in White, *The Middle Ground*.

27. Father Miguel Almela, in Opodepe, to Father Andrés Michel, in Ures, 15 March 1766. Archivo General de la Nación, Mexico (hereafter AGN), Jesuitas IV-10, exp. 176, folio 210; Marcial de Sossa to Father Michel, from Nacameri, 20 March 1766, AGN, Jesuitas IV-10, exp. 177, folio 211. A *fanega* is approximately 1.5 bushels or 55 liters of dry volume, used commonly to measure wheat and corn.

28. Father Miguel Almela to Father Andrés Michel, Opodepe, 22 June 1767, AGN, Jesuitas IV-10, exp. 230, folio 265.

29. See, for example, AGN, Jesuitas IV-10, exp. 69, folio 100 and exp. 71, folio 102, in which Gerónimo de Chave y Barretia, a merchant of Río Chico, first demanded of Father Michel payment of a debt of 450 pesos and threatened to involve the governor to force payment, but then assumed a more conciliatory tone by accepting the missionary's calculus of the amount owed, "for I value your friendship more than what the interest may be."

30. Archivo de la Mitra, Hermosillo, Archivo de la Parroquia del Sagrario (hereafter AMH, AS), vol. 1, folios 132–46, 1666–828. See Radding, *Wandering Peoples*, 66–99; and Radding, "From the Counting House to the Field and Loom."

31. AGN, Jesuitas IV-10, contains numerous expedients with correspondence among the missionaries and with local merchants that detail disputes over debts and the daily transactions revealed in the ledgers themselves.

32. Jesuit Eusebio Francisco Kino reported in 1710 that mission Dolores, the first of his *reducciones* in Pimería Alta, supplied livestock and grains valued at over three thousand pesos for the newer missions of this frontier *rectorado*. Kino, *Las misiones de Sonora y Arizona*, 366.

33. Nentvig, *Descripción geográfica*, 162–63.

34. Martin, *Governance and Society in Colonial Mexico*, 24–28.

35. AGN, Jesuitas IV-10, exp. 146, folio 189, 1746; exp. 158, folio 192, 1765; Biblioteca Nacional Fondo Franciscano (hereafter BNFF), vol. 32, exp. 659, 1803.

36. AMH, AS, vol. 1, 1666–1828.

37. Martin, *Governance and Society*, 57–58. Price histories for New Spain are well developed, using *alhóndiga* (public granary) and *diezmo* (tithe) figures. See especially Gibson, *The Aztecs under Spanish Rule*; Florescano, *Precios del maíz*; Morin, *Michoacán en la Nueva España*; Brading, *Haciendas and Ranchos*; Van Young, *Hacienda and Market*; Garner, *Economic Growth and Change in Bourbon Mexico*; Johnson and Tandeter, *Essays on the Price History*; and García Acosta, *Los precios del trigo*.

38. BNFF, vol. 35, exp. 761–62. Descriptive language in the rest of the paragraph comes from this Franciscan report.

39. Kessell, *Friars, Soldiers, and Reformers*, 174. Orden del Comandante Pedro de Nava sobre régimen de los pueblos, AMH, Archivo Diocesano de Hermosillo (hereafter AD), vol. 1, exp. 27, 1794. A half century earlier, Jesuit visitor Juan Antonio Baltasar had proposed secularizing a number of the missions of southern Sonora and Ostimuri (including Ónavas) for the purpose of advancing the mission frontier northward. Padre Juan Antonio Baltasar to Padre Provincial Cristóbal de Escobar y Llamas, 1744, Yale Beinecke Library, quoted in Burrus and Zubillaga, *El noroeste de México*, 163–64.

40. Indian justices of San Xavier to Fr. Thomas de Valencia, ABNB, Chiquitos, exp. 152, folio 23, 1771. This document was transcribed in Spanish and included in a file concerning a running conflict between Father Valencia and the governor of Chiquitos that was forwarded by the bishop of Santa Cruz to the *audiencia* of Charcas.

41. See, for example, the complaints brought before governor Barthelemí Verdugo by the entire *cabildo* of San Juan against the priest Domingo Viveños, ABNB, MYCH, vol. 25, folios 143–44.

42. ABNB, MYCH, vol. 24, exp. 7, folios 286–323, 1769.

43. ABNB, MYCH, vol. 35, folios 78–95.

44. ABNB, MYCH, vol. 26, exp. 29, folios 56–59; vol. 38, folios 272–337.

45. Knogler, "Relato sobre el país y la nación," 147–53, 156–57. Lathe work produced the wooden rosaries that formed an important part of the ceremonial life of the missions and were occasionally exported. Examples of inlaid wooden chests produced in the Chiquitos workshops are on display in the Museo de Historia maintained by the Universidad San Francisco Javier, Sucre, Bolivia. Pictorial evidence for weaving is more abundant for Moxos than for Chiquitos (María Mercado, *Album de paisajes*). It is not clear whether the women wove in their own homes or in mission workshops.

46. Libro de quentas de este Pueblo de Santhiago de Chiquitos, ABNB, MYCH, vol. 71, folios 15–18v, 1769.

47. Construction at the last of these, Santa Ana, was completed after the departure of the Jesuits.

48. Roth and Kühne, " 'Esta nueva y hermosa iglesia.' " These churches were restored under the direction of architect Hans Roth during the last quarter of the twentieth century.

49. "Visita eclesiástica y estatutos formulados por el Obispo de Santa Cruz de la Sierra, D. Francisco Ramón de Herboso, 1768–1769," ABNB, MYCH, vol. 24, folios 6–98; ABNB, MYCH, vol. 35, folios 1–36, 1768.

50. The quantitative evidence summarized in the following paragraphs comes from ABNB, MYCH, vol. 35, folios 78–95, 1771–76; ABNB, MYCH, vol. 38, folios 272–337, 1786–94.

51. Martínez de la Torre to the President of the Audiencia, San Xavier, 12 April 1768, ABNB, MYCH, vol. 23, exp. 9, folios 21–24.

52. ABNB, MYCH, vol. 23, exp. 28, folios 118–19; exp. 28, folios 129–50, 1772; MYCH, vol. 30, exp. 47, folios 181–189v, 1796.

53. ABNB, MYCH, vol. 23, exp. 5, folio 14v.

54. ABNB, MYCH, vol. 24, folios 354–355; ABNB, MYCH, vol. 25, folios 169–177v, 1777–1784; ABNB, MYCH, vol. 38, folios 272–337.

55. Corregidor Antonio Sesoane de los Santos to Governor Andrés Mentre, 14 April 1777, ABNB, MYCH, vol. 25, folio 169.

56. See, for example, the reports of governor Francisco Javier de Cañas to the *audiencia*, 1787 and 1788, in ABNB, MYCH 27, folios 46–85v.

57. Informe del Obispo de Santa Cruz al Presidente de la Audiencia, 1774, ABNB, MYCH, vol. 35, folios 40, 65–66; Block, *Mission Culture on the Upper Amazon*, 70–77, 132–71.

58. ABNB, MYCH, vol. 35, folios 1–26.

59. ABNB, MYCH, vol. 25, folios 91–93. Salt and saltpeter continued to be valued commodities for regional trade during the republican period. See Universidad Gabriel René Moreno (hereafter UGRM), Museo de Historia (hereafter MH), Fondo Prefectural, vol. 1, exp. 6, 1826.

60. "Comandante del pueblo de Santiago don Jph Lorenzo Chávez al Gobernador Verdugo," 21 April 1781, ABNB, MYCH, vol. 25, folio 138.

61. San Juan de Chiquitos, 31 May 1779, ABNB, MYCH, vol. 25, folios 139–141.

62. San Juan de Chiquitos, 1 June 1779, ABNB, MYCH, vol. 25, folios 141–144.

63. Río, *La aplicación regional*, 33–49; Polzer, *Rules and Precepts*, 23–24; González Ruiz, ed., *Etnología y misión en la Pimería Alta*, 125–87.

64. Radding, *Wandering Peoples*, 171–245; Río, *La aplicación regional*, 181–223.

65. Kessell, *Friars, Soldiers, and Reformers*, 19–21; McCarty, *A Spanish Frontier in the Enlightened Age*, 53–72.

66. Radding, *Wandering Peoples*, 256–63.

67. Radding, "Voces chiquitanas," 128–29.

3. Community and Property

1. On the use of social ecology, see the introduction. Soja, *Postmodern Geographies*, has challenged historians to incorporate space in their studies of long-term processes.

2. A *sitio* is a land measurement equivalent to 1,747 hectares when assigned for raising bovine cattle (*ganado mayor*), and equivalent to 776 hectares when used for smaller species of goats, sheep, and pigs (*ganado menor*). AGN, Archivo Histórico de Hacienda (hereafter, AHH), Temporalidades, vol. 1165; Radding, *Wandering*

Peoples, 177. *Sitios* designated in Sonoran land titles are nearly always for *ganado mayor*.

3. Archivo de Instrumentos Públicos de Jalisco (hereafter AIPJ), Ramo de Tierras y Aguas (hereafter RTA), first collection, vol. 42, exp. 2, folio 1–1v. The elder Salazar's 1746 title to Carrizalito is presented in folio 3 of the expedient.

4. AIPJ, RTA, first collection, vol. 42, exp. 2, folios 13–14, 28. *Mediannata* (half-year), used here, refers to a one-time payment of tax on securing a land title. Tomín, as used in this source, is equivalent to a *real*, or one-eighth of a peso.

5. Pfefferkorn, *Sonora*, 264.

6. Fontana and Matson, *Before Rebellion*, viii–xiii.

7. Officer, *Hispanic Arizona*, 36–37, 340.

8. See chapter 2 for an explanation of the repercussions of the Jesuit expulsion for mission economy.

9. AIPJ, RTA, first collection, vol. 41, exp. 33, folios 185–190v, 1795. The Ocuca ranch remained in the hands of the Redondo family until the mid-nineteenth century when, through marriage, it passed to the ownership of a German immigrant, Frederick Augustus Ronstadt. His descendants remained in Sonora and held Ocuca until the 1880s, when they moved to Tucson, Arizona. See Fontana, "Edward Ronstadt."

10. A villa was a Hispanic settlement distinguished from an Indian pueblo, with sufficient population and legal status to support a town council. See Lockhart and Schwartz, *Early Latin America*, 65–68, 102.

11. ABNB, MyCh, vol. 24, exp. 225, 1779.

12. D'Orbigny, *Viaje a la América Meridional*, 3:1195–202. D'Orbigny had no knowledge of the events of 1776–79, half a century before his visit, and tended to attribute good management of the missions uniformly to the Jesuits.

13. Regalsky, *Etnicidad y clase*.

14. See Radding, *Wandering Peoples*, chaps. 6, 7; Romero, *De las misiones a los ranchos y haciendas*; Lorenzana Durán, "Tierra, agua y mercado."

15. García Recio, *Análisis de una sociedad de frontera*, 277–310.

16. The strikingly different social ecologies and landscapes of Sonora and Chiquitos are reflected in the historiographies of both regions. Land tenure occupies an important place in the histories of Sonora and northern Mexico, while the theme proves less central to the history and anthropology of Chiquitos until the twentieth century. See Balzá Alarcón, *Tierra, territorio y territorialidad indígena*.

17. Radding, *Wandering Peoples*, 35–39; West, *Sonora*, 44–69.

18. Jesuit Nicolás de Acrera, "Informe sobre los Santos Reyes de Cucurpe, 1744," Bancroft Library (hereafter BL), M-M 1716, vol. 1–77. Sheridan, *Where the Dove Calls*, 53–88, describes in admirable detail contemporary farming and ranching practices in Cucurpe.

19. Tamarón y Romeral, *Descripción de la diócesis de Nueva Vizcaya*; Nentvig, *Descripción geográfica*; de los Reyes, *Observaciones*.

20. BNFF, vol. 34, exp. 759, folio 23, 1784.

21. "Los pimas de Xecatacari y Obiachi a Thomas de Esquivel, teniente de justicia mayor del Real de San Miguel Arcángel y su jurisdicción en la Provincia de Sonora," Biblioteca Pública del Estado de Jalisco (hereafter BPEJ), Archivo de la Real Audiencia de Guadalajara (hereafter ARAG), Ramos Civil (herafter RC) 27–9'359, 1716. See Radding, *Wandering Peoples*, 171–72, 337 n. 1.

22. "Título de confirmación de la merced y composición al pueblo común y naturales de Santa Anna de Nuri en el Reyno de la Nueva Vizcaya," AIPJ, Libros de Gobierno de la Audiencia (hereafter LGA), vol. 47, folios 218–20, 20v'222v.

23. Aconchi/Baviácora account ledgers, 1726, AMH, AS, vol. 1, 1666–1828.

24. Kessell, *Friars, Soldiers, and Reformers*, 206–14.

25. Documentary sources for land titles in Sonora are AGN, Tierras; Archivo Histórico del Gobierno del Estado de Sonora (hereafter AHGES), Títulos Primordiales (hereafter TP); AIPJ, LGA; and BPEJ, RC. See also Romero, *De las misiones a los ranchos y hacienda*.

26. AHGES TP, vol. 15, exp. 192, 1819–1821; AGN, Tierras, vol. 1423, exp. 8, 14, 15, 16.

27. Kessell, *Friars, Soldiers, and Reformers*, 272–73, 278–79 n. 4; Officer, *Hispanic Arizona*, 145–46.

28. AHGES TP, vol. 4, exp. 40, folios 899–947.

29. AHGES TP, vol. 4, exp. 1, 1831.

30. BNFF, vol. 32, exp. 659; "Plan de Pitic," AGN, Tierras, vol. 2773, exp. 22, 1785. Pitic subsequently became Hermosillo, the capital of the state of Sonora.

31. Fray Juan Felipe Martínez, "Padrón y inventario de la nueva misión de los seris" (1796), AMH. University of Arizona Special Collections (hereafter UASC), microfilm 811, roll 2; Molina Molina, *Historia de Hermosillo antiguo*, 120–23.

32. Archivo General de Notarías, Hermosillo (hereafter A NO); Radding, *Wandering Peoples*, 214–29.

33. A NO, vol. 17, folios 37–39, 1809.

34. Decree of Visitor-General José de Gálvez, BNFF, vol. 34, exp. 740–41; Pedro de Corbalán, "Informe" (unsigned draft), AHGES 1772; and Pedro de Corbalán, "Informe Reservado," BNFF, vol. 34, exp. 738, 1778; Pedro de Nava, Archivo General de Indias, Sevilla (hereafter AGI), Guadalajara, vol. 586, 1794.

35. BNFF 35/722, 1790; Núñez Fundidor, *Bacerác en 1777*, 67–91; Radding, *Wandering Peoples*, 188–93. See chapter 1 for a full description of floodplain farming methods in Sonora.

36. Decreto No. 89 del Estado de occidente, Alamos, 30 September 1828, "Ley para el repartimiento de los pueblos indígenas, reduciéndolos a propiedad particular," in Pesqueira, "Documentos para la historia de Sonora" (unpublished collection), Univer-

sidad de Sonora, Hermosillo, series 2, vol. 1, 1821–45. Colonial antecedents cited in the decree refer especially to the Recopilación de Leyes de los Reynos de las Indias (1680) and to the Constitution of Cádiz, 1813. The Recopilación comprised the standard reference to colonial legislation, and the Constitution of Cádiz provided the legal antecedent to liberal legislation in the Latin American republics.

37. Governor Manuel Escalante y Arvizu to President Anastacio Bustamante, 4 June 1836, AGN, Gobernación, vol. 3; Sonoran state decrees in AHGES vol. 36–1, exp. 1052. Numerous land claims and disputes, similar to the cases summarized above, are found in AHGES TP, vol. 4.

38. Prov. de Chiquitos, 1828, Pueblo de la Concepción, "Estado de cuentas que rinde D. José Ramón Baca," UGRM, MH, Fondo Prefectural, vol. 37, 1828. Chapter 7 discusses state management of the corporate economy of the missions as recorded by the governorship of Chiquitos and the prefecture of Santa Cruz.

39. D'Orbigny, *Viaje a la América Meridional*, 3:1085–213, 4:1241–300.

40. Prefect of Santa Cruz to the Governor of Chiquitos, 30 August 1847, UGRM, MH, Fondo Prefectural, vol. 37, exp. 81.

41. Castelnau, *En el corazón de América del Sur*, 45–70; Castelnau, *Expédition dans les parties centrales de l'Amérique du Sud*, 3:205–65.

42. Castelnau, *En el corazón de América del Sur*, 52–62; Castelnau, *Expédition dans les parties centrales de l'Amérique du Sud*, 3:214–24.

43. "Denuncia de tierras por C. Demetrio Manuel Soruco, del vecindario de Santa Cruz, en el lugar llamado 'de Espejo' ubicado al frente de 'la Calera' en las inmediaciones del camino principal de S.C. a los Departamentos del interior, desde una legua de la rivera occidental del Piraí," UGRM, MH, Fondo Prefectural, vol. 38, folios 1–12, 1847–48. The document does not summarize the space that was delimited, but if the lands were measured in linear leagues, the property would have comprised approximately three and a half *sitios de ganado mayor*. Descriptive passages and quotations in the following paragraph come from this document.

44. Ibid., folios 1–2.

45. "Prefectura y Comandancia General del Departamento de Santa Cruz al Ministro de Estado del Interior, Secretario del Culto," 28 January 1852, Archivo Arquidiócesis de Santa Cruz (hereafter AASC), Provisorato, folios 1–2, 1852. National legislation dealing with the rental and sale of Indian lands in the Andean highlands under the rubric of consolidating *sayañas* (scattered peasant plots) was codified by decree in 1866 and 1867. ABNB, Anuario Administrativo, 143–48.

46. José Ramón Suárez to the Bishop of Santa Cruz, Porongo, 1 March 1852, AASC, Provisorato, folios 2–3, 1852.

47. José Lorenzo Sambies to the Bishop of Santa Cruz, Abajo, 3 April 1852, AASC, Provisorato, folios 4v–5, 1852. The *legua*, or league, measured approximately three

miles and was the customary linear measurement of lands assigned to Indian pueblos. See note 36 on the Recopilación.

48. Juan Miguel Montero to the Bishop of Santa Cruz, Gutierrez, 4 April 1852, AASC, Provisorato, folio 5v, 1852.

49. Ibid., AASC Provisorato, folio 6, 1852.

50. The drought probably occurred during 1843–47; see note 40 above.

51. José Lorenzo Suárez to the Bishop of Santa Cruz, Santa Cruz, AASC, Provisorato, folios 6v–7, 1852; UGRM, MH, Fondo Prefectural, vol. 1, exp. 6, 1826.

52. Governor of Chiquitos to the Prefect, Letter no. 35, 12 April 1834, UGRM, MH, Fondo Prefectural, vol. 1, exp. 16.

53. J. Gregorio Basan to the Bishop of Santa Cruz, San Rafael, 16 August 1852, AASC, Provisorato, folio 7v, 1852.

54. ABNB, Ministro de Hacienda, vol. 151, exp. 67, folios 2, 22, 32, 1858; vol. 154, exp. 63, folios 1–5v, 1859.

55. Land rental bid by Manuel Velarde, San Ignacio, 11 November 1859, ABNB, Ministro de Hacienda, vol. 154, exp. 63.

56. C. Domingo Riberto, natural y cacique del pueblo de San José, to the president of the Republic of Bolivia, ABNB *Ministro de Hacienda*, vol. 154, exp. 63, folios 6–7.

57. Jefe Político de Chiquitos y Guarayos al Ministro de Hacienda, San Miguel, 9 August 1861, ABNB, Ministro de Hacienda 1861, vol. 160, exp. 48, folios 1–2.

58. Bernard Lepetit, "Remarques sur la contribution de l'espace à l'analyse historique," paper presented at EHESS, Paris, 1990, quoted in Cariño Olvera, *Historia de las relaciones hombre-naturaleza*, 22–24.

4. Ethnic Mosaics and Gendered Identities

1. García Canclini, *Hybrid Cultures*, 2–11.

2. Collier, *Marriage and Inequality in Classless Societies.*

3. Clifford, *Routes*, 25, eschews "the organic, naturalizing bias of the term 'culture' " in favor of the " 'chronotope' . . . of travel encounters" that comprehends both time and space. My rendering of what I call ethnic mosaics in this chapter follows the emphasis on change found in both Clifford and Mathewson, "Tropical Riverine Regions," in their departure from the classic definitions of ethnicity commonly based on Barth, *Ethnic Groups and Boundaries.*

4. Lockhart, *We People Here.*

5. Bernard Fontana translated *Tohono O'odham* poetically as "people of earth and little rain" (Fontana, *Of Earth and Little Rain*, 1981). Likewise, Chantell Cramaussel noted that the Tepehuanes of Nueva Vizcaya, and today of northern Mexico, refer to

themselves as Ódami, "the people" (Cramaussel, "De cómo los españoles clasificaban a los indios"). For eastern Bolivia, Tomichá, *La primera evangelización*, 152 n. 35, quoted an anonymous colonial document, "Gramática de la lengua de los indios llamados Chiquitos, pertenecientes al govierno de Chuquisaca en el reyno del Perú," in which the author asserted that the Chiquitos preferred to call themselves M'oñeyca, "the men" (*oñei-s*, in the singular, means "man").

6. Valiñas Coalla makes this point cogently in "Lo que la lingüística yutoazteca podría aportar."

7. Métraux, "The Guaraní"; Tomichá, *La primera evangelización*, 178–80; Fischermann, "Los pueblos indígenas," 36–38. The Guaraní were known also as Carijó or Cainguá, "people of the forest."

8. Cramaussel, "De cómo los españoles clasificaban a los indios," 280–84; Pennington, *Vocabulario en el Lengua Nevome*, 49; Pennington, *The Tepehuan of Chihuahua*, 1–44.

9. James Scott has referred to this kind of classificatory exercise in different contexts as rendering societies "legible." Scott, *Seeing Like a State*, 11, 22–33.

10. Sauer, *The Distribution of Aboriginal Tribes and Languages*; Sauer, *Aboriginal Population of Northwestern Mexico*; Beals, *The Comparative Ethnology of Northern Mexico*; Bandelier, *Final Report*; Lumholtz, *Unknown Mexico*. Robert West, in *Sonora*, and Carroll Riley, in *The Frontier People*, continued to build on that tradition, and Peter Gerhard ably synthesized much of this literature in *The North Frontier of New Spain*.

11. Hervás y Panduro, *Catalogo de las lenguas de las naciones conocidas*. This edition in Spanish constitutes a portion of the author's encyclopedic *Idea dell'universo*, published in Italian in 1778–87. The linguistic part of his twenty-one volume work comprised the last five volumes, published 1784–87. Hervás, an ex-Jesuit, corresponded directly with his correligionists who had served in the Chiquitos missions, such as Joaquín Camaño, probable author of the *Grammática de la lengua chiquita* (undated, in the University of Jena) (Tomichá, *La primera evangelización*, 158–59). Hervás's linguistic studies included native language families of both North and South America, including Tarahumara, Opata, Pima, Eudeve, Iroquois, and Algonquian; Tupí-Guaraní, Guarayo, Sirionó, and the wide-ranging network of Arawakian languages. See Hervás y Panduro, *El lingüista español Lorenzo Hervás*, especially 13–65.

12. Adam and Henry, *Arte y vocabulario de la lengua chiquita*. The Jesuit manuscripts published in this volume are: the *Grammática de la lengua chiquita*, attributed to Joaquín Camaño; *Arte de la lengua chiquita* (1718); *Pláticas para el uso de la lengua chiquita: Vocabulario de la lengua chiquita, parte segunda, chiquito-español, y parte tercera, de sus raíces*. Tomichá, *La primera evangelización*, 152 n. 36.

13. D'Orbigny, *Viaje a la América Meridional*, vol. 4; d'Orbigny, *L'homme américain*. Nordenskiöld published widely on the associations among different South Ameri-

can native peoples based on his observations of material culture, cosmologies, and languages. See, for example, Nordenskiöld, *An Ethno-geographical Analysis*; and Nordenskiöld, *Exploraciones y aventuras en Sudamérica*. Métraux, "Tribes of Eastern Bolivia"; Parejas Moreno, "Chiquitos," 260–66.

14. Mathewson, "Tropical Riverine Regions."

15. Whitehead, "Ethnic Transformation and Historical Discontinuity." Jean and John Comaroff argue, in similar terms, that in the "long conversation" between the tribal peoples of South Africa and the British Nonconformist missionaries, the "natives" reworked signifiers of identity and moral truth into their own values and "traditions," for which the colonized peoples created a name: *setswana*. See Comaroff and Comaroff, *Of Revelation and Revolution*, 17–18, 39.

16. *Blacks* is the translation for the Spanish term *Negros*, referring to enslaved persons or freed slaves of African origin or descent. The category *castas* comprehended the in-between populations whose ontological and cultural existence arose from *mestizaje*: mestizos, mulattoes, *castizos* (mestizo and Spaniard), *moriscos* (mulatto and Spaniard; in Spain, it implied Muslims who remained in the Christian portions of Spain after the reconquest), and the like. In Spanish, terms such as *negro* and *mestizo* are not normally capitalized. The term did not carry the pejorative implications that it did in the English-speaking United States, although enslaved persons and freed slaves suffered social discrimination in colonial societies. The literature on the *castas* as a system of classification is voluminous, encompassing the disciplines of history, demography, art history, and literature. Recent outstanding contributions include Cope, *The Limits of Racial Domination*; Twinam, *Public Lives, Private Secrets*; McCaa, "Marriageways in Mexico and Spain"; Kuznesof, "Ethnic and Gender Influences"; Klor de Alva, "The Postcolonization of the (Latin) American Experience"; and Rabasa, *Writing Violence on the Northern Frontier*. *Casta* paintings by wealthy Spaniards and produced by a number of artistic workshops in New Spain depicted stylized racial categories (Patricia Seed, "Caste and Class Structure," 10).

17. Carolyn Martin Shaw's construction of "part societies" is instructive for our discussion of colonial ethnic polities: Shaw, *Colonial Inscriptions*, 6–9.

18. Gutiérrez, *When Jesus Came, the Corn Mothers Went Away*, 227–97; Martin, *Governance and Society in Northern Mexico*, 125–83; García Recio, *Análisis de una sociedad de frontera*, 417–31.

19. The Chiquitano language and a number of languages spoken by indigenous peoples of the Chaco are gendered. See Knogler, "Relato sobre el país y la nación," 143; Saeger, *The Chaco Mission Frontier*, 77–79.

20. Radding, *Wandering Peoples*, 22–33; Spicer, *Cycles of Conquest*, 8–15.

21. Santamaría, "Fronteras indígenas"; Block, *Mission Culture on the Upper Amazon*; Saeger, *The Chaco Mission Frontier*; Fernández, *Relación historial*, 45, 125–28. See note 8 above, as well as the discussion of *parcialidades* below.

22. Merrill, "Conversion and Colonialism."

23. See chapter 2.

24. Bartolomé de las Casas and Juan Ginés de Sepúlveda are the two most famous figures associated with the written treatises over "just war," the enslavement of the Indians, and their evangelization. Las Casas, *Del único modo de atraer*; Las Casas, *Brevíssima relación*; Sepúlveda, *Democrates alter*; Sepúlveda, *Historia del nuevo mundo*. See also Bakewell, *A History of Latin America*, 79–80, 122. Rabasa, *Writing Violence on the Northern Frontier*, 67–72, denounces the inherent contradiction in the notion of a "peaceful conquest" and the threat of violence in the reading of the *requerimiento*.

25. Cuello, "The Persistence of Indian Slavery"; Deeds, "Rural Work in Nueva Vizcaya"; Sheridan, *Anónimos y desterrados*, 90–97.

26. Cramaussel, "De cómo los españoles clasificaban a los indios," 288–300, argues convincingly that many of these so-called nations were creations of the *encomienda* grants, on paper, and had little intrinsic ethnic meaning.

27. Ortega Noriega, *Un ensayo de historia regional*, 44–49; Radding, *Wandering Peoples*, 31–38.

28. West, *Sonora*, 44–69; Radding, *Wandering Peoples*, 151–52; AGN, Jesuitas IV-10, exp. 116, 119. These workers were called *tapisques*.

29. Dobyns et al., "What were Nixoras?"; Montané, "De *nijoras* y 'españoles a medias' "; Brooks, *Captives and Cousins*. Athapaskan-speaking bands were known as Apaches.

30. See chapter 1, notes 50 and 51.

31. Garavaglia, "The Crises and Transformations of Invaded Societies," 2:9–14; Whigham, "Paraguay's *Pueblos de Indios*," 159–61; Saeger, *The Chaco Mission Frontier*, 9–14.

32. García Recio, *Análisis de una sociedad de frontera*, 177–93. The *indios de servicio* were also called *yanaconas*, using the Andean term that referred to Indians who did not pay tribute to the crown but were held in personal service to individual colonists.

33. Ibid., 194–209. *Malocas* (from Portuguese and Araucanian languages) refers both to slaving raids and to the protective "long houses" in Indian villages; *caça a o indio* may translate to hunting "like the Indians" for the purpose of taking human captives, or simply "hunting Indians."

34. Ibid., 156–59, 217.

35. Ibid., 161–63, 176–79, 207–9.

36. Provisión de la Audiencia de Charcas, *La Plata*, 9 November 1700. Real Academia de la Historia, Madrid, Collection Mata Linares, vol. 56, folios 138–52.

37. García Recio, *Análisis de una sociedad de frontera*, 206–7 n. 230, 186–200. The crown's directive prohibiting future raids dated from 1720.

38. Ibid.. 240 n. 109. The archival reference is "Memoria de los indios encomendados en Santa Cruz de la Sierra," San Lorenzo AGI, Charcas 158, 1717.

39. García Recio, *Análisis de una sociedad de frontera*, 190–93; Denevan, "Stone versus Metal Axes."

40. ABNB, MyCh, exp. 12, 1707. The term *natural child* referred to the offspring of two persons who were free to marry, but had not done so. They were not spurious, such as the products of an adulterous union or a liaison that violated religious vows of chastity.

41. The Recopilación de Leyes de los Reinos de las Indias (1681) was frequently cited by *audiencia* judges, plaintiffs, and defendants.

42. ABNB, MyCh, vol. 23, exp. 7, folios 32–33. Thomasa's successful bid for freedom occurred a full generation after the capture of as many as two thousand Itonamas by Santa Cruz governor José Cayetano Hurtado Dávila, an act protested by the Jesuits and censured by the *audiencia*. See note 36 above.

43. "Eugenio y Jacinto Masavi, naturales del Pueblo de Chiquitos," ABNB, MyCh, vol. 23, exp. 7, folios 28–29, 1751–53.

44. ABNB, MyCh, vol. 23, exp. 7, folio 23. On the *protector de indios*, see García Recio, *Análisis de una sociedad de frontera*, 255–56. An authoritative source for the officio in northern New Spain is Cutter, *The Protector de indios in Colonial New Mexico, 1659–1821*.

45. These cases and their significance for the wider province of Santa Cruz are discussed in greater detail in Radding, "Voces chiquitanas."

46. Lovell and Lutz, *Demography and Empire*, 77, 97, 130, 173, 175; Haskett, *Indigenous Rulers*, 9–10; Restall, *The Maya World*, 30; Stavig, *The World of Túpac Amaru*, 3–8.

47. Saeger, *The Chaco Mission Frontier*, 131–33, makes this point cogently in reference to the Mocobí-Abipon wars of the 1770s and 1780s. The localized pattern of revolts in Chiquitos and the ways in which they portray ethnic identity will be discussed in chapter 5.

48. Archivo General de la Nación, Argentina (hereafter AGNA) 6467/10, 12, Biblioteca Nacional (hereafter BN), Chiquitos Annúas, 367; AGNA 6127/10, 14, BN, Chiquitos Annúas, 355.

49. *Gramática de la lengua de los indios llamados Chiquitos*, prologue folio 3, quoted in Tomichá, *La primera evangelización*, 152 n. 35.

50. Tomichá, *La primera evangelización*, 150. Tapuymiri was more probably a specific ethnic designation for groups of people living in the extreme eastern portion of the province near the Laguna de los Xarayes.

51. Díez Astete and Murillo, *Pueblos indígenas de tierras bajas*, 75–84; Sieglinde Falkinger, personal communication, 25 April 2002, Santa Cruz de la Sierra.

52. Tomichá, *La primera evangelización*, 154–97.

53. Ibid., 366–72; Fischermann, "Viviendo con los paí."

54. Reff, *Disease, Depopulation, and Culture Change*, 97–179; Jackson, *Indian Population Decline*, 167. On the smallpox epidemic of 1780–81 in New Mexico, see Frank, *From Settler to Citizen*, 55–62; Molina del Villar, *La Nueva España y el matlazáhuatl*, espe-

cially 53–134; Elsa Malvido, "¿El arca de Noé o la caja de Pandora?" Historians and medical specialists continue to debate the pathology of *matlazáhuatl*; its name in Nahuatl (*matlatl* plus *zahuatl*, meaning "red eruptions") refers to the external manifestation of symptoms and was used generically to signal plague.

55. Registers of baptisms, marriages, and burials for the Pimería Baja, 1798–1820, 1804–6; for the Pimería Alta, 1794–96; 1804–6, AMH, AS, vol. 22; BNFF, vol. 36, exp. 815, vol. 37, exp. 829, vol. 36, exp. 802. Radding, *Wandering Peoples*, 112–25. Bishop Francisco Rouset de Jesús ordered the diocesan census in 1796; parish priests compiled it over the next half decade. It constitutes the only comprehensive household census known for colonial Sonora and is the basis for tables 8, 9, and 10 below.

56. Sheridan, *Anónimos y desterrados*; Griffen, *Culture Change and Shifting Populations*; Griffen, *Indian Assimilation*; Cramaussel, "De cómo los españoles clasificaban a los indios"; Río, *Conquista y aculturación*; Cariño Olvera, *Historia de las relaciones hombre-naturaleza en Baja California Sur*; Magaña, *Población y misiones de Baja California*.

57. Radding, *Wandering Peoples*, 155–56; Rodríguez Gallardo, *Informe sobre Sinaloa y Sonora*, 44–55, 59–61, 102–4; UASC microfilm 370; AGN, AHH, vol. 278, exp. 18.

58. Radding, *Wandering Peoples*, 146–50; AGN, Jesuitas II-29, exp. 19. Fathers Joseph Roldán, Thomas Pérez, and Manuel Aguirre to Padre Ignaci Lizassoaín, 1762, Documentary Relations of the Southwest, University of Arizona (hereafter DRSW), roll III-C-4.

59. On Sobaípuris, see Officer, *Hispanic Arizona*, 4; Kessell, *Friars, Soldiers, and Reformers*, 7, 35, 40, 78; Rodríguez Gallardo, *Informe sobre Sinaloa y Sonora*, 85, 104; AGI, Guadalajara, vol. 135, exp. I.2, exp. I.3a; on Tohono O'odham in the missions, among many sources, see Matson and Fontana, *Friar Bringas Reports to the King*; Barbastro, *Sonora hacia fines del siglo* XVIII; on *nijoras*, see note 29 above.

60. Radding, *Wandering Peoples*, 270–74; Captain Lorenzo Cancio to Marqués de Croix, 1766, Presidio of San Carlos de Buenavista, AGN, Provincias Internas (hereafter PI), vol. 48, exp. 1, folios 27–36.

61. AGN, Jesuitas IV-10, exp. 231, folio 266. *Gente de razón* ("people of reason") was the designation given to Spaniards and mestizos—people of mixed ethnic origins who could claim parentage by a Spaniard—with privileges of exemption from obligatory labor and tribute payment.

62. Bishop Reyes to Viceroy Matías de Gálvez, 1784, BNFF, vol. 34, exp. 759.

63. Early Jesuit records for Tepache come from the visit made by Father Juan de Almonacir, 1685, AGN, AHH, Temporalidades, vol. 1126, exp. 4, quoted in Polzer, *Rules and Precepts*, 37; Tamarón y Romeral, *Descripción de la diócesis de Nueva Vizcaya*, 1013; Nentvig, *Descripción geográfica*, 174; Gerhard, *The North Frontier of New Spain*, 282.

64. *Laboríos* were Indian laborers in a Spanish *labor* (hacienda). See Gerhard, *The North Frontier of New Spain*, 370.

65. Radding, *Wandering Peoples*, 187, 191, 288–90, 353 n. 53; BNFF, vol. 35, exp. 722, 767.

66. Tamarón y Romeral, *Descripción de la diócesis de Nueva Vizcaya*.

67. Le Roy Ladurie, *Montaillou*, made the residential *domus* a central feature of his analysis of late fourteenth-century peasant society viewed through the church's documented interrogation of the religious heretics of Montaillou.

68. The ages listed in the census are not exact, but rather constitute estimates. Historical demographers working with early modern censuses in both Europe and the Americas have pointed out that enumerators tended to "round off" ages to the nearest decade or five-year interval. Nevertheless, a comparison of age ranges is a valid approach to analyzing probable household formation in a context of high fertility and high mortality.

69. Gutiérrez, *When Jesus Came, the Corn Mothers Went Away*, 315–18, 385 n. 46; Lavrín, *Sexuality and Marriage in Colonial Latin America*.

70. Radding, *Wandering Peoples*, 185–87; AGN, Tierras, vol. 474, exp. 2, folios 1–161; AMH, AD, vol. 1, exp. 1–30, 1765; BNFF, vol. 37, exp. 833.

71. Estado de Ónavas, 5 February 1802, AMH, AD, vol. 3, exp. 7.

72. AMH, AD, vol. 1; UASC microfilm 811, roll 2; BNFF, vol. 35, exp. 762.

73. Kraniauskas, "Hybridity in a Transnational Frame," 236.

74. Comaroff and Comaroff, *Of Revelation and Revolution*; Todorov, *La conquista de América*; Clendinnen, *Ambivalent Conquests*.

75. See note 3 above on Clifford's use of *chronotope* to theorize processes of cultural change across space and time. Jean and John Comaroff make a similar observation concerning the frontier between the northern and southern Sotho-Tswana peoples of South Africa. Comaroff and Comaroff, *Of Revelation and Revolution*, 40–42.

5. Power Negotiated, Power Defied

1. Furlong, *José Sánchez Labrador*, 37–52. Furlong published the works of Sánchez Labrador as *Enciclopedia Rioplatense* in twenty volumes in 1932. Sánchez Labrador arrived in the Americas in 1734, leaving when the Jesuits were expelled, in 1768. His writings contributed substantially to geography, ethnography, and languages of the greater Río Plata region, extending through the modern nations of Argentina, Paraguay, and Bolivia. He lived thirty years after the expulsion in the Papal States, in Ravenna, where he died in 1798. During that time he wrote prolifically about his own experiences as a missionary among the Guaraní, Toba, and Mbaya (Guarcurúan) peoples.

2. On *minca*, see chapter 1, note 42.

3. Hardt and Negri, *Empire*, 23–30; Foucault, *Power/Knowledge*.

4. Gibson, *Tlaxcala en el siglo* XVI, 70–123; Gibson, *The Aztecs under Spanish Rule*, 166–93; Lockhart, *The Nahuas after the Conquest*, 14–58; Haskett, *Indigenous Rulers*, 13–17; Stern, *Peru's Indian Peoples*, 114–37; Restall, *The Maya World*, 51–83.

5. Benton, *Law and Colonial Cultures*, 7–15, 39–45.

6. See Williams, Jr., *Linking Arms Together*, 20–39, on the heterogeneous and multicultural societies of early contact among Native Americans and Europeans on the Atlantic coast and woodlands of eastern North America.

7. This discussion is informed by my reading of Gramsci, *The Prison Notebooks*; Foucault, *The History of Sexuality*; Foucault, *Discipline and Punish*; Burchell, Gordon, and Miller, *The Foucault Effect*; Bourdieu, *Outline of a Theory of Practice*; Benton, *Law and Colonial Cultures*, 258; Genovese, *Roll, Jordan, Roll*; Gaventa, *Power and Powerlessness*; Lears, "The Concept of Cultural Hegemony"; Scott, *Weapons of the Weak*; Scott, *Domination and the Arts of Resistance*; Scott, *Seeing like a State*; Joseph and Nugent, *Everyday Forms of State Formation*; Mallon, *Peasant and Nation*; Vaughan, *Cultural Politics in Revolution*; Jacobsen, "Montoneras"; and Sahlins, *Islands of History*.

8. Cutter, "The Legal System as a Touchstone of Identity"; Martin, *Governance and Society*, 125–48.

9. Haskett, *Indigenous Rulers*, 95–99. Haskett wisely counsels, however, that the Nahuatl terms do not necessarily translate in function from their pre-Hispanic referents to their colonial usages.

10. West, *Sonora*, 62–66, appendix D: *Sellos*, 1684, 1714, 140.

11. *Diccionario de la lengua española*, 19th ed., s.v. "*alguacil*." Haskett, *Indigenous Rulers*, 98–99.

12. Nentvig, *Descripción geográfica*, 164–65.

13. Gibson, *Tlaxcala in the Sixteenth Century*; Gibson, *The Aztecs under Spanish Rule*; Haskett, *Indigenous Rulers*, 60–85, 91–94; Lockhart, *The Nahuas after Conquest*.

14. Pfefferkorn, *Sonora*, 266–67.

15. Nentvig, *Descripción geográfica*, 165.

16. Fernández, *Relación historial*, 125–26; Métraux, "The Social Organization," 20–21, relies on the same Jesuit chronicles for his description of Manasí social organization. See chapter 4, table 5.

17. Knogler, "Relato sobre el país y la nación," 178. In chapter 4, we saw that *M'oñeyca*,. meaning "the men," stood for the combined Chiquitos "nation" of linguistically related *parcialidades* (see Tomichá, *La primera evangelización*, 152–53 nn. 35 and 38; and *Gramática de la lengua de los indios llamados Chiquitos*).

18. Saeger, *The Chaco Mission Frontier*, 113–18; see comparative descriptions of the limited powers of "alliance chiefs" among the Algonquian peoples of the Great Lakes region in White, *The Middle Ground*.

19. Block, *Mission Culture on the Upper Amazon*, 86.

20. In New Spain, *corregidores* were appointed officers of the Spanish colonial bu-

reaucracy who oversaw tribute collection and *repartimiento* labor drafts. It is interesting that in the mission provinces of Moxos, Chiquitos, the Chaco, and Paraguay, the *corregidor* was an Indian officer. Parejas Moreno, "Organización misionera," 276; Saeger, *The Chaco Mission Frontier*, 122–23; Whigham, "Paraguay's *Pueblos de Indios*," 161–63.

21. D'Orbigny, *Viaje a la América Meridional*, 4:1257–60.

22. Hausberger, "La violencia"; Radding, "From the Counting House to the Field and Loom," 80–82.

23. Nentvig, *Descripción geográfica*, 165; d'Orbigny, *Viaje a la América Meridional*, 4: 1258.

24. Haskett, *Indigenous Rulers*, 27–59; Spicer, *The Yaquis*, 39–40, summarized documented complaints of the missionaries' manipulation of council elections as a source of the Yaqui rebellion of 1740.

25. Marcelino de la Peña, Governor of Chiquitos, to the Prefect of Santa Cruz, 1832, UGRM MH Fondo Prefectural, vol. 1, exp. 12, folios s/n.

26. D'Orbigny, *Viaje a la América Meridional*, 3:1147–213, 4:1241–79. D'Orbigny further codified a racializing lexicon to classify Native Americans in *L'homme américain*. The use of science to create categories in which to place different species, languages, and peoples was noted in chapter 4 in relation to the linguistic studies by Lorenzo Hervás y Panduro, Lucien Adam, and Victor Henry, as well as to the ethnographic surveys by Alfred Métraux.

27. Ramón Conrado Morón Tomichá, interview by Justa Morón Macoñó, San Ignacio de Velasco, August–November 2001, transcribed by Cynthia Radding, April 2002. Justa Morón adapted the spelling of these seven section names from her father's pronunciation and diagrammed their location. *Piokó* resembles the documentary name of one of the *parcialidades*, Piococa (pl.), compiled by Roberto Tomichá, meaning "fish" and included in table 4. The Piococa were associated with both the Tao and Piñoco dialects.

28. Ibid.

29. Whigham, "Paraguay's *Pueblos de Indios*," 162, comments on the native elite of the Guaraní missions constituted in the *cabildo*.

30. Radding, *Wandering Peoples*, 146–50.

31. Ibid., 256–63, 276–79; Kessell, *Friars, Soldiers, and Reformers*, 137–38.

32. Los pimas de Xecatacari y Obiachi solicitan a Tomás de Esquibel, teniente de justicia mayor del Real de San Miguel Arcángel, de la Provincia de Sonora, para fundar un pueblo en Buenavista, 26 May 1716, BPEJ; ARAG, RC, exp. 27–9'359. Quoted in Radding, *Entre el desierto y la sierra*, 139–53. In 1742, a presidio was established in Buenavista, as is shown in map 1. See also Radding, *Wandering Peoples*, 170–75.

33. Representación del Padre comisario de las misiones, fray Juan Prestamero, sobre quejas de los ópatas contra el padre ministro de Opodepe, BNFF, vol. 34, exp. 735, 1777.

Correspondence among Manuel Grijalba, Opata governor of the pueblo of Opodepe, Pedro Tueros, the commander of the presidio of San Miguel de Horcasitas, and brigadier Theodoro de Croix, first commandant general of the Internal Provinces, BNFF, vol. 40, exp. 912, 1778–1779. Opodepe, the head village, and Nacameri, its *visita*, are located northwest of the presidio of Horcasitas, shown on map 1.

34. The *común* is a ubiquitous term in colonial Spanish America that takes on a variety of meanings. In the Andean world, it served to distinguish the peasant base from the elite stratum of *kurakas* (lords), see Stavig, *The World of Túpac Amaru*; Penry, "The Rey Común." The *común* in northwestern Mexico encompassed the communal lands of the missions and the people who inhabited them. See Sheridan, *Where the Dove Calls*; Radding, *Wandering Peoples*, 171–75. Deeds, *Defiance and Deference*, argues that the *común* appears only in the eighteenth-century documentation for Nueva Vizcaya.

35. Frank, *From Settler to Citizen*, 70–75.

36. Kessell, *Friars, Soldiers, and Reformers*, 110; see Font and Eixarch, *Fray Pedro Font*.

37. An important historical and anthropological literature on the pampas regions of South America has developed these same arguments concerning the savvy movements of raiders and the ritualized exchanges of prisoners. See, for example, Saeger, *The Chaco Mission Frontier*, 6–13; Jones, "Comparative Raiding Economies"; Socolow, "Spanish Captives in Indian Societies"; Brooks, " 'This Evil Extends Especially."

38. I have argued in *Wandering Peoples*, 153–58, that colonial policies of forced *reducción*, together with Spanish landownership, resulted in the dispersal of native pueblos into smaller *rancherías*.

39. Don Garate, Tumacácori National Monument Database, National Park Service, 17 August 2002.

40. The episodes related here are based on AGN, PI, vol. 86, exp. 1.

41. "Diario de . . . P. Francisco Pimentel, capellán de la expedición encabezada por el Coronel Diego Ortiz Parrilla," AGN, Jesuitas, leg. 2–5, exp. 2, translated in Sheridan, *Empire of Sand*, 177–231; "Relación de la Expedición de las Provincias de Sinaloa, Ostimuri y Sonora en el Reino de Nueva España," AGI, Guadalajara, vol. 416, translated in ibid., 274–402. Radding, *Wandering Peoples*, 260–61, 275–77.

42. Governor Tienda del Cuervo to Viceroy Marqués de Cruillas, AGN, PI, vol. 86, exp. 1, folios 108–11, 1762.

43. Ibid., folios 115–116.

44. Cap. Anza to Governor Tienda de Cuervo, S. José de Pimas, 10 July 1765, ibid., folios s/n.

45. Ibid., folio 15.

46. P. Juan Mariano Blanco to P. Lorenzo Joseph García, 7 May 1765, ibid., folio 16. The letter was copied in the document forwarded to the Viceroy Cruillas by Governor Juan de Pineda.

47. Gov. Juan de Pineda to Viceroy Marqués de Cruillas, San Miguel de Horcasitas, 21 May 1765, ibid., folio 12.

48. Captain Gabriel Antonio Vidósola to Governor Tienda de Cuervos, Fronteras, 26 July 1762, ibid., folios 132–140v; direct quotations on folios 137–38. Gabriel Antonio de Vildósola was the brother-in-law of Juan Bautista de Anza; both men had military careers and represented families who built their patrimony and prestige on military service in the presidios.

49. Gov. Tienda de Cuervo to the Viceroy, San Miguel de Horcasitas, 7 May 1762, ibid., folios 105–6.

50. ACSC, exp. 6-2C, 1779. The Manapeca ethnic designation does not appear in the lists of tribal and linguistic groups compiled by Roberto Tomichá, based on Jesuit catalogues, nor in the ethnic designations indicated by Alcides d'Orbigny (see chapter 4). It is probable that the Manapecas were so named in the late eighteenth century as a subgroup of the Manasí (Manasica) family of related Chiquitano languages.

51. Porongo is located outside of the Chiquitanía in the province of Cercado, west of Santa Cruz.

52. ACSC, exp. 6-2C, 1779.

53. ACSC, 6-1, A3 1790.

54. ABNB, MYCH, vol. 29, folios 113–206, 1790.

55. ABNB, MYCH, vol. 29, exp. 21, folios 163–165, 27 July 1790. The governor's message to the *cabildo* officers and their replies are copied in folios 175–182.

56. Fischermann, "Los Rojas."

57. This is the guiding principle for the comparative and collaborative volume edited by Donna Guy and Thomas E. Sheridan, *Contested Ground*, which comprehends the expansive frontiers of northern New Spain and the greater Río de la Plata.

58. White, *The Middle Ground*; Hämäläinen, "The Comanche Empire"; Jones, "Comparative Raiding Economies."

6. Spiritual Power, Ritual, and Knowledge

1. *Shaman* (German, *Schamane*; Russian, *shaman*; Sanskrit, *sramanás*, sometimes interpreted as "ascetic") in anthropological literature commonly refers to religious intermediaries who, in different circumstances, exercise beneficent healing or maleficent, destructive powers. *The American Heritage Dictionary of the English Language*, ed., s.v. "Shaman." On the antiquity of shamanism in America, see Lynch, "Earliest South American Lifeways," 1:221.

2. Salomon, "Testimonies," 80; Roosevelt, "The Maritime, Highland, Forest Dynamic," 334; Shimada, "Evolution of Andean Diversity," 444–58; Monteiro, "Invaded Societies," 985.

3. *Priest* (Old English, *prêost*; Latin, *presbyter*, meaning "elder"). Funk & Wagnalls, *New International Dictionary of the English Language, Comprehensive Edition* (Chicago: Ferguson Publishing Company, 1997), 2:1000.

4. MacCormack, "Ethnography in South America," 1:159.

5. Fischermann, "Un pueblo indígena interpreta su mundo."

6. Ibid., 25; Fischermann, "Viviendo con los paí," 53–54. On consensus and disagreement in the reproduction of knowledge, see Merrill, *Rarámuri Souls*, 13–15.

7. Fernández, *Relación historial*, 191–203. San Ignacio de Zamucos was established in 1724 and maintained for a little over two decades (see chapter 4, note 53).

8. Vicente Rafael explains the expansive meanings of *nono* (spirit or soul) in Tagalog by asserting that native peoples of Southeast Asia and the Philippines believed that spirits could move between life and death, thus eluding boundaries and clear definitions that would translate easily into the Christian lexicon. Rafael, *Contracting Colonialism*, 111–15.

9. Fernández, *Relación historial*, 129–33. The *isituús* probably correspond to the *jichis*, spirits of forests and waterways in the natural world that figure importantly in contemporary Chiquitano cosmologies (Riester and Fischermann, *En busca de la loma santa*, 150). See chapter 1, note 33, and this chapter below.

10. García Recio, *Análisis de una sociedad de frontera*, 88, 180; Antonio Menacho, *Por tierras de Chiquitos*, 36–43.

11. Fernández, *Relación historial*, 128. On the widespread belief that Saint Thomas had visited the native peoples of the Americas in ancient times, see Cañizares-Esguerra, *How to Write the History of the New World*, 312–19; Jacques Lafaye, *Quetzalcóatl et Guadalupe*, 253–88.

12. Jesuit visitor Juan Patricio Fernández summarized and reworked the letters and reports of his fellow missionaries Lucas Caballero and Juan Bautista de Zea in his hagiographic history of the missions of Chiquitos, representing their actions as heroic and self-sacrificing, explicitly in search of the crown of martyrdom. See Caballero, *Relación de las costumbres y religión*, quoted in Hoffmann, *Las misiones jesuíticas entre los Chiquitanos*, 12.

13. Fernández, *Relación historial*, 131.

14. Ibid., 132–33.

15. Ibid., 132, 142.

16. Ibid., 140–41.

17. Ibid., 140; on epidemics, 123, 137, 141, 159.

18. Nabhan, *Cultures of Habitat*, 157.

19. The O'odham or Pima represent different cultural communities, related linguistically and associated with a broad geographical range running from northwestern Sonora, in the heart of the desert, to the central mountainous valleys of the upper Yaqui drainage (see chapter 1, note 27). The Cahitan-speaking villagers known historically and ethnographically as the Yaquis and Mayos of southern Sonora and northern

Sinaloa conserve richly elaborated, mixed religious traditions. The Rarámuri (Tarahumara) of the eastern valleys and barrancas of the Sierra Madre in the colonial province of Nueva Vizcaya (present-day Chihuahua) share in the broadly defined Uto-Aztecan languages and cultures of northwestern Mexico.

20. Nabhan, *Cultures of Habitat*, 180–82; Nabhan, *Counting Sheep*; Szuter, Crosswhite, and Crosswhite, "The Sonoran Desert," appendix 5a, "Mammals of the Sonoran Desert," 296.

21. Nabhan, *Cultures of Habitat*, 166–82.

22. Kozak and López, *Devil Sickness and Devil Songs*, 81, 108–9.

23. Songs 17 and 18 in the song set by José Manol, in ibid., 140.

24. Kozak and López, *Devil Sickness and Devil Songs*, 98–100. Michel Foucault explicitly couples knowledge and pleasure in "The Will to Knowledge" and "Sex, Power, and the Politics of Identity" in *Ethics, Subjectivity, and Truth* (13, 165–66, respectively).

25. Bahr, "Pima-Papago Christianity"; Kozak and López, *Devil Sickness and Devil Songs*, 41–50, 76–79.

26. Griffith, *Beliefs and Holy Places*, 69–76, argues convincingly that *sa:nto himdag* took shape during the latter nineteenth century, distinguishing its core of beliefs and practices from Roman Catholicism "in form and intent." On the plurality of religious beliefs held by many indigenous communities, especially in reference to the Ayoreóde of Bolivia and the Rarámuri of Mexico, see note 8 above.

27. Pérez de Ribas, *History of the Triumphs of Our Holy Faith*, 96–97.

28. Pérez de Ribas referred to the *monte* in terms of his own efforts to reduce the Batucaris to the Ahome mission in the lower Fuerte river valley (see chapter 2, note 14).

29. Pérez de Ribas, *History of the Triumphs of Our Holy Faith*, 151–52. This miracle was attributed to the Feast of the Nativity of Mary, 8 September 1600.

30. Ibid., 177–79.

31. Ibid., 188–90. Father Clerici began working among the Guasave in 1609. The reference to forty years is metaphorical, with biblical overtones. The Israelites wandered during forty years in the wilderness before settling in the land of Canaan; Jesus fasted for forty days in the desert, where Satan visited him and tempted him with worldly wealth and power, thus establishing the liturgical season of Lent. Forty years, as recorded in many colonial documents referring to Indian communities in New Spain, seems to stand for a generation; in this case, it is a marker between a pagan past and a Christian present.

32. Kozak and López, *Devil Sickness and Devil Songs*, 86.

33. Underhill, *Papago Woman*, 50–52, 79–81. The saguaro is the one of the most common plants of the Sonoran Desert, and it forms an integral part of O'odham culture (see chapter 1).

34. Sigal, ed., *Infamous Desire*; Gutiérrez, *When Jesus Came, the Corn Mothers Went Away*, 33–35.

35. Zepeda, "Where the Wilderness Begins," quoted in Nabhan, *Cultures of Habitat*, 162–63.

36. Nabhan, *Cultures of Habitat*, 170–71; Nabhan suggests that the myth of *ne:big* is associated with Pleistocene fauna whose fossil remains are found in Quitovac. Kozak and López, *Devil Sickness and Devil Songs*, 85. On Quitovac and *wi:gita*, see Underhill, *Papago Indian Religion*; and Bahr, "Papago Ocean Songs and the Wi:gita." On *i'itoi*, see Griffith, *Beliefs and Holy Places*, 14–22.

37. Nabhan, *Cultures of Habitat*, 152–54; Spicer, *The Yaquis*, 64.

38. Merrill, *Rarámuri Souls*, 73–75.

39. Kozak and López, *Devil Sickness and Devil Songs*, 64–67.

40. Merrill, *Rarámuri Souls*, 73–78. *Rarámuri* signifies "humans," including both Indians and non-Indians. Its more restricted meanings refer to Native Americans (distinguished from Chabóchi) and, specifically, to the Rarámuri ethnic community. The term can also connote gender, identifying Rarámuri men, while *igómele* stands for Rarámuri women.

41. Ibid., 81.

42. The four plants are *bakánawi* (*Scirpus* sp.), *hícuri* (*Lophophora williamsii*), *uchurí* (*Echinocereus*, *Coryphanta*, and *Mammillaria*), and *rikúhuri* (*Datura* spp.). Ibid., 73–75; Bye, "Hallucinogenic Plants of the Tarahumara." *Hícuri* (peyote) is known as a hallucinogenic plant among many native peoples of North and Central America. Among the Rarámuri, its ceremonial powers are carefully guarded. Merrill, *Rarámuri Souls*, 123.

43. Nabhan, *Gathering the Desert*, 11–15.

44. Griffith, *Beliefs and Holy Places*, 4–13. Griffith explores the linguistic and legendary linkages of stories about water snakes and serpents in the wider Mesoamerican and Uto-Aztecan cultural spheres. Basso, *Wisdom Sits in Places*, 14–17, recounts Charles Henry's eloquent story of the drying of Snakes' Water spring near Cibecue, Arizona, because "Water had been offended by acts of disrespect" (16).

45. The concept of *jichis* was first introduced in chapter 1 and above, in note 9. Riester and Fischermann, *En busca de la loma santa*, 150–53, 161; see Fischermann, "Contexto social y cultural," 34; Quiroga and Fischermann, "Pueblo viejo'puerto nuevo."

46. Riester and Fischermann, *En busca de la loma santa*, 151; Mario Arrien and Rosa María Quiroga, personal communication, Santa Cruz, Bolivia, August 1999.

47. I have noted above that Jesuit missionaries Fernández and Pérez de Ribas observed the accusation of women as sorcerers and even reported missionaries' exorcisms to release them from the devil's power. Rarámuri contemporary belief asserts that men and women, Indians and non-Indians, can be shamans, but that the majority of sorcerers, and those most feared, are Rarámuri men (Merrill, *Rarámuri Souls*, 75). See, however, Deeds, "Double Jeopardy: Indian Women in Jesuit Missions of Nueva Vizcaya."

48. Riester and Fischermann, *En busca de la loma santa*, 161–70; Freyer, *Los Chiquitanos*, 50, 70–71; Fischermann, "Contexto social y cultural," 36; Knogler, "Relato sobre el país," 154. Knogler reported that healing shamans were called *bazübos*, "those who suck." See Tomichá, *La primera evangelización*, 230–32.

49. Significantly, Vicente Rafael notes that in Tagalog, *loob*, literally meaning "inside," became entangled with the concepts of "shame" (*hiya*) and a network of indebtedness (*utang na loob*) that described the relationship between the sinning believer and a forgiving god. Missionary texts employed *loob* to translate the notion of the Christian soul. Rafael, *Contracting Colonialism*, 121–35. "The significance of *loob* lies in the fact that it marks out the space within which objects and signs from the outside can be accumulated and from which and toward which they can be issued in payment of a debt" (125).

50. Barbastro, *Sonora hacia fines del siglo* XVIII, 31–32, 56–57. The archaeological and documentary evidence for the provinces of Sonora and Sinaloa points to the accommodation of family dwellings in the missions, in contrast to the construction of compulsory separate dormitories for men and women that have been described for the Dominican and Franciscan missions of Baja and Alta California. Jackson, *Indian Population Decline*; Haas, *Conquests and Historical Identities*.

51. This summary is based on the excellent historical photographs, architectural diagrams, and descriptions in Pickens, *The Missions of Northern Sonora*; Officer, Schuetz-Miller, and Fontana, *The Pimería Alta*, 61–94; Buschiazzo, "San Javier del Bac"; Radding, "Crosses, Caves, and *Matachinis*," 189–91.

52. Griffith, *Beliefs and Holy Places*, 82–86; Fontana, "The O'odham," 25. O'odham ceremonial houses called *vahki* (rain house), round house (*olas ki*), or big house (*gu ki*) fulfill functions suggestive of the *kiva* (underground ceremonial room) that figure so prominently among the Hopis and Pueblos. See Ortiz, *The Tewa World*; and Griffith, *Beliefs and Holy Places*, 78–93.

53. Griffith, *Beliefs and Holy Places*, 82–86. O'odham chapels always open to the east, in contrast to Catholic churches, which do not have a uniform orientation to the cardinal points.

54. Kühne, "Semana Santa y Fiesta Patronal," 558. See also Luke 19:29, "And it came to pass, when he was come nigh to Bethphage and Bethany, at the mount called the mount of Olives, he sent two of his disciplines."

55. Gisbert, "El arte misional," 81–82.

56. Knogler, "Relato sobre el país," 170–72; Gutiérrez da Costa and Viñuales, "Territorio, urbanismo y arquitectura"; Mariluz Urquijo, "Las escuelas"; Fischermann, "Los Rojos." The exception to this model is the church of San José, in the southern Chiquitanía, which is built of stone and fired brick, with vaulted roofs, in the style of urban baroque churches in Spanish America.

57. Kühne, "Semana Santa y Fiesta Patronal," 558.

58. Strack, " 'De tradiciones españolas y costumbres raras,' " photos 102, 124–25.

59. Strack, *Frente a Dios y los pozokas*; Ramón Conrado Morón Tomichá, interview by Justa Morón Macoñó, San Ignacio de Velasco, August–November 2001, transcribed by Cynthia Radding, April 2002.

60. Radding, "Crosses, Caves, and *Matachinis*," 194; Spicer, *The Yaquis*, 59–113; Kolaz, "Tohono O'odham *Fariseos*"; Merrill, *Rarámuri Souls*, 26, 29, 169–71, 177–81; Abercrombie, *Pathways of Memory and Power*, 109–12.

61. Falkinger and Kühne, " 'Que para aprenderla no basta muchos años,' " 31–33; Fernández, *Relación historial*.

62. Fray Francisco Antonio Barbastro al Virrey Conde de Revillagigedo sobre las misiones de la Pimería Alta y Baja, 1793, AGN, PI, exp. 2, folios 529–44.

63. Outstanding published Jesuit grammars and vocabularies for Sonora include Smith, *A Grammatical Sketch of the Heve Language*; Pennington, *Vocabulario en la Lengua Névome*; Pennington, *Arte y vocabulario*; and Aguirre, *Doctrina christiana*. For Chiquitos, similar primary sources and grammatical studies include Joaquín Camaño, author of a *Grammática de la lengua chiquita*, quoted in Tomichá, *Primera evangelización*, 158–59; Hervás y Panduro, *Catalogo de las lenguas de las naciones conocidas*; and Adam and Henry, *Arte y vocabulario de la lengua chiquita*. See chapter 4, notes 11 and 12.

64. "Sermones, confesionario breve, catechismo breve, oraciones, vocabulario breve en la lengua Opata," anonymous, undated bound volume attributed by Alphonse Pinart to Francisco Barbastro, who served in the Sonora missions under the auspices of the Colegio de Propaganda Fide de la Santa Cruz de Querétaro for over a quarter of a century, 1772–1800. BL, Hubert Howe Bancroft Collection (hereafter HHB), M-M 483.

65. Francisco Barbastro, "Plática del nacimiento de Jesuchristo, año de 1792," BL, HHB, M-M 483.

66. Ibid.; Radding, "Crosses, Caves, and *Matachinis*," 189.

67. Lockhart, *The Nahuas after the Conquest*; MacCormack, " 'The Heart Has Its Reasons' "; Burkhart, *The Slippery Earth*; Gruzinski, *The Conquest of Mexico*; and López Austin, *Hombre-dios*.

68. Rafael, *Contracting Colonialism*, esp. chaps. 2–4. See note 49 above.

69. Falkinger and Kühne, " 'Que para aprenderla no basta muchos años,' " 35–36; Knogler, "Relato sobre el país," 175.

70. Ibid., 174.

71. Ibid., 178. *Ma onyeica atonie* translates into "I am fully a man." See chapter 4, note 49.

72. See chapter 5, note 56.

73. Sumaria recibida por Dn José Miguel Hurtado, administrador del pueblo de San Rafael, contra el Gobernador de la Prov. de Chiquitos, Dn Juan Bautista de Altolaguirre, ABNB, Rück exp. 291, 1811, citation from folios 73–76.

74. Falkinger and Kühne, " 'Que para aprenderla, no basta muchos años,' " 36. Fal-

kinger, whose research on the sermons constitutes an invaluable repository for the culture and language of the Chiquitano peoples, has shown that the sermons made for an instrument of literacy in their own language at a time when Chiquitanos were largely excluded from public education in Bolivia. I am indebted to her for sharing her insights with me about the importance and the eloquence of the sermons. Patricio Bariquí Méndez, religious leader of Comunidad Palmarito, part of the parish and subprefecture of Concepción, in an interview by Cynthia Radding and Bernardo Fischermann, Palmarito, 14 August 1997, emphasized the importance of reciting sermons from memory in Chiquitano. Abercrombie, *Pathways of Memory and Power*, 8–9, and Rappaport, *The Politics of Memory*, emphasize the importance of local archives for indigenous communities of the Andes, even when their contents can no longer be read.

75. Kozak and López, *Devil Sickness and Devil Songs*, 127–52. Shamans dream the songs, but they arrange them into sets that then become a standardized sequence to effect a cure.

76. Merrill, *Rarámuri Souls*, 65–66.

77. Ibid., 62–84.

78. Fernández, *Relación historial*, 38–40; Knogler, "Relato sobre el país," 149; Justa Morón, interview by the author, San Ignacio de Velasco, 12 August 2001.

79. Spicer, *The Yaquis*, 59–113; Radding, "Crosses, Caves, and *Matachines*," 177–201; Radding, *Wandering Peoples*, 54–5; Pennington, *The Pima Bajo of Central Sonora*, 149–50; Merrill, *Rarámuri Souls*, 60–61, 108–11.

80. Nentvig, *Descripción geográfica*, 164–165; Mann, "The Power of Song," 217–29; on Chiquitos, Edward Kühne, personal communication, Santa Cruz de la Sierra, Bolivia, May 2002.

81. Gisbert, "El arte misional," 80; Pacquier, *Las chemins du baroque*, 185–202.

82. Nawrot, "Producciones de opera."

83. Knogler, "Relato sobre el país," 180–81.

84. The beauty of this music and the breadth of its musicology deserve a comprehensive study. This short treatment, in the context of spiritual landscapes, is based primarily on Nawrot, "Producciones de opera"; Nawrot, *Indígenas y cultura musical*; Waisman, " 'Soy misionero porque canto, taño, y danzo' "; Seoane Urioste, "Música en las misiones jesuíticas"; and the author's observance of musical festivals in Santa Cruz de la Sierra and the restored mission pueblos of Chiquitos in 2000 and 2002.

85. See chapter 5, note 51. Rabasa, *Writing Violence on the Northern Frontier*, 280–82, elaborates on the theme of "writing violence" as "an inverted allegory," referring to the violence of colonial historicity, its documentation of physical coercion, and the threat of terror.

86. Mirafuentes Galván, "Agustin Ascuhul."

87. Testimony of Juan Matheo Pinto, maestro y governador que ha sido deste pueblo [Ráhum] 1 July 1739, AGN, *Californias*, vol. 64, exp. 8, folios 136–56, and folios 148–49;

Spicer, *The Yaquis*, 32–57; Hu-Dehart, *Missionaries, Miners, and Indians*, 59–86; Radding, "Peasant Resistance."

88. Bell, *Ritual*, 166–71.

7. Transitions from Colony to Republic

1. Eric Van Young, most recently, has published a masterful revisionist history of peasant revolt and the Mexican struggle for independence in *The Other Rebellion;* Walker, *Smoldering Ashes*, opens new perspectives on the commonly accepted sequence of events from the Tupac Amarú rebellions of the 1780s and the independence movements in the Andes.

2. ABNB, MyCh, exp. Ad 159, 1801; Administrador General de las Misiones al Ministro Protector de Misiones, La Plata, 22 January 1803, ABNB, Escribanía de Cámara (hereafter EC), exp. 1805. These two files provide the documentary basis for the summary and interpretation of Sibilat's failed enterprise in Chiquitos. I have published a previous interpretation of this material in Radding, "Historical Perspectives on Gender, Security, and Technology."

3. ABNB, EC, exp. 36, folio 2.

4. ABNB, EC, exp. 36, folios 3–4.

5. ABNB, MyCh, exp. Ad 159, folios 4–5, 22–23. Concerning the Indian justices' anger over whipping and their sensitivity to the presence of married women in the *colegio*: these were issues that had galvanized the uprisings of 1779 in Concepción and of 1790 in San Ignacio, as discussed in chapter 5.

6. ABNB,MyCh, exp. Ad 159, folio 12. A *vara* is a linear measurement approximately equal to 0.835 meters. Fires were frequently noted, occurring in the linear rows of houses where Indian families lived.

7. Larson, *Cochabamba*, 245–58.

8. Weiner and Schneider, *Cloth and Human Experience*; Strack, "Forasteros matizando recuerdos."

9. UGRM, MH, Fondo Prefectural, vols. 9, 10, 11. Governor Marcelino de la Peña of Chiquitos, in Santa Ana, to the prefect of Santa Cruz, on 25 March 1831, bargained with a Brazilian merchant, Felicino de la Contaliste (married and living in Santa Cruz), to acquire metal files in exchange for salt. An arroba generally was considered twenty-five pounds in weight.

10. Tarifa que propone el Gob. de la Prov. de Chiquitos Marcelino de la Peña al Prefecto del Dept. de los precios de efectos de producciones y manufacturas de ella, bajo que deben ser abonados los sueldos de los empleados y demás ventas particulares para el año de 1833. Santa Ana, 11 December 1832, UGRM, MH, Fondo Prefectural, vol. 12.

11. Governor of Chiquitos to the Prefect of Santa Cruz, 17 January 1829, UGRM, MH,

Fondo Prefectural 1829, vol. 9, folio 87. *Acta* for the meeting of 7 January 1829, and the formal statement of the conditions for securing loans from each administrator.

12. Marcelino de la Peña to the Prefect de Santa Cruz, 18 July 1831, UBRM, MH, Fondo Prefectural 1831, vol. 11.

13. Estado General del Tesoro Público de Santa Cruz, 1830, UGRM, MH Fondo Prefectural, vol. 10; UGRM MH Fondo Prefectural, vol. 9, 1829 provides monthly statements of revenues according to category, with specific references to mission products.

14. Ministerio del Interior January–February 1830, Circular 22, 10 August 1830, UGRM, MH, Fondo Prefectural, vol. 10.

15. Platt, *La persistencia de los ayllus*, 34–35; Platt, "Ethnic Calendars and Market Interventions." Indian tribute remained an important rubric of Bolivian national tax revenues until the 1870s, estimated at 54 percent in 1846 (Jorge A. Ovando Sanz, *El tributo indígena en las finanzas bolivianas del siglo* XIX, quoted in García Jordán, *Cruz y arado*, 249). Thomas Abercrombie described the tax-paying festivals (*Kawiltu Kupraña*) that he witnessed in Santa Bárbara de Culta in *Pathways of Memory and Power*, 388–89.

16. Exp. Obrado a instancia de la Contaduría de Retasas de la Capital de Buenos Ayres para que la Rl Audiencia de la Plata informe el motivo porque los indios de las Provincias de Moxos y Chiquitos no pagan los reales tributos a Su Magestad, ABNB, EC, exp. 121, 1805.

17. Governor Josef Ramón de la Roca to the Bishop Don Francisco Xavier Adazábal, concerning "el nuevo plan de gobierno y libre comercio que su Magestad tiene mandado se entable en esta provincia," ACSC, exp. 6-1 A-1, San Xavier, 10 April 1811.

18. D'Orbigny, *Viaje a la América Meridional*, 3:1147–49.

19. Cuenta que rinde el administrador de este pueblo, Teniente del Batallón de Milicias Provinciales de Santa Cruz, José Aguilar y Flores, de las entradas, salidas, y existencias de su temporalidad, desde el día 31 diciembre 1822 hasta el 31 diciembre 1829; Pueblo de Santa Ana, 1828. Cuenta que rinde el ex adm. C. Francisco Xavier Cardoso de los bienes de temporalidad datados desde el 29 abril 1828 hasta el 18 diciembre 1828, UGRM, MH, Fondo Prefectural, vol. 9, Pueblo de San Francisco Xavier, 1829.

20. D'Orbigny, *Viaje a la América Meridional*, 3:1140. D'Orbigny identified the disease as *viruela boba*.

21. Gov. Peña of Chiquitos to the Prefect of Santa Cruz, 16 January, 3 March, and 25 March 1831, written from the pueblos of Santa Ana and San Miguel, UGRM, MH, Fondo Prefectural, vol. 11.

22. Prefect of Santa Cruz to Gov. of Chiquitos, UGRM, MH, Fondo Prefectural, vol. 37, exp. 71, 29 July 1847, and exp. 81, 30 August 1847.

23. AGN, Misiones, vol. 3, exp. 3, folios 46–47; exp. 36, folios 142–43.

24. Kessell, *Friars, Soldiers, and Reformers*, 237; Jackson, *Indian Population Decline*, 167.

25. Kessell, *Friars, Soldiers, and Reformers*, 246, presents slightly different figures for 1819–20, but in the same proportions of Indians and Spaniards and *castas*.

26. Fray José Pérez, Comisario Prefecto of the Missions of Pimería Alta, AGN, Oquitoa, 31 December 1818, Misiones, vol. 3, exp. 3, folios 46–47.

27. Mexican Hispanophobia was not misplaced. In July 1829, an expeditionary force of three thousand troops left Havana, Cuba, to attack the port of Tampico, Tamaulipas, in an attempt to reconquer Mexico. Spanish forces initially took control of the port, but they were forced to surrender in October, having sustained heavy losses to yellow fever and the siege that Mexican forces imposed, cutting off their invaders' supply lines. The Franciscan Diego Bringas, a fervently royalist Creole, served as chaplain to the invading army. Kessell, *Friars, Soldiers, and Reformers*, 269; Flores Caballero, *Counterrevolution*, 81–133; Costeloe, *Response to Revolution*, 52–116.

28. José María Pérez Llera, "Apuntes sobre los acontecimientos acaecidos en este colegio de la Santísima Cruz de Querétaro desde el año de 1821 y sus misiones hasta el mes de diciembre de 1844," Archivo del Colegio de la Santa Cruz de Querétaro (hereafter ACQ), quoted in Kessell, *Friars, Soldiers, and Reformers*, 277–82. Escalante y Arvizu would later become governor of the independent state of Sonora.

29. Kessell, *Friars, Soldiers, and Reformers*, 270, 279.

30. Fray José María Pérez Llera to the Governors of Occidente and Sonora, 1830–31, Archivo Histórico del Estado de Sonora (hereafter AHES), 1–2: 93, 94; ACQ, Doc 24. Fr. Pérez Llera, "Estado en que recibí las misiones de la Alta Pimería en el año de '30," transcribed by Kieran R. McCarty.

31. AHES, 1–2: 95; ACQ, Doc 36, 37, 44. Francisco Redondo and Francisco Mendoza administered the temporalities of Tubutama and Oquitoa, 1828–30. Leonardo Escalante al Ministro de Justicia y Negocios Eclesiásticos: Informe sobre la situación actual en que se hallan las temporalidades de las misiones de la Pimería Alta, 6 August 1831, ACQ, Doc 87.

32. Radding, *Wandering Peoples*, 277; Officer, *Hispanic Arizona*, 62–66; Griffen, *The Apaches at War and Peace*; Velasco, *Noticias estadísticas*, 81–111; Zúñiga, *Rápida ojeada*, 61–92.

33. Velasco, *Noticias estadísticas*, 100; "Pueblo de Bacoachi," *Boletín de la Sociedad Mexicana de Geografía y Estadística* (1861): 54–61, reprinted in Escudero, *Noticias estadísticas de Sonora y Sinaloa*, 343. We learn here that Bacoachi had two periods of growth stemming from peace accords reached with the Apaches in 1804 and 1806.

34. See chapter 5, note 49.

35. Zúñiga, *Rápida ojeada*.

36. Hardy, *Travels in the Interior of Mexico*, 88–91; Escudero, *Noticias estadísticas de Sonora y Sinaloa*, 288. Population data from Velasco, *Noticias estadísticas*, 55.

37. Radding, *Wandering Peoples*, 217–29, based on an analysis of notarial records in the Hermosillo Archivo de Notarías; Velasco, *Noticias estadísticas*, 55. The combined

population of the presidio and villa of Pitic and the Seri mission located on the south bank of the river opposite Pitic was just over one thousand souls in 1796. Fray Juan Felipe Martínez, "Padrón y inventario de la nueva mission de los seris," AMH, 1796, UASC microfilm 811, roll 2.

38. Hardy, *Travels in the Interior of Mexico*, 96. *Upper Sonora* refers to the highlands extending east and north of the lower San Miguel and Sonora river valleys that converge in Hermosillo.

39. Ibid., 109–10.

40. Ibid., 120–21.

41. Ibid., 126–45.

42. See chapter 4, tables 8–10, and my discussion of them; Hardy, *Travels in the Interior of Mexico*, 150.

43. Velasco, *Noticias estadísticas*, 55. Escudero, *Noticias estadísticas de Sonora y Sinaloa*, 235.

44. Hardy, *Travels in the Interior of Mexico*, 168; Escudero, *Noticias estadísticas de Sonora y Sinaloa*, 241–42.

45. Escudero, *Noticias estadísticas de Sonora y Sinaloa*, 235–36.

46. Ibid., 236–37; Hardy, *Travels in the Interior of Mexico*, 171–79. At the time Hardy traveled through Mexico, the peso and dollar were equivalent in value. A *marco* of silver was worth seven pesos, two reales.

47. Velasco, *Noticias estadísticas*, 178, 190.

48. Velasco, *Noticias estadísticas*, 173–87. Velasco was the lieutenant (*teniente general*) to the subdelegate of San Francisco, Leonardo Escalante, who reported to the intendant governor Alejo García Conde.

49. Ibid., 187–89. Eight reales equaled one peso. *Maquila* refers to the performance of one stage of a manufacturing or industrial process, in this case refining pieces of mineral ore "on speculation" from the *rescatadores* who had obtained them from trading with the Indian workers in the mines.

50. Ibid., 191.

51. Escudero, *Noticias estadísticas de Sonora y Sinaloa*, 110–11. Escudero's comment tells us less about the work habits of mulattoes than it does about his own racialized thinking about Mexican society.

52. Radding, *Wandering Peoples*, 295–96; Juan Ysidro Bohórquez al Presidente de la República, and José Anrríquez a las Cámaras Generales, 1836, AGN, Gobernación, Caja 4.

53. González Navarro, "Instituciones indígenas del México independente," 1:217–313; Mallon, *Peasant and Nation*, 65–72; Guardino, *Peasants, Politics, and the Formation of Mexico's National State*, 91–94; Ducey, "Indios liberales."

54. Guardino, *Peasants, Politics, and the Formation of Mexico's National State*, 189–91; Mallon, *Peasant and Nation*, 157–61; Thomson, "Bulwarks of Patriotic Liberalism";

Thomson, "Los indios y el servicio militar"; Hernández Chávez, *Anenecuilco*, 48–78; Escobar, "¿Qué sucedió."

55. On ethnogenesis, see Hill, *History, Power, and Identity*; on hegemony, see chapter 5, note 7.

56. Household censuses for Tepache and Oposura and the community-wide census for Ónavas were discussed in chapter 4.

57. Escobar Ohmstede, Falcón, and Buve, "Presentación." The whole volume from which this article is taken proves helpful. Van Young, *The Other Rebellion*, and Van Young, "La otra rebelión," conceptualize the local nature of peasant politics and massive rural mobilization during the wars for independence.

58. Florencia Mallon, in *Peasant and Nation*, has argued forcefully for multiple layers of contested meanings and practices of hegemony, reaching into the internal politics of Indian communities and the gendered relations of peasant households. Guardino, *Peasants, Politics, and the Formation of Mexico's National State*, and Thomson, "Movilización conservadora," demonstrate the importance of both interclass alliances and popular liberalism in shaping the nation-state. Antonio Annino explores these issues in "El primer constitucionalismo mexicano." Carmagnani and Romano, "Componentes sociales," relates the restructuring of territories to the reconfiguring of social hierarchies in the postcolonial republics of Latin America.

59. Van Young, "La otra rebelión," 54–55; Escobar Ohmstede and Falcón, "Los ejes de la disputa."

60. Hardy, *Travels in the Interior of Mexico*, 164–65; Escudero, *Noticias estadísticas de Sonora y Sinaloa*, 319; Velasco, *Noticias estadísticas*, 137–39.

61. Hu-Dehart, *Yaqui Resistance and Survival*, 21, quoting Archivo de la Defensa Nacional, Expediente, 271.

62. These were the laws numbered 88, 89, and 92, issued in September 1828. See the discussion of their impact on indigenous lands in chapter 3.

63. Hu-Dehart, *Yaqui Resistance and Survival*, 20–55; Spicer, *The Yaquis*, 130–40; Figueroa, *Los que hablan fuerte*, 72–82. This section is based on Evelyn Hu-Dehart's carefully documented study and Edward H. Spicer's culturally sensitive reading of the historical record, as well as my own research in the Archivo Histórico del Estado de Sonora on these military operations, correspondence with Juan María Jusacamea, and the movements of Yaqui laborers in Sonora.

64. Hu-Dehart, *Yaqui Resistance and Survival*, 56–74; Spicer, *The Yaquis*, 139–45. *Caudillismo* is a dominant theme of nineteenth-century Latin American history. Indigenous, mestizo, and Creole leaders took on the role of caudillo in different times and places, negotiating local alliances into broader regional movements. See Mallon, *Peasant and Nation*, 143–46.

65. José Velasco Toro, "Los Yaquis: Panorama de una resistencia activa," unpublished manuscript, 16, quoted in Figueroa, *Los que hablan fuerte*, 82.

66. Spicer, "People on the Desert," 29–30; Mariano Salcido, interview by the author, Quitobac, Sonora, February 1985.

67. Río, "Las efímeras 'ciudades,' " 678–79.

68. AHES, 4–1, Cuaderno 4, 27 March and 1 June 1839.

69. AHES, 4–1, Cuaderno 5, 30 April 1839.

70. Szuter, Crosswhite, and Crosswhite, "The Sonoran Desert," 254–55.

71. AHES, 4–2, Cuaderno 16, Governor Gándara from Potam, 22 May 1840.

72. Informe del Teniente Coronel José María Parra, 25 May 1840, AHES, 4–2, Cuaderno 15; Cuaderno 18, correspondence 7 July 1840.

73. AHES, 4–2, Cuadernos 15 and 16.

74. AHES, 4–2, Cuaderno 26, 1841.

75. Contemporary developments beginning in 1847 marked the first stages of armed conflict in Yucatán between ladinos (mestizos/*vecinos*) and Mayas, confounded with the political separatism among the ladino elite that became known as the "caste war." Hostilities continued for over half a century. The classic study of these events published in English is Reed, *The Caste War of Yucatán*. Reed began his history with the presentation of two opposing worlds: the ladinos and the *macehuales* (Mayan peasants), introducing each of them according to their perceived relationship with the forest and the landscapes they sought to create. The duality of ladino/*macehual* is, in part, replicated in the language of Yaqui resistance that developed during the nineteenth century in Sonora: *yoris*, or non-Indians, versus *yoremes*, Cahitan-speaking Yaquis and Mayos (Spicer, *The Yaquis*, 134). I argue here that these two worlds were not separate but interdependent, even as they were antagonistic.

76. García Jordán, *Cruz y arado*, 274–77.

77. These themes are developed at length in chapters 5 and 6.

78. ABNB, Gabriel René Moreno (hereafter GRM), M416.VIII. Constitución de la República Boliviana [1826] reimpresa de orden del Supremo Gobierno en Chuquisaca, 30 May 1827; ABNB, GRM, M405.XXII. Constitución Política de la República Boliviana, 1831, reformada en algunos de sus artículos por el Congreso Constitucional de 1834. This second constitution retained the distinctions between Bolivians and citizens established in the first document, but eliminated the explicit requirement of literacy. See Barragán Romano, *Indios, mujeres y ciudadanos*, 14–23.

79. Ibid., 48–54.

80. Langer, "El liberalismo."

81. The erosion of communal resources in the Chiquitanía was discussed in chapter 3.

82. Barragán Romano, *Indios, mujeres y ciudadanos*, 9–13, argues convincingly that the liberal notion of the modern nation-state was an unfulfilled project that translated into a history of social and political strife. The following works are representative of the debates concerning Creole visions of republican citizenship: Regalsky, *Etnicidad y clase*,

54–55; Guerra, *Modernidad e independencias*; Anderson, *Imagined Communities*; Guerrero, "El proceso de identificación." Marta Irurozqui has developed the intellectual and political connections between the concepts of *vecino* and *citizen* in "La vecindad y sus promesas."

83. Irurozqui, "El bautismo de la violencia."

84. D'Orbigny, *Viaje a la América Meridional*, 3:1175; Ovando Sanz, *La invasión brasileña a Bolivia*, 73–77. This episode has been repeated in a number of secondary sources and comprises an important part of the regional historiography for Chiquitos; I have not found independent archival sources to corroborate the event.

85. Ovando Sanz, *La invasión brasileña a Bolivia*, 50–53, 130–33. The treaty was probably signed in Villa Bella de Mato Grosso. Villa Bella and Casalbasco, a Brazilian military post, figured repeatedly in reports of border incidents between Chiquitos and Mato Grosso.

86. Ibid., 40–68, based on Lecuna, *Documentos*; O'Leary, *Memorias*; and Grondona *Descripción sinóptica de la provincia de Chiquitos*.

87. Ovando Sanz, *La invasión brasileña a Bolivia*, 70–73; Lecuna, *Documentos*, 1:184–85. Somewhat remarkably, Sebastián Ramos salvaged his life and political career from this treasonous initiative and played an important role in the frontier between Chiquitos and Mato Grosso during the 1840s, as will be discussed in the concluding chapter.

88. Prefectura de Santa Cruz. Actas de Gobierno. Correspondencia, Ramo de Hacienda. Legajo sobre Chiquitos, 1837: elecciones de los cantones de provincia. UGRM, MH, Fondo Prefectural, vol. 16, 1834.

89. *Vecino* implied local residence and commitment to the community, while *citizen* denoted an active concept, a status to be earned. Irurozqui, "La vecindad y sus promesas," 210–12; Guerrero, "El proceso de identificación," 32–35.

90. UGRM, MH, Fondo Prefectural, vol. 28, 1839.

91. Ibid.

92. Canedo, "Movimientos indígenas," on the Chiriguano frontier; Fischermann, "Historia chiquitana."

93. UGRM, MH, Fondo Prefectural, vol. 28, 1839. These letters are dated in Santa Cruz, 5 November 1840, 22 March 1841, and 30 March 1841. *Fianzas* for administrators are recorded in the Fondo Prefectural, vol. 8, 1828.

94. José Lorenzo Suárez, 14 August 1840; 11 November 1840; 2 January 1841; and 5 April 1841, adjudicación de terrenos a Miguel Santos Rivero y Ramón Antonio Paz, denominados de la Concepción del pueblo de San José de esa Provincia, UGRM, MH, Fondo Prefectural, vol. 29, 1841. This Miguel Santos Rivero may well be the same Rivero who was appointed *juez de letras* (teacher) and voted as first elector for the canton of San Ignacio, 24 April 1839.

95. Monthly account formed by Priest José Manuel Ayala, administrator of the

construction of this Holy Cathedral Church, showing the expenses that are legally justified by the accompanying receipts, UGRM, MH, Fondo Prefectural, vol. 30, 1841. I am indebted to Angel Yandura, of the Isosog Guaraní nation, professor of Guaraní culture and language at the Universidad Gabriel René Moreno, for his help in identifying different ethnic surnames in the archival lists.

96. Fischermann, "Historia chiquitana," 78.

8. Contested Landscapes

1. "Staying earth" is referenced in Kozak and López, *Devil Sicknesss and Devil Songs*, 65. Griffith, *Beliefs and Holy Places*, captures this quality admirably for the Pimería Alta and the contemporary O'odham villages of Arizona and Sonora. Similarly, Keith Basso's ethnographically sensitive reading of Apache toponyms in the vicinity of Cibecue, in eastern Arizona, meshes the moral content of Apache cautionary tales with the landscapes of "narrated place-worlds" in *Wisdom Sits in Places*.

2. Santamaría, "Fronteras indígenas," emphasizes the extensive internal frontiers formed, in part, by the river systems of the interior of South America. Northern New Spain was designated the *provincias internas*, and institutionalized under that name by the Military Commandancy of the Interior Provinces, created in 1779.

3. José Manuel Herrera to the Governor of Chiquitos, transcribed to the Prefect of Santa Cruz, from Santa Ana, 31 October 1835, UGRM, MH, Fondo Prefectural, vol. 17, 1835. Villa María lay in Brazil, northeast of Santo Corazón (Greever, *José Ballivián*, maps B and C).

4. Informe del Gobernador de la Provincia de Chiquitos al Prefecto de Santa Cruz, enciso #7, UGRM, MH, Fondo Prefectural, vol. 6, 1832; Governor of Chiquitos, Marcelino de la Peña, to Prefect of Santa Cruz, UGRM MH Fondo Prefectural, vol. 17, 1835.

5. See chapter 7, table 16. The description he sent to Governor Peña, probably knowing that it would be forwarded to the prefecture of Santa Cruz, reflects the language of nineteenth-century promotional literature issued for midwestern and western territories in the United States. See Lewis, "Rhetoric of the Western Interior."

6. See chapter 3 and chapter 5; d'Orbigny, *Viaje a la América Meridional*, 3:1195–202; chapter 7, table 13; and UGRM, MH, Fondo Prefectural, vol. 31, 1842.

7. UGRM, MH, Fondo Prefectural, vol. 31, folios 1, 10, 1842. The *actas* celebrated in San Ignacio and San Miguel, 3–4 January 1842, which followed instructions received from the prefecture and notice of the national ceremony of 9 December 1841, identified Ballivián's victory at Ingavi with the "campos de Viacha," in the department of La Paz.

8. Felipe Bertrés, Mapa corográfico de la República de Bolivia con la topografía de las fronteras limítrofes, 1843; Mapa de la República de Bolivia, executed by military officers

Juan Ondarza and Juan Mariano Mujía and the cartographer Lucio Camacho, published in 1859. García Jordán, *Cruz y arado*, 252 n.12, 257–58 n.26, 271–80; Greever, *José Ballivián*, 123. Cinchona bark from the Peruvian and Bolivian Amazonian forests was a preferred source of quinine for treating malaria, becoming a prized commodity on the nineteenth-century international markets. Balick and Cox, *Plants, People, and Culture*, 27–31. On the impact of cinchona gathering and its commercialization in the province of Moxos, see Block, *Mission Culture on the Upper Amazon*, 159–60.

9. García Jordán, *Cruz y arado*, 261–67; Fifer, *Bolivia*, 172–73. The Empresa Nacional de Bolivia and the Sociedad Progresista de Bolivia were both commercial companies for land development.

10. García Jordán, *Cruz y arado*, 270–71 n. 70; Fifer, *Bolivia*, 170–74; Greever, *José Ballivián*, 113–18. The exact dimensions of the Otúquis concession vary according to different authors and sources.

11. Governor of Chiquitos to the Prefect, Santa Ana, 8 January 1834, UGRM, MH, Fondo Prefectural, vol. 16, exp. 4; Greever, *José Ballivián*, 114–17.

12. Fifer, *Bolivia*, 171; Greever, *José Ballivián*, 119. Paraguayan military forces occupied the colonial Spanish Fort Borbón as their northernmost outpost along the contested waterway of the Paraguay river.

13. See chapter 1; Lahmeyer Lobo, "A importância estratégica," 11–13.

14. Bach, *Descripción de la nueva provincia*, appendix 1. Bolivia issued this edition of Bach's history of Olíden's province of Otúquis in order to substantiate its claims against Paraguay on the eve of the Chaco War of 1932–35.

15. See the passage quoted from *ecónomo* José Manuel Herrera of Santo Corazón above, and the strongly worded recommendation from the prefect of Santa Cruz to the governor of Chiquitos to order the administrators to oversee the planting of cacao and coffee in the pueblos, UGRM, MH, Fondo Prefectural, vol. 37, exp. 71, 29 July 1847.

16. UGRM, MH, Fondo Prefectural, vol. 12, exp. 65. Bach sent an impassioned complaint to Marcelino de la Peña of Chiquitos, dated 5 October 1832, from Brazil, directed to the Bolivian Congress, the president, and the minister of the interior, in which he accused the priest of Asunción de Guarayos, Manuel Rodríguez, of setting the fire and violently opposing Bach's settlement.

17. UGRM, MH, Fondo Prefectural, vol. 12, exp. 50, 1832, 14, exp. 27, 1833.

18. UGRM, MH, Fondo Prefectural, vol. 17, exp. 86, 1835; 29, 1841. The governor of Chiquitos advised the prefect that five Brazilian slaves had eluded their convoy and taken five Indian youth [*muchachos*] from San Rafael, presumably headed for the villa of Santa Cruz. The governor forwarded a notice (in Portuguese) from the administrative officer of Cuiabá, Manoel do Espíritu Santo, dated 8 August 1835, in which he described the five runaway slaves. Brazilian slaves had good reasons for seeking their freedom in Bolivian national territory: in 1831, Bolivia declared all slaves born since

1825 to be free; in 1838, all foreign slaves who entered Bolivia were legally free and, in 1851, Bolivia abolished slavery altogether (Klein, *African Slavery*, 251).

19. Bororós were identified on maps and drawings of the mid-nineteenth century, possibly related to the colonial *parcialidades* of Zamuco- and Tao-speakers known as Borós and Orobedas (see chapter 4, tables 6 and 7). Bororós have been studied ethnographically more consistently in Brazil than in Bolivia. See Maybury-Lewis, *Dialectical Societies*.

20. José Miguel Peinado to the Governor of Chiquitos, UGRM, MH, Fondo Prefectural, vol. 32, Juan Felipe Baca, Salinas del Marco, 19 October 1842; Bolivian Santiago Baca to the Governor of Chiquitos, 2 January 1843; Sebastián Ramos to the Governor of Chiquitos, San Ignacio, 21 January 1843.

21. Campo de las Salinas. Sebastián Ramos, juez territorial al cura encargado de la economía del Pueblo de Santo Corazón, Don José Gabriel Merino Valde Iglesias, UGRM, MH, Fondo Prefectural, vol. 29, 1841. On Ramos's petition to return to Bolivia in 1832, see, Governor Marcelino de la Peña to the Prefect of Santa Cruz, UGRM, MH, Fondo Prefectural, vol. 12, exp. 67, 81.

22. Sebastián Ramos, Teniente Coronel Jefe de la Frontera Oriental al Prefecto de Santa Cruz, Manuel Rodríguez Magasinos, UGRM, MH, Fondo Prefectural, vol. 36. Santo Corazón, 30 June 1847.

23. Communications to the Governor of Chiquitos, no. 38, 7 May 1847, and no. 48, 31 May 1847, UGRM, MH, Fondo Prefectural, vol. 37.

24. Decree of 27 January 1853, signed by President Manuel I. Belzú and submitted to U.S. secretary of state Edward Everett. Notes from the Bolivian legation in the United States to the Department of State, General Records of the Dept. of State, Record Group 39, quoted in Fifer, *Bolivia*, 177.

25. Fifer, *Bolivia*, 178–85; 212–21.

26. Lahmeyer Lobo, "A importância estratégica," 15–20. In 1867–68, Bolivian president, general Mariano Melgarejo, simultaneously negotiated a treaty of navigation and territorial concessions with Brazil, the opening of a Bolivian consulate in Asunción, Paraguay, and a treaty with Argentina over boundaries in the Chaco Boreal (ibid., 20; Fifer, *Bolivia*, 179–80).

27. Sociedad Progresista de Bolivia, *Empresa de navegación á vapor, caminos carreteros, colonización, ganadería, agricultura, industria y comercio*, promotional pamphlet, Sucre, 30 September 1874; contract between the Bolivian government in thirty-three articles, proposed 31 January 1864 and signed 1 February 1864, Bolivian Political Pamphlet Collection, UIUC Rare Book and Special Collections Library, 1–21. On the significant differences between Bolivians and citizens, see chapter 7, 268, 279–81.

28. On the disbandment of Chiquitano herds, see chapter 3, 111–13; on the Ley de Exvinculación de Tierras Indígenas (Law of Division of Indigenous Landholdings) of

1874, see Langer, "El liberalismo," esp. 79–82; García Jordán, *Cruz y arado*, 291–94; Langer, "Missions and the Frontier Economy."

29. Taussig, *Shamanism, Colonialism, and the Wild Man*; Stanfield, *Red Rubber, Bleeding Trees*; Block, *Mission Culture on the Upper Amazon*, 161–63.

30. Adelman and Aron, "From Borderlands to Borders," 814–16; Truett, "Neighbors by Nature," 164–65. An excellent treatment of the social and cultural development of Tucson and southern Arizona, focused on the Mexican population, is Sheridan, *Los Tucsonenses*.

31. Ruibal Corella, "*¡Y Caborca se cubrió de glora!*"

32. Bartlett, *Personal Narrative*, 1:423–24.

33. Ibid., 1:275–77.

34. Ibid., 1:265–66, 170–71, 290–92.

35. Ibid., 1:283–84.

36. Ibid., 1:460–62. Ronstadt later purchased the hacienda of Ocuca in the Altar valley. See chapter 3, note 9.

37. Ibid., 1:445–46. Juan and Refugio Tánori were Opata leaders who, when Bartlett visited Sonora, fought with government troops in campaigns against the Apaches. By 1858, however, they had joined Yaquis in open rebellion against the Liberal regime of Ignacio Pesqueira; in 1865, Refugio Tánori joined Creole and indigenous adversaries of Pesqueira to support the French occupation of Sonora. See Hu-Dehart, *Yaqui Resistance and Survival*, 76–85.

38. Bartlett, *Personal Narrative*, 442–43.

39. Emory, *Report on the United States and Mexican Boundary Survey*, 1:123.

40. Ibid.

41. See chapter 6, note 24; Kozak and López, *Devil Sickness and Devil Songs*, 81, 108–9.

42. Emory, *Report on the United States and Mexican Boundary Survey*, 1:94. The powerful Escalante family had registered numerous *sitios* of grazing land in Cuchuta, Teuricachi, Mututicachi, and Santa Rosa, in the territory bounded by the presidios of Fronteras, Bacoachi, and San Bernardino during the final years of colonial rule (Radding, *Wandering Peoples*, 200–202). Lieutenant Ignacio Pérez obtained title to the ranch of San Bernardino, spreading over 73,000 acres, in 1822 (Officer, *Hispanic Arizona*, 106–8). My interpretation of Emory's and Bartlett's reading of the Sonoran landscape is informed by a pathbreaking work of environmental history in Africa, Fairhead and Leach, *Misreading the African Landscape*, esp. 237–95, where they critique colonial discourses of degradation that had characterized native land-use practices.

43. Emory, *Report on the United States and Mexican Boundary Survey*, 1:124–25. Emory may have anticipated by over a century the electrically powered, deep well irrigation of the coastal desert west of Hermosillo that has transformed the ecology and the economy of the Sonoran capital city and its environs.

44. Voss, *On the Periphery of Nineteenth-Century Mexico*; see chapter 7.

45. Kino, *Las misiones de Sonora y Arizona*.

46. Turner, *The Frontier in American History*; Bolton, *Rim of Christendom*.

47. To cite selected works, see White, *The Roots of Dependency*; White, *The Middle Ground*; Limerick, *The Legacy of Conquest*; Cronon, Miles, and Gitlin, *Under an Open Sky*; Weber, *The Spanish Frontier in North America*; Weber, *The Mexican Frontier*; John, *Storms Brewed in Other Men's Worlds*; Hickerson, *The Jumanos*; Hämäläinen, "The Western Comanche Trade Center"; and Gutiérrez and Orsi, *Contested Eden*.

48. Adelman and Aron, "From Borderlands to Borders," launched a provocative forum essay in the *American Historical Review* in which they attempted to intersect the political dimensions of *borderlands* with the cultural dimensions of *frontiers* as spaces of liminality. Their construction of *bordered lands* in reference primarily to the borders of the United States, Mexico, and Canada as modern nation-states, implying a kind of narrative closure to the "shift from inter-imperial struggle to international coexistence" (816), raised problems of conceptualization, in reference especially to indigenous peoples, and of empirical interpretation in their comparisons of Spanish, French, and British imperial policies in North America. John R. Wunder and Pekka Hämäläinen wrote a spirited challenge to Adelman and Aron, criticizing their vision of "sanitized" borderlands (1229) in which Indian agency had only a minimal role (Wunder and Hämäläinen, "Of Lethal Places and Lethal Essays").

49. Lehmann, ed., *Ecology and Exchange in the Andes*; Saignes, "De la filiation à la residence"; Masuda, Shimada, and Morris, *Andean Ecology and Civilization*; Larson, *Cochabamba*, 13–50. Karl Zimmerer, in "Discourses on Soil Loss in Bolivia," has challenged the notion of vertical archipelagos with his observations of "patchwork" mixed cropping patterns in two highland regions of Peru and Bolivia.

50. Santamaría, "Fronteras indígenas"; Roosevelt, "The Maritime, Highland, Forest Dynamic"; Renard Casevitz, Saignes, and Taylor, *El este de los Andes*.

51. Adelman and Aron, "From Borderlands to Borders," 815–16; Weber and Rausch, introduction.

52. The bibliography is far too vast to cite here, but noteworthy contributions include: on the fur trade, Trigger, *Natives and Newcomers*; Brown, *Strangers in Blood*; Van Kirk, *"Many Tender Ties"*; White, *The Middle Ground*; on captives and human commodity trade, Brooks, *Captives and Cousins*; one late nineteenth-century capitalist borderlands, Truett, "Neighbors by Nature"; for the Andean world, Larson and Harris, *Ethnicity, Markets, and Migration in the Andes*; Saignes, *Los Andes orientales*.

53. Martinez, *Border People*; Truett, "Neighbors by Nature," 163–65, argues that the "corridors of steel" that brought the railroad through southern Arizona and Sonora in the 1880s and the web of industrial mines opened in both states linked the United States and Mexico as neighbors, but that "after 1854, the stories of the American West and the Mexican North take their leave of one another." Nevertheless, it can also be shown that the two border regions of Sonora and Arizona grew together across the binational

boundary, both in networks of trade and migration and in rising spirals of conflict and violence. Stern, "Buildings, Boundaries, and Blood," documents graphically the ways in which medical screening at the Ciudad Juárez–El Paso border became an exercise in humiliation and intimidation invoking racialized and gendered stereotypes of disease and contamination.

54. The sense of exceptionalism and difference in northern Mexico and the southwestern United States has produced a rich literature and repertoire in the plastic and performing arts, analyzed sensitively by García Canclini, *Hybrid Cultures*, esp. 206–41. See, among many literary and historical treatments of border cultures and identities, Saldívar, *The Dialectics of Our America*; Romero, Hondagneu-Sotelo, and Ortiz, *Challenging Fronteras*; Bonilla, Melendez, and Los Angeles Torres, *Borderless Borders*; Limón, *Dancing with the Devil*.

55. Fifer, *Bolivia*, 205–21.

Glossary

Terms particular to Sonora (So) and to Chiquitos (Ch) are marked accordingly. (Q) refers to Quechua.

acequias: irrigation ditches
adelantado: frontier governor; has strong militaristic overtones
aguadas permanentes: watering holes sufficient to water and graze livestock
alcabala: colonial sales tax
alcalde: judge and *cabildo* officer
alcalde de primer voto: first magistrate
alcalde de segundo voto: second magistrate
alcalde mayor: district magistrate
aldea: village, fixed settlement
alférez: second lieutenant
alguacil: administrative officer overseeing community resources
alhóndiga: public granary
almas: souls
almud: a dry measure, sometimes one-twelfth of a *fanega*, sometimes one-quarter
annuas: annual reports issued by the Jesuits
apoderado: person holding power of attorney over another
arriero: muleteer
arroba: a weight of twenty-five pounds
arroyo: brook
asojna (Ch): small bird; in Chiquitano mythology, it was once endowed with shamanic powers, marking the changing of the seasons
audiencia: colonial court governing institution presided over by four judges
ayllu (Q): lineages claiming to descend from a common ancestor and sharing territorial use-rights
baldíos: "vacant lands"; also referred to as *tierras realengas*
bañados: bogs
bandeir: see *maloca*
bandeirantes: Portuguese explorers and slave hunters
bárbaro: barbarian

barbecho: fallow lands
barrancas: raised earthworks
batán: wooden implements for grinding rice, maize, and other seeds
bayeta: imported cloth
besiro (Ch): "correct speech" or "the right way"; possible base for the name Chiquitano
bosque: uncultivated forestland for gathering or for clearing garden plots
bosque alto: high forest
bretaña: a sort of fine linen
bronco y alzado: branded and wild
bufón: clown
caballería: measure of land of 2,663.5 hectares or 6,579 acres
cabecera: head village of a mission
cabildo: town council
caça a o indio: see *maloca*
cacicazgo: office held by cacique
cacique: local (indigenous) lord or chief
caja real: provincial treasury
calandria: clothier's press
calentura tersiana: disease resembling malaria
calpul: Maya term for social divisions in specific sections of indigenous towns; also *cah*
calpulli: Nahuatl term for social divisions in specific sections of indigenous towns
campo cerrado: forested savannah
campo húmedo: humid savannah
capillas de posa: chapels along pilgrimage routes
carga: measure of volume that varies by product; equal approximately to 111 liters or 3.1 bushels
cargo: products shipped from the missions to La Plata and Potosí
carpir: clearing of undergrowth following burning; occurs before the onset of the rains in late September and October
carta anúa: annual report submitted by missionaries
casa de asiento: permanent earthen dwelling
casa de tierra: permanent earthen dwelling
casados/*casadas*: married men/women
Casas Grandes: pre-Columbian population center in present-day Chihuahua
casta: mixed-race castes, like mestizo or mulatto; often used to refer to the mixed-blood masses of the colonial era
catálogo: catalog
caudillo: patriarch of notable landholding families
chabóchi (So): humans in Rarámuri cosmology
chácara (Q): small cultivated plot; also *chacra*

chaco: seasonal cultivated plot; swidden plot
chapapa: loft for storage and hanging implements
chapayeka (So): masked demonic assistant in religious rituals
cheeserúx (Ch): honored and revered healers and sorcerers
chicha (Q): corn beer brewed in the Andean region
chico/chiquito: small; possible base for the name Chiquitano
chininises: ritual dancers in Chiquitano religious celebrations
chiqui-s: "egg" or "testicles"; possible base for the name Chiquitano
chunguperedatei (Ch): large white swan, associated with outsiders
ciudadano: citizen
ciudadanos en ejercicio: citizens in full exercise of their rights eligible to vote or be elected to public office
cofradía: lay Catholic brotherhood dedicated to charity work
cojñone (Ch): outsiders
colegio: school and home for Jesuits
comisario: member of the *cabildo*
compadrazgo: ritual sponsorship; similar to the bonds of godparentage
composición: payment for legal title to lands already occupied
común: mission communal lands; also refers to common wealth
contribución: personal contribution tax on all non-Indian property holders
convento: convent
cordillera: mountain range
corregidor: highest male officer of the mission, with both administrative and judicial authority
corúa (So): water-dwelling snake that protects the flow of water
coyote: indigenous and African racial mixture; in Sonora it meant mestizo
cruceño: adjective for someone from Santa Cruz de la Sierra
cruceros de día: minor Indian officials who notified priests of emergencies
cura: rector priest in charge
curandero: healer
curezei (Ch): cross
da:m kacim (So): in O'odham cosmology, up above laying
data: trade goods sent to the missions including freight costs, clerical stipends, and lay salaries paid to the administrators and priests
denuncia: claim to public land
despoblado: area without livestock or people
diezmo: tithe
doajkam (So): O'odham concept of wilderness, signifying health, wholeness, and the renewal of life
ecónomo: lay administrator providing day-to-day supervision of the mission pueblos

en borrasca: mine that does not yield
encomendero: one who received an *encomienda*
encomienda: grant of Indians held in service
en mancomún: kind of communal regime recognizing de facto use and occupation more than formal property demarcations
entrada: overture—both peaceful and coercive—made to native *rancherías* to accept mission life in the uncertain climate created by the fearful sequence of epidemic diseases and by Spanish demands for forced labor under the *encomienda*
escrutador: electoral examiner
español: Spanish
estancia: ranch
existentes: total population in the missions; translated as "souls" who "existed" in the pueblos from one year to the next
fanega: volume of grain equal to 55 liters or 1.6 bushels; also, land measurement equal to area of land sown with one *fanega*, approximately 3.5 hectares or 8.5 acres
fariseos: pharisees
fianza: bond
fiesta-cargo: system of officeholding establishing a hierarchy of interlocking posts between *cofradías* and indigenous village governance
fiscal: indigenous council official
fomento: merchandise distributed to the Indians
fundo legal: minimal allotment to Indian pueblos and Spanish towns as municipal property; conventionally a square league
gambusino (So): mine prospector
gamuza: tanned leather hide
ganado de las secciones: pueblos' livestock
generación: identity based on ethnic or kinship alliances
gente de razón: "people of reason"; category used to separate the indigenous (legally considered children) from Europeans
gentil: adult Indian not converted to Christianity; someone who lived outside the organizational structure of the missions
guedé (Ch): sun; in Chiquitano mythology, it redistributes power among the strong and the weak
hacienda de beneficio: large benefiting patio for processing ore
hanansaya (Q): lower subdivision of an *ayllu*
hechicero: shaman
henaxíx-tí (Ch): *jichi* of the spirits of all animals
hijo de pueblo: resident of indigenous communities
horcon: timber column sunk into the ground
huduñig (So): in O'odham cosmology, sunset place

huérfana: orphan
hurinsaya (Q): upper subdivision of an *ayllu*
huya aniya (So): Cahitan notion of "forest world"; like Spanish, *monte*
i:bui (So): spiritual world; conceptualized as breath
i'itoi (So): O'odham's man-god creator
indio de servicio: Indian who did not pay tribute to the crown, but was held in personal service to an individual colonist
indio encomendado: Indian subject to serving *encomienda*
indio laborío: Indian working for contract and separated from the missions
indistintamente: informal land possession; without distinguishing among Indians and *vecinos*
isituús (Ch): lesser order of divine spirits guarding the rivers and lagoons and keeping them stocked with fish
jatokaáx (Ch): humanoid dwarfs held responsible for the disappearance of children
jiawul himdag (So): the devil way
jichi: spirit associated with water, wind, and different kinds of landforms—hills, woods, savannas (*pampa*, *monte*)—as well as with the human activities in nature such as hunting and cultivation
juez: judge
juez comisario: local judge with authority delegated from the *alcalde* mayor
juez de paz: justice of the peace
Junta Provincial de Temporalidades: authority established by the Audiencia of Charcas to oversee the properties and economic administration of the missions after the expulsion of the Jesuits
justicias: magistrates and native councilmembers
ka:cim jewed (So): in O'odham, "staying earth"; opposite of *i:bui*
kawichí: Rarámuri (Tarahumara) meaning "of wildness"
labores: communal fields
labores de los vecinos: small farm plots
lienzo: linen or canvas cloth
llanos: flatlands, grasslands
macana: club
mador: missionary's direct assistant and ecclesiastical notary
maloca: aggressive slaving expedition, not unlike the Portuguese *bandeira* or with the purpose of enslaving Indian captives
mameluco: Portuguese contrabandist
manta: common cloth, typically shawls or blankets
mapono (Ch): most knowledgeable and powerful shamans who transported souls
maquila: small, ore-processing patios
marco: measurement of silver equal to 8.5 pesos

marqueta: crude cake of wax

matachines: ritual dancers in religious festivals of Cahitan and Tarahumara peoples of northwestern Mexico

matlazáhuatl: plague that attacked New Spain in the 1730s; name means "red eruptions" in Nahuatl

mayordomo: foreman

mediannata: tax

mehi weco (So): in O'odham cosmology, fire below

memoria: list of purchases for the missions

mesa: electoral board of five electors

mescaleras: women that gathered agave roots

Mesoamerica: Central America and present-day Mexico

mestizaje: racial mixture; also applied to cultural mixing

mestizo: racial mixture of indigenous and European people

metate: grinding stone used for lava

milicia: militia

milpa: family plot, cultivated whether irrigated or swidden

milpa de temporal: rainfall-dependent small plot for growing maize

minca/*minga* (Quechua: *minc'ay*; Chiquitano: *metósh*): work exchanged through reciprocal obligations

mocoé (Ch): fallow lands

mojonera: wooden and masonry landmark

montaña: mountains

monte: wooded place or mountain

Moxos/Mojos: province in eastern Bolivia

naciones: ethnic identities or chiefdoms

naturales: Spanish term for the native people

nawait: ceremonial fermenting and drinking of wine fermented from the juice of the *sahuaro* by the Tohono O'odham of Sonora

ne:big (So): mythical O'odham water beast

negros: Africans

nijoras: captives separated from their communities of origin

noria: well

Nueva Galicia: province of New Spain; corresponds to present-day western Mexico

Nueva Vizcaya: province of New Spain; corresponds to present-day north-central Mexico

oboíx (Ch): feared and reviled healers and sorcerers

obraje: factory, often for textile production

oidag: fields

o'odham himdag (So): the people's way
oquipaú (Ch): souls of the dead
padrón: official house-to-house census
pagaduría: exchequer
pampa: flatlands
pampa arbolada: wooded flatlands
pantanal: swamp
parcialidades: distinct language and kin groups that retained ethnic and cultural significance
parientes: kinspeople
párvulos/párvulas: children under the age for catechism, confession, and communion
pasados: older men and former officials constituting a village elite
pascola (So): Cahitan ritual clown appearing in religious dances for the deer
paurú (Ch): natural waterhole or spring
peones: laborers, often on haciendas
piezas de rescate: "ransomed pieces"; Indians captured directly from their communities or seized from enslavers
pinole: basic corn meal ground from roasted kernels
poblado: small settlement, like a ranch
polity: political representation and authority of local governance in Indian villages
potrero: pasture or small ranch dedicated to breeding horses
pozoka: *cabildo* officers ritually visiting different pueblos during saint's day celebrations
prestación vial: municipal tax in kind
principales: indigenous chiefs
puesto: geographic location of a small settlement
putugútoi (Ch): jaguar; in Chiquitano mythology, it symbolizes the power of nature
quipoci (Ch): maternal deity, like the Christian Mary
ramada: one-room adobe structure
ranchería: hamlet smaller than a village, often seminomadic
ranchero: small private landholder using extrafamilial labor
realenga: public land
real: monetary unit, eight *reales* to one peso
reales (*de minas*): royal mining grants and mines
rebajas: losses
receptoría: mission receiving office
recepturía: mission central receiving station
rectorado: Jesuit administrative district in northwestern Mexico
reducciones: mission towns intended to enforce permanent settlement in nomadic frontiers

regidores: local officials
repartimiento: forced labor recruited from native villages and sent to Spanish estates and mines for stipulated periods of time
república de españoles: system of laws and customs, rights and privileges applied to Europeans
república de indios: system of laws and customs, rights and privileges applied to indigenous people
requerimiento: initial declaration of Spanish dominion pronounced on landfall or encounter with a new ethnic group
rescatadores: itinerant peddlers
retablo: altar ornament, often composed of religious paintings in wooden frames
roza: cycle of cutting, burning, clearing, planting and harvesting, combining the rotation of crops within each plot with the rhythm of fallowing; also known as *roce y quema*
rozada: heavy cutting (beginning in May) and burning (in the dry months of July and August)
sacerdote: priest
salineras: salt mines
salvaje: barbarian
sánti (So): God's helpers (Rarámuri souls)
sa:nto himdag (So): the saint way
saraó (Ch): dance of European origin performed in religious ceremonies in Chiquitos
serranía: piedmont region of Sonora
serrano: someone from the highlands or piedmont
si'alig weco (So): in O'odham cosmology, beyond the eastern horizon
síndico: member of the *cabildo*
sínodo: missionaries' stipend
sitio: land measurement equivalent to 1,747 hectares when assigned for raising bovine cattle (*ganado mayor*); or equivalent to 776 hectares when used for smaller species of goats, sheep, pigs (*ganado menor*)
solteros/ solteras: unmarried adolescents or young adults
sosegado: living in peace
subdelegado: local Spanish administrator
suerte: small plot of farm land equivalent to 10.5 hectares or 26.3 acres
sukrísto (So): God's song in Rarámuri cosmology; from Spanish *Jesucristo*
swidden agriculture: see *roza*
tacú: wooden implements for cooking and grinding
tacuara: arrow shot
tambos (Q): roadside hostels for travelers
tapisque: worker drafted for harvest labor on private haciendas

tapuymiri: slaves of small things; possible base for the name Chiquitano
temastianes: catechism teachers and sacristans for Mass
teniente: local Spanish official with administrative and magisterial duties
teniente de alcalde mayor: local Spanish administrator
teniente de justicia mayor: judicial and administrative post
ternero: veal calf
terreno: titled properties with fixed boundaries
tesgüino (So): corn beer
tierras realengas: nominally vacant lands not yet reduced to private ownership under the terms of Spanish law
tinaja: natural reservoir filled with water seepage from hillside springs
tipoi/ tipoy: women's dress
tocuyo: plain, homespun cotton cloth
tohono (So): desert
toírr (Ch): *jichi* of the spirit of the *chacos* and their produce
toldos: canvas for carrying burdens
tomín (pl. *tomines*): unit of currency, a subdivision of the *real*
topiles (So): indigenous council official; similar to a constable
torno: spinning wheel
trincheras: defensive terraces on hillsides
tuúx (Ch): *jichi* of water; associated especially with the natural springs or openings in the earth with flowing fresh water
usos y costumbres: local customs
vaquero: cowboy or cattle guard
vara: linear measure equal to 0.8 meters or 2.7 feet
vecino: Hispanic landholding resident of a given place
villa: legally recognized Spanish town, distinct from an Indian pueblo
visita: mission visiting station
viudos/ viudas: widowed men and women
wahia: well or natural water source near hills in the Sonoral Desert
wi:gita (So): annual ritual of dance, song, and storytelling
yanaconas: see *indios de servicio*
yarituses: birdlike figures in saint's day rituals of San Xavier in Chiquitos
yuca: manioc or cassava
zacate: grass or hay
zapata: bracket for a beam
zona serrana: highlands zone

Bibliography

Primary Sources

Adam, Lucien, and Victor Henry. *Arte y vocabulario de la lengua chiquita con algunos textos traducidos y explicados compuestos sobre manuscritos inéditos del XVIII siglo*. Paris: Maisonneuve et Cie, 1880.

Aguirre, Manuel. *Doctrina christiana y pláticas doctrinales traducidas en lengua opata por el P. Rector Manuel Aguirre de la compañía de Jesús, quien las dedica al Illmo. Sr. doctor d[on] Pedro Tamarón, Obispo de Durango*. Mexico City: Colegio de San Ildefonso, 1767.

Bach, Moritz. *Descripción de la nueva provincia de Otuquis en Bolivia*. 2d rev. and enl. ed. La Paz: "Unidas," 1929.

Bandelier, Adolph. *Contributions to the History of the Southwestern Portion of the United States*. Cambridge, MA: John Wilson and Son, 1890.

——. *A History of the Southwest: A Study of the Civilization and Conversion of the Indians in Southwestern United States and Northwestern Mexico from the Earliest Times to 1700*. Ed. Ernest J. Burus. 1887. Rome: Jesuit Historical Institute, 1969.

——. *Final Report of Investigations among the Indians of the Southwestern United States: Carried on Mainly in the Years from 1880 to 1885*. 2 vols. Cambridge, MA: John Wilson and Son, 1890–92.

Barbastro, Francisco Antonio. *Sonora hacia fines del siglo XVIII: Un informe del misionero franciscano fray Francisco Antonio Barbastro con otros documentos complementarios*. Ed. Lino Gómez Canedo. Guadalajara: Librería Font, 1971.

Bartlett, John Russell. *Personal Narrative of Exploration and Incidents in Texas, New Mexico, California, Sonora, and Chihuahua, Connected with the United States and Mexican Boundary Commission during the Years 1850, '51, '52, and '53*. 2 vols. New York: D. Appleton, 1854.

Burrus, Ernest J., and Félix Zubillaga, eds. *El noroeste de México: Documentos sobre las misiones jesuíticas, 1600–1769*. Mexico City: Universidad Nacional Autónoma de México, 1986.

Caballero, Lucas. *Relación de las costumbres y religión de los indios Manasicas*. 1706. Ed. Manuel Serrano y Sanz. Madrid: Librería General de V. Suárez, 1933.

Castelnau, Francis de. *En el corazón de América del Sur, 1843–1847*. Ed. Ana María Lema and Carlos Ostermann. Trans. María Teresa Bedoya de Ursic. La Paz: Editorial Los Amigos del Libro, 2001.

——. *Expédition dans las parties centrales de l'Amérique du Sud, de Río de Janeiro a Lima, et de Lima au Pará: Exécutée por ordre du goubernement français pendant les années 1843 a 1847*. 6 vols. Paris: P. Bertrand, Libraire-Editeur, 1851.

De los Reyes, Antonio. *Observaciones sobre el obispado de Sonora*. 1784. Valencia: Generalitat Valenciana, 1989.

D'Orbigny, Alcide Dessalines. *El hombre americano, considerando sus aspectos fisiológicos y morales*. Trans. Alfredo Cepeda. 1839. Buenos Aires: Editorial Futuro, 1944.

——. *L'homme américain de l'Amérique Méridionale considéré sous ses rapports physiologiques et morau*. 2 vols. Paris: Pitoit-Levrault, 1839.

——. *Viaje a la América Meridional . . . Realizado de 1826 a 1833*. 4 vols. Trans. Alfredo Cepeda. 1845. Buenos Aires: Editorial Futuro, 1945.

Emory, William H. *Report on the United States and Mexican Boundary Survey*. 7 vols. Washington, DC: U.S. Senate, 1857.

Escudero, José Agustín de. *Noticias estadísticas de Sonora y Sinaloa*. 1849. Ed. Héctor Cuauhtémoc Hernández Silva. Hermosillo, Mexico: Universidad de Sonora, 1997.

Fernández, Juan Patricio. *Relación historial de las misiones de indios chiquitos que en el Paraguay tienen los padres de la Compañía de Jesús*. 1726. Ed. Daniel Santamaría. Jujuy, Argentina: Universidad Nacional de Jujuy, Centro de Estudios Indígenas y Coloniales, 1994.

Font, Pedro, and Tomás Eixarch. *Fray Pedro Font, diario íntimo, y diario de Fray Tomás Eixarch*. Ed. Julio César Montané Martí. Hermosillo, Mexico: Universidad de Sonora, 2000.

Furlong S. J., Guillermo, ed. *José Sánchez Labrador, S. J. y su "yerba mate."* 1774. Buenos Aires: Librería del Plata, S. R. L., 1952.

Gramática de la lengua de los indios llamados Chiquitos. 17?

Grondona, José Esteban. *Descripción sinóptica de la provincia de Chiquitos*. 1833.

Hammond, George P., and Agapito Rey, eds. *Narratives of the Coronado Expedition, 1540–1542*. Albuquerque: University of New Mexico Press, 1940.

Hardy, R. W. H. *Travels in the Interior of Mexico, in Baja California, and around the Sea of Cortes, 1825, 1826, 1827, 1828*. 1829. Glorieta, NM: Rio Grande Press, 1977.

Hervás y Panduro, Lorenzo. *Catálogo de las lenguas de las naciones conocidas, y numeración, división, y clases de éstas según la diversidad de sus idiomas y dialectos*. 6 vols. Madrid: Administración Real Arbitrio de Beneficios, 1800–1805.

——. *El lingüista español Lorenzo Hervás: Estudio y selección de obras básicas*. 1874. Sel. Antonio Tovar. Ed. Jesús Bustamante. Madrid: Sociedad General Española de Librería, S. A., 1986.

Humboldt, Alexander von, and Aimé de Bonpland. *Personal Narrative of Travels to the*

Equinoctial Regions of the New Continent during the Years 1799–1804. Trans. Helen Maria Williams. 1818–29. Amsterdam: Theatrum Orbis Terrarum, 1971.

Kino, Eusebio Francisco. *Las misiones de Sonora y Arizona: Comprendiendo la crónica titulada "Favores celestiales" y "Relación diaria de la entrada al noroeste."* Ind. Francisco Fernández del Castillo. 1702. Mexico City: Editorial Porrúa, 1989.

Knogler, P. Julián. "Relato sobre el país y la nación de los chiquitos." In *Las misiones jesuíticas entre los Chiquitanos*, ed. Werner Hoffmann, 121–85. Buenos Aires: Fundación para la Educación, la Ciencia y la Cultura, 1979.

Las Casas, Bartolomé de. *Brevíssima relación de la destruyción de las Indias*. 1552. Seville, Spain: Er Revista de Filosofía, 1991.

——. *Del único modo de atraer a todos los pueblos a la verdadera religión. Advertencia preliminar y edición y anotación del texto latino*. 1536. Ed. Agustín Millares Carlo. Trans. Atenógenes Santamaría. Mexico City: Fondo de Cultura Económica, 1942.

Lecuna, Vicente, ed. *Documentos relativos a la creación de Bolivia*. 2 vols. Caracas: Litografía del Comercio, 1924.

Linné, Carl von (Linnaeus). "Specimen academicum de Oeconomia Naturae." *Amoenitatis Academicae* II (1751): 1–58.

Lumholtz, Carl. *Unknown Mexico: Explorations in the Sierra Madre and Other Regions, 1890–1898*. 2 vols. New York: Charles Scribner's Sons, 1902.

Nentvig, Juan. *Descripción geográfica, natural y curiosa de la provincia de Sonora*. 1764. Ed. Germán Viveros. Mexico City: Archivo General de la Nación, 1971.

Nordenskiöld, Erland. *An Ethno-geographical Analysis of the Material Culture of Two Indian Tribes in the Gran Chaco*. Göteborg: Elanders boktryckeri aktiebolag, 1919.

——. *Exploraciones y aventuras en Sudamérica*. Trans. Gudrun Birk and Angel E. García. La Paz: Apoyo Para el Campesinado Indígena del Oriente Boliviano, 2001.

——. *Forschungen und Abenteuer in Südamerika*. Stuttgart: Strecker und Schröder, 1924.

Núñez Cabeza de Vaca, Alvar. *Comentarios de Alvar Núñez Cabeza de Vaca, adelantado, y governador del Rio de la Plata, escritos por Pedro Hernández, escribano y secretario de la provincia*. 5th ed. Madrid: Espasa-Calpe, 1971.

——. *Relación*. In *Alvar Núñez de Cabez de Vaca: His Account, His Life, and the Expedition of Pánfilo de Narváez*, vol. 1. Ed. Rolena Adorno and Patrick Charles Pautz. Lincoln: University of Nebraska Press, 1999.

Núñez Fundidor, Fray Angel Antonio. *Bacerác en 1777: Carta edificante de Fray Angel Antonio Núñez Fundidor*. Ed. Julio César Montané Martí. Hermosillo, Mexico: Contrapunto 14, 1999.

Obregón, Baltasar de. *Obregón's History of 16th Century Explorations in Western America Entitled Chronicle, Commentary, or Relation of the Ancient and Modern Discoveries in New Spain and New Mexico, Mexico*. Ed. and trans. Agapito Rey and George P. Hammond. 1584. Los Angeles: Wetzel Publishing, 1928.

Oviedo y Valdés, Gonzalo Fernández de. *Historia general y natural de las Indias, islas y tierra firme del mar océano.* 4 vols. Ed. José Amador de los Rios. Madrid: Real Academia de la Historia, 1851–55.

Pérez de Ribas, Andrés. *History of the Triumphs of Our Holy Faith amongst the Most Barbarous and Fierce Peoples of the New World.* 1645. Trans. Daniel T. Reff, Maureen Ahern, and Richard K. Danford. Tucson: University of Arizona Press, 1999.

——. *Triunfos de nuestra santa fé entre gentes las mas bárbaras y fieras del nuevo orbe, por el padre Andres Pérez de Ribas. Procedida de los naufragios de Alvar Núñez Cabeza de Vaca.* 1645. Mexico City: Editorial "Layac," 1944.

Pesqueira, Fernando. "Documentos para la historia de Sonora." Unpublished collection, Universidad de Sonora, Hermosillo, series 2, vol. 1, 1821–45.

Pfefferkorn, Ignaz. *Sonora: A Description of the Province.* 1795. Ed. Bernard Fontana. Trans. Theodore E. Treutlein. Tucson: University of Arizona Press, 1989.

Recopilación de Leyes de los Reinos de las Indias, Mandadas imprimir y publicar por la majestad católica del Rey Carlos II. 5th ed. 1681. Madrid, 1841.

Rodríguez Gallardo, José Rafael. *Informe sobre Sinaloa y Sonora, 1750.* Ed. Germán Viveros. Mexico City: Archivo General de la Nación, 1975.

Sepúlveda, Juan Ginés de. *Democrates alter: Tratado sobre las causas justas de la guerra contra los indios.* 1547. Mexico City: Fondo de Cultura Económica, 1941.

——. *Historia del nuevo mundo.* 1562. Ed. and trans. Antonio Ramírez de Verger. Madrid: Alianza Editorial, 1987.

Smith, Buckingham. *A Grammatical Sketch of the Heve Language Translated from an Unpublished Spanish Manuscript.* 1861. New York: Cramoisy, 1974.

Tamarón y Romeral, Pedro. *Demostración del vastísimo obispado de la Nueva Vizcaya.* 1765. Mexico City: Antigua Librería Robredo, 1937.

——. *Descripción de la diócesis de Nueva Vizcaya.* 1765. Madrid: Aguilar, 1958.

Turner, Frederick Jackson. *The Frontier in American History.* Foreword by Ray Allen Billington. New York: Holt, Rinehart, and Winston, 1962.

Velasco, José Francisco. *Noticias estadísticas del Estado de Sonora.* 1850. Hermosillo, Mexico: Gobierno del Estado de Sonora, 1985.

Zúñiga, Ignacio de. *Rápida ojeada al estado de Sonora.* 1835. Hermosillo, Mexico: Gobierno del Estado de Sonora, 1985.

Secondary Sources

Abercrombie, Thomas A. *Pathways of Memory and Power: Ethnography and History among an Andean People.* Madison: University of Wisconsin Press, 1998.

Ackerly, Neal W., Jerry B. Howard, and Randall H. McGuire. *La Ciudad Canals: A Study of Hohokam Irrigation Systems at the Community Level*. 7 vols. Tempe: Arizona State University Anthropology Field Studies, 1985.

Adelman, Jeremy. *Frontier Development: Land, Labour, and Capital on the Wheatlands of Argentina and Canada, 1890–1914*. Oxford: Clarendon, 1994.

——, ed. *Colonial Legacies: The Problem of Persistence in Latin American History*. New York: Routledge, 1999.

Adelman, Jeremy, and Stephen Aron. "From Borderlands to Borders: Empires, Nation-States, and the Peoples in between in North American History." *American Historical Review* 104.3 (1999): 814–41.

Adorno, Rolena. "The Negotiations of Fear in Alvar Núñez Cabeza de Vaca's *Naufragios*." *Representations* 33 (1991): 163–99.

Adorno, Rolena, and Patrick Charles Pautz, eds. *Alvar Núñez Cabeza de Vaca: His Account, His Life, and the Expedition of Pánfilo de Narváez*. 3 vols. Lincoln: University of Nebraska Press, 1999.

Afzal-Khan, Fawzia, and Kalpana Seshadri-Crooks, eds. *The Pre-occupation of Postcolonial Studies*. Durham, NC: Duke University Press, 2000.

Ahern, Maureen. "The Cross and the Gourd: The Appropriation of Ritual Signs in the *Relaciones* of Alvar Núñez Cabeza de Vaca and Fray Marcos de Niza." In *Early Images of the Americas: Transfer and Invention*, ed. Jerry M. Williams and Robert E. Lewis, 215–44. Tucson: University of Arizona Press, 1993.

Alvarez, Ana María. "Sociedades agrícolas." In *Historia general de Sonora*, vol. 1, *Período prehistórico y prehispánico*, ed. Sergio Calderón Valdés, 225–60. Hermosillo, Mexico: Gobierno del Estado de Sonora, 1985.

Amith, Jonathan D. "The Möbius Strip: A Spatial History of a Colonial Society; Central Guerrero, Mexico, from the Sixteenth to Nineteenth Centuries." PhD diss., Yale University, 2001.

Anderson, Benedict. *Imagined Communities: Reflections on the Origin and Spread of Nationalism*. 2d ed. London: Verso, 1991.

Andrien, Kenneth J., and Lyman L. Johnson, eds. *The Political Economy of Spanish America in the Age of Revolution, 1750–1850*. Albuquerque: University of New Mexico Press, 1994.

Annino, Antonio. "El primer constitucionalismo mexicano, 1810–1830." In *Para una historia de América*, vol. 3, *Los nudos* (2), ed. Marcello Carmagnani, Alicia Hernández Chávez, and Ruggiero Romano, 140–89. Mexico City: Fondo de Cultura Económica, El Colegio de México, Fideicomiso Historia de las Américas, 1999.

——, ed. *America Latina: Dallo stato coloniale allo stato nazione*. Milan: Franco Angeli, 1987.

Arrien, Mario. *Cacería y pesca en dos comunidades chiquitanas de Lomerío*. Santa Cruz de la Sierra, Bolivia: Informe BOLFOR, 1997.

Arrien, Mario, Pedro Salvatierra, and Andrés Pessoa. "El kúyuri ó el origen de los insectos que pican." Posthumo Hans Roth, *Presencia en la cultura* 1.5 (2000).

Aston, T. H., and C. H. E. Philpin, eds. *The Brenner Debate: Agrarian Class Structure and Economic Development in Pre-industrial Europe*. Cambridge: Cambridge University Press, 1985.

Bahr, Donald. "Papago Ocean Songs and the Wi:gita." *Journal of the Southwest* 33.4 (1991): 539–56.

——. "Pima-Papago Christianity." *Journal of the Southwest* 30.2 (1988): 133–67.

Bakewell, Peter. *A History of Latin America: Empires and Sequels*. London: Blackwell, 1997.

Balick, Michael J., and Paul Alan Cox, *Plants, People, and Culture: The Science of Ethnobotany*. New York: Scientific American Library, 1996.

Ballantyne, Anthony. *Orientalism and Race: Aryanism in the British Empire*. Houndmills, Basingstoke, UK: Palgrave, 2002.

Balzá Alarcón, Roberto. *Tierra, territorio y territorialidad indígena: Un estudio antropológico sobre la evolución en las formas de ocupación del espacio del pueblo indígena chiquitano de la ex-reducción jesuita de San José*. Santa Cruz de la Sierra, Bolivia: Apoyo Para el Campesinado Indígena del Oriente Boliviano, 2001.

Barragán Romano, Rossana. *"¿Indios de arco y flecha?" Entre la historia y la arqueología de las poblaciones del norte de Chuquisaca (siglos XV–XVI)*. Sucre, Bolivia: Antropólogos del Surandino, Inter-American Foundation, 1994.

——. *Indios, mujeres y ciudadanos: Legislación y ejercicio de la ciudadanía en Bolivia (siglo XIX)*. La Paz: Fundación Diálogo, 1999.

Barth, Fredrik. *Ethnic Groups and Boundaries: The Social Organization of Culture Difference*. Boston: Little, Brown, 1969.

Basso, Keith H. *Wisdom Sits in Places: Language and Landscape among the Western Apache*. Albuquerque: University of New Mexico Press, 1996.

Batres, Leopoldo. "Visit to the Archaeological Remains of La Quemada, Zacatecas, Mexico." In *The North Mexican Frontier: Readings in Archaeology, Ethnohistory, and Ethnography*, ed. Basil C. Hedrick, J. Charles Kelley, and Carroll L. Riley, 1–20. Carbondale: Southern Illinois University Press, 1971.

Beals, Ralph L. *The Comparative Ethnology of Northern Mexico before 1750*. Berkeley: University of California Press, 1932.

Becker, Marc. "Comunas and Indigenous Protest in Cayambé, Ecuador." *Americas* 55.4 (1999): 531–59.

Beezley, William H., Cheryl E. Martin, and William E. French, eds. *Rituals of Rule, Rituals of Resistance: Public Celebrations and Popular Culture in Mexico*. Wilmington, DE: Scholarly Resources, 1994.

Bell, Catherine. *Ritual: Perspectives and Dimensions*. New York: Oxford University Press, 1997.

Bell, Stephen. *Campanha Gaúcha: A Brazilian Ranching System, 1850–1920*. Stanford, CA: Stanford University Press, 1998.

Bender, Gordon L., ed. *Reference Handbook on the Deserts of North America*. Westport, CT: Greenwood, 1982.

Benton, Lauren. *Law and Colonial Cultures: Legal Regimes in World History, 1400–1900*. Cambridge: Cambridge University Press, 2002.

Birk, Gudrun. *Plantas útiles en bosques y pampas chiquitanas: Un estudio etnobotánico con perspectiva de género*. Santa Cruz de la Sierra, Bolivia: Apoyo para el Campesinado Indígena del Oriente Boliviano, 1995.

Block, David. *Mission Culture on the Upper Amazon: Native Tradition, Jesuit Enterprise, and Secular Policy in Moxos, 1660–1880*. Lincoln: University of Nebraska Press, 1994.

Bolton, Herbert Eugene. *Rim of Christendom: A Biography of Eusebio Francisco Kino, Pacific Coast Pioneer*. New York: Macmillan, 1936.

Bonilla, Frank, Edwin Melendez, and Maria de Los Angeles Torres, eds. *Borderless Borders: U.S. Latinos, Latin Americans, and the Paradox of Interdependence*. Philadelphia: Temple University Press, 1998.

Bonnell, Victoria E., and Lynn Hunt, eds. *Beyond the Cultural Turn: New Directions in the Study of Society and Culture*. Berkeley: University of California Press, 1999.

Bourdieu, Pierre. *Outline of a Theory of Practice*. Trans. Richard Nice. Cambridge: Cambridge University Press, 1977.

Bowen, Thomas. *Unknown Island: Seri Indians, Europeans, and San Esteban Island in the Gulf of California*. Albuquerque: University of New Mexico Press, 2000.

Brading, David. *Haciendas and Ranchos in the Mexican Bajío: León, 1700–1860*. Cambridge: Cambridge University Press, 1978.

Brah, Avtar, and Annie E. Coombes, eds. *Hybridity and Its Discontents: Politics, Science, and Culture*. London: Routledge, 2000.

Braniff, Beatriz. *La frontera protohistórica Pima-Ópata en Sonora, México: Proposiciones arqueólogicas preliminares*. 3 vols. Mexico City: Instituto Nacional de Antropología e Historia, Colección Científica, 1992.

——, ed. *La Gran Chichimeca: El lugar de las rocas secas*. Mexico City: Consejo Nacional para la Cultura y las Artes, 2001.

Braudel, Fernand. *Civilization and Capitalism, 15th–18th Century*. 3 vols. Trans. Siân Reynolds. 1979. Berkeley: University of California Press, 1992.

Brooks, James F. *Captives and Cousins: Slavery, Kinship, and Community in the Southwest Borderlands*. Chapel Hill: University of North Carolina Press, 2002.

——. " 'This Evil Extends Especially . . . to the Feminine Sex': Negotiating Captivity in the New Mexico Borderlands." *Feminist Studies* 22.2 (1996): 279–309.

Brown, Jennifer S. H. *Strangers in Blood: Fur Trade Company Families in Indian Country*. Vancouver: University of British Columbia Press, 1980.

Burchell, Graham, Colin Gordon, and Peter Miller, eds. *The Foucault Effect: Studies in Governmentality; With Two Lectures by and an Interview with Michel Foucault*. Chicago: University of Chicago Press, 1991.

Burkhart, Louise. *The Slippery Earth: Nahua-Christian Moral Dialogue in Sixteenth-Century Mexico*. Tucson: University of Arizona Press, 1989.

Burton, Antoinette. *At the Heart of the Empire: Indians and the Colonial Encounter in Late-Victorian Britain*. Berkeley: University of California Press, 1998.

Buschiazzo, Mario J. "San Javier del Bac." *Anales del Instituto de Arte Americano e Investigaciones Estéticas* 6 (1953): 67–73.

Bye, Robert. "Hallucinogenic Plants of the Tarahumara." *Journal of Ethnopharmacology* 1 (1979): 23–48.

Calderón Valdés, Sergio, ed. *Historia general de Sonora*. Vol. 1, *Período prehistórico y prehispánico*. Hermosillo, Mexico: Gobierno del Estado de Sonora, 1985.

Canedo, Isabel. "Movimientos indígenas en el Oriente boliviano: Caracterización de las tensiones en la frontera del siglo XIX." In *Santa Cruz en el siglo XIX: Ponencias presentadas en el II ciclo de historia cruceña*, ed. Loreto Correa Vera, 59–73. Santa Cruz de la Sierra, Bolivia: Museo de Historia, Universidad Gabriel René Moreno, 1997.

Cañizares-Esguerra, Jorge. *How to Write the History of the New World: Histories, Epistemologies, and Identities in the Eighteenth-Century Atlantic World*. Stanford, CA: Stanford University Press, 2001.

Cárdenas de la Peña, Enrique, ed. *Temas médicos de la Nueva España*. Mexico City: Instituto Mexicano del Seguro Social, Instituto Cultural Domecq, 1992.

Cariño Olvera, Martha Micheline. *Historia de las relaciones hombre-naturaleza en Baja California Sur, 1500–1940*. Mexico City: Secretaría de Educación Pública, 2000.

Carmagnani, Marcello. *El regreso de los dioses: El proceso de reconstitución de la identidad étnica en Oaxaca, siglos XVII y XVIII*. Mexico City: Fondo de Cultura Económica, 1988.

Carmagnani, Marcello, and Ruggiero Romano. "Componentes sociales, siglos XIX y XX." In *Para una historia de América*, vol. 1, *Las estructuras*, ed. Carmagnani, Alicia Hernández Chávez, and Rugierro Romano, 363–83. Mexico City: Fondo de Cultura Económica, Colegio de México, Fideicomiso Historia de las Américas, 1999.

Carmagnani, Marcello, Alicia Hernández Chávez, and Rugierro Romano, eds. *Para una historia de América*. 2 vols. Mexico City: Fondo de Cultura Económica, Colegio de México, Fideicomiso Historia de las Américas, 1999.

Carney, Judith A. *Black Rice: The African Origins of Rice Cultiviation in the Americas*. Cambridge, MA: Harvard University Press, 2001.

Castetter, Edward F., and Willis H. Bell. *Pima and Papago Indian Agriculture*. Albuquerque: University of New Mexico Press, 1942.

Centurión, Teresa R. de, and Ivo J. Kraljevic, eds. *Las plantas útiles de Lomerío*. Santa Cruz de la Sierra, Bolivia: Proyecto de Manejo Forestal Sostenible, Herbario, CICOL, 1996.

Chakrabarty, Dipesh. *Provincializing Europe: Postcolonial Thought and Historical Difference*. Princeton, NJ: Princeton University Press, 2000.

Chambers, Iain, and Lidia Curti, eds. *The Post-colonial Question: Common Skies, Divided Horizons*. London: Routledge, 1996.

Chassen-Lopez, Francie. "Maderismo or Mixtec Empire? Class and Ethnicity in the Mexican Revolution, Costa Chica of Oaxaca, 1911." *Americas* 55.1 (1998): 91–128.

Clendinnen, Inga. *Ambivalent Conquests: Maya and Spaniard in Yucatan, 1517–1740*. Cambridge: Cambridge University Press, 1987.

Clifford, James. *Routes: Travel and Translation in the Late Twentieth Century*. Cambridge, MA: Harvard University Press, 1997.

Clifford, James, and George Marcus, eds. *Writing Culture: The Politics and Poetics of Ethnography; A School of American Research Advanced Seminar*. Berkeley: University of California Press, 1986.

Collier, George A. *Basta! Land and the Zapatista Rebellion in Chiapas*. With Elizabeth Lowery Quaratiello. Rev. ed. Oakland, CA: Food First, 1999.

Collier, George A., Renato I. Rosaldo, and John D. Wirth, eds. *The Inca and Aztec States, 1400–1800: Anthropology and History*. New York: Academic Press, 1982.

Collier, Jane Fishburne. *Marriage and Inequality in Classless Societies*. Stanford, CA: Stanford University Press, 1988.

Comaroff, Jean, and John Comaroff. *Of Revelation and Revolution*. Vol. 1, *Christianity, Colonialism, and Consciousness in South Africa*. Chicago: University of Chicago Press, 1991.

Cope, Douglas R. *The Limits of Racial Domination: Plebian Society in Colonial Mexico City, 1660–1720*. Madison: University of Wisconsin Press, 1994.

Correa Vera, Loreto, ed. *Santa Cruz en el siglo XIX: Ponencias presentadas en el II ciclo de historia cruceña*. Santa Cruz de la Sierra, Bolivia: Museo de Historia, Universidad Gabriel René Moreno, 1997.

Cosgrove, Denis E. *Social Formation and Symbolic Landscape*. London: Croom Helm, 1984.

Costa, Maria de Fátima. *História de um país inexistente: O pantanal entre os séculos XVI e XVIII*. São Paulo: Livreria Kosmos Editora, 1999.

Costeloe, Michael P. *Response to Revolution: Imperial Spain and the Spanish American Revolutions, 1810–1840*. Cambridge: Cambridge University Press, 1986.

Cramaussel, Chantell. "De cómo los españoles clasificaban a los indios: Naciones y encomiendas en la Nueva Vizcaya Central." In *Nómadas y sedentarios en el norte de México: Homenaje a Beatriz Braniff*, ed. Marie-Areti Hers et al., 218–84. Mexico City: Universidad Nacional Autónoma de México, 2000.

Cronon, William. *Changes in the Land: Indians, Colonists, and the Ecology of New England*. 1983. New York: Hill and Wang, 1997.

——. "A Place for Stories: Nature, History, and Narrative." *Journal of American History*, 78.4 (1992): 1347–76.

——, ed. *Uncommon Ground: Toward Reinventing Nature*. New York: Norton, 1995.

Cronon, William, George Miles, and Jay Gitlin, eds. *Under an Open Sky: Rethinking America's Western Past*. New York: Norton, 1992.

Crosby, Alfred W. *The Columbian Exchange: Biological and Cultural Consequences of 1492*. Westport, CT: Greenwood, 1972.

——. *Ecological Imperialism: The Biological Expansion of Europe, 900–1900*. Cambridge: Cambridge University Press, 1986.

Crosswhite, Frank S. "Desert Plants, Habitat, and Agriculture in Relation to the Major Pattern of Cultural Differentiation in the O'odham People of the Sonoran Desert." *Desert Plants* 3 (1981): 47–76.

Cuello, José. "The Persistence of Indian Slavery and Encomienda in the Northeast of Colonial Mexico, 1577–1723." *Journal of Social History* 21.4 (1988): 683–700.

Custodio, Bolcato, and Juan Carlos Ruiz, eds. *Las misiones de ayer para los días de mañana*. Santa Cruz de la Sierra, Bolivia: Colegio de Arquitectos de Santa Cruz, 1993.

Cutter, Charles R. "The Legal System as a Touchstone of Identity in Colonial New Mexico." In *The Collective and the Public in Latin America: Cultural Identities and Political Order*, ed. Luis Roniger and Tamar Herzog, 57–70. Brighton, UK: Sussex Academic Press, 2000.

——. *The Protector de Indios in Colonial New Mexico, 1659–1821*. Albuquerque: University of New Mexico Press, 1986.

Daniels, Stephen, and Denis Cosgrove. "Introduction: Iconography and Landscape." In *The Iconography of Landscape: Essays on the Symbolic Representation, Design, and Use of Past Environments*, ed. Daniels and Cosgrove, 1–10. Cambridge: Cambridge University Press, 1988.

Davis, Natalie Zemon. "Iroquois Women, European Women." In *Women, "Race," and Writing in the Early Modern Period*, ed. Margo Hendricks and Patricia Parker, 243–58. London: Routledge, 1994.

——. *Women on the Margins: Three Seventeenth-Century Lives*. Cambridge, MA: Harvard University Press, 1995.

Dean, Warren. *With Broadax and Firebrand: The Destruction of the Brazilian Atlantic Forest*. Berkeley: University of California Press, 1995.

Deeds, Susan. *Defiance and Difference in Mexico's Colonial North: Indians under Spanish Rule in Nueva Vizcaya*. Austin: University of Texas Press, 2003.

——. "Double Jeopardy: Indian Women in Jesuit Missions of Nueva Vizcaya." In *Indian Women of Early Mexico*, ed. Susan Schroeder, Stephanie Wood, and Robert Haskett, 255–72. Norman: University of Oklahoma Press, 1997.

——. "Rural Work in Nueva Vizcaya: Forms of Labor Coercion on the Periphery." *Hispanic American Historical Review* 69.3 (1989): 425–49.

Denevan, William M. "Stone versus Metal Axes: The Ambiguity of Shifting Cultivation in Prehistoric Amazonia." *Journal of the Steward Anthropological Society* 20.1–2 (1992): 153–65.

Diamond, Jared. *Guns, Germs, and Steel: The Fates of Human Societies*. New York: Norton, 1997.

Díez Astete, Alvaro, and David Murillo. *Pueblos indígenas de tierras bajas: Características principales*. La Paz: Ministerio de Desarrollo Sostenible y Planificación, Viceministerio de Asuntos Indígenas y Pueblos Originarios, Programa Indígena-Programa de las Naciones Unidas para el Desarrollo, 1998.

Di Peso, Charles C. *Casas Grandes: A Fallen Trading Center of the Gran Chichimeca*. Ed. Gloria J. Fenner. Dragoon, AZ: Amerind Foundation, 1974.

Dobb, Maurice. *Studies in the Development of Capitalism*. London: Routledge and K. Paul, 1946.

Dobyns, Henry F., et al. "What Were Nixoras?" *Southwestern Journal of Anthropology* 16.2 (1960): 230–68.

Doolittle, William E. *Cultivated Landscapes of Native North America*. Oxford: Oxford University Press, 2000.

——. *Pre-Hispanic Occupance in the Valley of Sonora, Mexico: Archaeological Confirmation of Early Spanish Reports*. Tucson: University of Arizona Press, 1988.

Dore, Elizabeth, and Maxine Molyneux, eds. *Hidden Histories of Gender and the State in Latin America*. Durham, NC: Duke University Press, 2000.

Doyel, David E. "The Transition to History in Northern Pimería Alta." In *Columbian Consequences*, vol. 1, *Archaeological and Historical Perspectives on the Spanish Borderlands West*, ed. David H. Thomas, 139–58. Washington, DC: Smithsonian Institution Press, 1989.

Ducey, Michael T. "Indios liberales y liberales indigenistas: Ideología y poder en los municipio rurales de Veracruz, 1821–1890." In *El siglo XIX en las Huastecas*, ed. Antonio Escobar Ohmstede and Luz Carregha Lamadrid, 111–36. Mexico City: Centro de Investigaciones y Estudios Superiores en Antropología Social, 2002.

Escobar Ohmstede, Antonio. "¿Qué sucedió con la tierra en las huastecas decimonónicas?" In *El siglo XIX en las Huastecas*, ed. Antonio Escobar Ohmstede and Luz Carregha Lamadrid, 137–66. Mexico City: Centro de Investigaciones y Estudios Superiores en Antropología Social, 2002.

Escobar Ohmstede, Antonio, and Romana Falcón. "Los ejes de la disputa: Introducción." In *Los ejes de la disputa: Movimientos sociales y actores collectivos en América Latina, siglo XIX*, ed. Antonio Escobar Ohmstede and Falcón, 9–24. Madrid: Iberoamericana, 2002.

Escobar Ohmstede, Antonio, Romana Falcón, and Raymond Buve. "Presentación: Las sociedades locales frente a las tendencias modernizadoras de los estados nacionales del siglo XIX latinoamericano." In *Pueblos, comunidades y municipios frente a los proyectos modernizadores en América Latina, siglo XIX*, ed. Antonio Escobar

Ohmstede, Falcón, and Buve, 1–8. Amsterdam: Centrum voor Studie en Documentatie van Latijns-Amerika, 2002.

Escobar Ohmstede, Antonio, and Luz Carregha Lamadrid, eds. *El siglo xix en las Huastecas*. Mexico City: Centro de Investigaciones y Estudios Superiores en Antropología Social, 2002.

Escobar Ohmstede, Antonio, and Patricia Lagos Preisser, eds. *Indio, nación y comunidad en el México del siglo xix*. Mexico City: Centro de Investigaciones y Estudios Superiores en Antropología Social and Centro Francés de Estudios Mexicanos y Centroamericanos, 1993.

Escobar Ohmstede, Antonio, and Romana Falcón, eds. *Los ejes de la disputa: Movimientos sociales y actores collectivos en América Latina, siglo xix*. Madrid: Iberoamericana, 2002.

Escobar Ohmstede, Antonio, Romana Falcón, and Raymond Buve, eds. *Pueblos, comunidades y municipios frente a los proyectos modernizadores en América Latina, siglo xix*. Amsterdam: Centrum voor Studie en Documentatie van Latijns-Amerika, 2002.

Escobar, Arturo. "Constructing Nature: Elements for a Poststructural Political Ecology." In *Liberation Ecologies: Environment, Development, Social Movements*, ed. Richard Peet and Michael Watts, 46–68. London: Routledge, 1996.

Fairhead, James, and Melissa Leach. *Misreading the African Landscape: Society and Ecology in a Forest-Savanna Mosaic*. Cambridge: Cambridge University Press, 1996.

Falkinger, Sieglinde, and Eckart Kühne. " 'Que para aprenderla no basta muchos años': La lengua chiquitana." In *Las misiones jesuíticas de Bolivia: Martin Schmid, 1694–1772; Misionero, músico y arquitecto entre los Chiquitanos*, ed. Kühne, 31–38. Santa Cruz de la Sierra, Bolivia: Asociación Suiza por la Cultura "Pro Helvetia," 1996.

Falkinger, Sieglinde. "Porque somos cristianos no somos salvajes." In *Mujeres, negros y niños en la música y sociedad colonial iberoamericana: iv reunión científica*, ed. Victor Rondón, 144–59. Santa Cruz de la Sierra, Bolivia: Asociación Pro-Arte y Cultura (APAC), 2002.

Farriss, Nancy. *Maya Society under Colonial Rule: The Collective Enterprise of Survival*. Princeton, NJ: Princeton University Press, 1984.

Feierman, Steven. "Colonizers, Scholars, and the Creation of Invisible Histories." In *Beyond the Cultural Turn: New Directions in the Study of Society and Culture*, ed. Victoria E. Bonnell and Lynn Hunt, 182–216. Berkeley: University of California Press, 1999.

Felger, Richard S., and Mary Beth Moser. *People of the Desert and Sea: Ethnobotany of the Seri Indians*. Tucson: University of Arizona Press, 1985.

Fifer, J. Valerie. *Bolivia: Land, Location, and Politics since 1825*. Cambridge: Cambridge University Press, 1972.

Figueroa, Alejandro. *Los que hablan fuerte, desarrollo de la sociedad Yaqui*. Hermosillo,

Mexico: Instituto Nacional de Antropología e Historia, Centro Regional del Noroeste, 1985.

Fischermann, Bernardo. "Campesino e indígena: La cultura chiquitana actual." In *Las misiones jesuíticas de Bolivia: Martin Schmid, 1694–1772; Misionero, músico y arquiteco entre los chiquitanos*, ed Eckart Kühne, 103–10. Santa Cruz de la Sierra, Bolivia: Asociación Suiza por la Cultura "Pro Velvetia," 1996.

——. "Contexto social y cultural." In *Las plantas útiles de Lomerío*, ed. Teresa R. Centurión and Ivo J. Kraljevic, 25–34. Santa Cruz de la Sierra, Bolivia: Proyecto de Manejo Forestal Sostenible, Herbario, CICOL, 1996.

——. "Historia chiquitana en la segunda mitad del siglo XIX." In *Santa Cruz en el Siglo XIX: Ponencias presentadas en el II ciclo de historia cruceña*, ed. Loreto Correa Vera, 75–86. Santa Cruz de la Sierra, Bolivia: Museo de Historia, Universidad Gabriel René Moreno, 1997.

——. "Un pueblo indígena interpreta su mundo: La cosmovisión de los Ayoreóde del Oriente Boliviano." *Fundación Cultural* 4.13 (2000): 14–37.

——. "Los pueblos indígenas." In *Santa Cruz: Tiempo y espacio*, ed. Alcides Parejas Moreno and Fischermann, 25–38. Santa Cruz de la Sierra, Bolivia: Cooperativa Rural de Electrificación, 2000.

——. "Los Rojos: Artesanos y sacerdotes cruceños en la Chiquitanía." In *Memoria: III Festival Internacional de Música Renacentista y Barroca Americana Misiones de Chiquitos 2000; Del 26 de julio al 13 de agosto de 2000*, 141–50. Santa Cruz de la Sierra, Bolivia: Asociación Pro Arte y Cultura, 2000.

——. "Viviendo con los paí." In *Las misiones jesuíticas de Bolivia: Martin Schmid, 1694–1772; Misionero, músico y arquitecto entre los Chiquitanos*, ed. Eckart Kühne, 47–54. Santa Cruz de la Sierra, Bolivia: Asociación Suiza por la Cultura "Pro Helvetia," 1996.

Flint, Richard. *Great Cruelties Have Been Reported: The 1544 Investigation of the Coronado Expedition.* Dallas: Southern Methodist University Press, 2002.

Flint, Richard, and Shirley Cushing Flint, eds., *Documents of the Coronado Expedition, 1539–1542.* Dallas: Southern Methodist University Press, 2005.

Flores Caballero, Romeo R. *Counterrevolution: The Role of the Spaniards in the Independence of Mexico, 1804–38.* Trans. Jaime E. Rodríguez O. 1969. Lincoln: University of Nebraska Press, 1974.

Florescano, Enrique. *Precios del maíz y crisis agrícolas en México, 1708–1860: Ensayo sobre el movimiento de los precios y sus consecuencias económicas y sociales.* Mexico City: El Colegio de México, 1969.

Fontana, Bernard L. "Edward Ronstadt: A Life Well-Lived." *Southwestern Mission Research Center Newsletter* 35.128–29 (2001): 1–16.

——. "History of the Papago." In *Handbook of North American Indians*, vol. 10, *Southwest*, ed. Alfonso Ortiz, 137–48. Washington, DC: Smithsonian Institution Press, 1983.

——. *Of Earth and Little Rain: The Papago Indians.* Flagstaff, AZ: Northland, 1981.

——. "Pima and Papago: Introduction." In *Handbook of North American Indians*, vol. 10, *Southwest*, ed. Alfonso Ortiz, 125–36. Washington, DC: Smithsonian Institution Press, 1983.

——. "The O'odham." In *The Pimería Alta, Missions and More*, ed. James E. Officer, Mardith Schuetz-Miller, and Bernard L. Fontana, 19–28. Tucson: Southwestern Mission Research Center, 1996.

Fontana, Bernard L., and Daniel S. Matson, eds. *Before Rebellion: Letters and Reports of Jacobo Sedelmayr, S.J.* Tucson: Arizona Historical Society, 1996.

Foucault, Michel. *Discipline and Punish: The Birth of the Prison*. Trans. Alan Sheridan. 1975. New York: Vintage, 1979.

——. *Ethics: Subjectivity and Truth*. Ed. Paul Rabinow. Trans. Robert Hurley. New York: New Press, 1998.

——. *The History of Sexuality*. Trans. Robert Hurley. 1976. New York: Pantheon, 1978.

——. *Power/Knowledge: Selected Interviews and Other Writings, 1972–77*. Ed. and trans. Colin Gordon. New York: Pantheon, 1980.

Frank, Ross. *From Settler to Citizen: New Mexican Economic Development and the Creation of Vecino Society, 1750–1820*. Berkeley: University of California Press, 2000.

Frank, Zephyr Lake. "The Brazilian Far West: Frontier Development in Mato Grosso, 1870–1937." PhD diss., University of Illinois–Urbana-Champaign, 1999.

Freyer, Bärbel. *Los Chiquitanos: Descripción de un pueblo de las tierras bajas orientales de Bolivia según fuentes jesuíticas del siglo XVIII*. Santa Cruz de la Sierra, Bolivia: ATLANTIDA, 2000.

Gade, Daniel W. *Nature and Culture in the Andes*. Madison: University of Wisconsin Press, 1999.

Galloway, Patricia, ed. *The Hernando de Soto Expedition: History, Historiography, and "Discovery" in the Southeast*. Lincoln: University of Nebraska Press, 1997.

Ganson, Barbara. *The Guaraní under Spanish Rule in the Río de la Plata*. Stanford, CA: Stanford University Press, 2003.

Garate, Don. "1751 Smallpox Epidemic." Mission 2000 Database, Tumacácori National Monument Database, National Park Service, *http://data2.itc.nps.gov/tuma/Results.cfm*. Accessed 17 August 2002.

Garavaglia, Juan Carlos. "The Crises and Transformations of Invaded Societies: The La Plata Basin (1535–1650)." In *The Cambridge History of the Native Peoples of the Americas*, vol. 3, *South America*, ed. Frank Salomon and Stuart B. Schwartz, 2:1–58. Cambridge: Cambridge University Press, 1999.

García Acosta, Virginia. *Los precios del trigo en la historia colonial de México*. Mexico City: Centro de Investigaciones y Estudios Superiores en Antropología Social, 1988.

García Canclini, Néstor. *Hybrid Cultures: Strategies for Entering and Leaving Modernity*. Trans. Christopher L. Chiaparri and Silvia L. López. 1990. Minneapolis: University of Minnesota Press, 1995.

García Jordán, Pilar. *Cruz y arado, fusiles y discursos: La constitución de los orientes en el Perú y Bolivia, 1820–1940*. Lima: Instituto Francés de Estudios Andinos, Instituto de Estudios Peruanos, 2001.

García Martínez, Bernardo. *Los pueblos de la sierra: El poder y el espacio entre los indios del norte de Puebla hasta 1700*. Mexico City: El Colegio de México, 1987.

García Martínez, Bernardo, and Alba González Jácome, eds. *Estudios sobre historia y ambiente en América*. Vol. 1. Mexico City: El Colegio de México, Instituto Panamericano de Geografía e Historia, 1999.

García Recio, José María. *Análisis de una sociedad de frontera: Santa Cruz de la Sierra en los siglos XVI y XVII*. Seville, Spain: Diputación Provincial de Sevilla, 1988.

Garner, Richard L. *Economic Growth and Change in Bourbon Mexico*. With Spiro E. Stefanou. Gainesville: University Press of Florida, 1993.

Gaventa, John. *Power and Powerlessness: Quiescence and Rebellion in an Appalachian Valley*. Urbana: University of Illinois Press, 1980.

Geertz, Clifford. *Agricultural Involution: The Process of Ecological Change in Indonesia*. Berkeley: Published for the Association of Asian Studies by University of California Press, 1963.

——. *The Interpretation of Cultures*. New York: Basic, 1973.

Genovese, Eugene. *Roll, Jordan, Roll: The World the Slaves Made*. New York: Pantheon, 1974.

Gerhard, Peter. *The North Frontier of New Spain*. Rev. ed. Norman: University of Oklahoma Press, 1993.

Gibson, Charles. *The Aztecs under Spanish Rule: A History of the Indians of the Valley of Mexico, 1519–1810*. Stanford, CA: Stanford University Press, 1964.

——. *Tlaxcala en el siglo XVI*. Trans. Agustín Bárcena. Mexico City: Fondo de Cultura Económica, 1991.

——. *Tlaxcala in the Sixteenth Century*. New Haven, CT: Yale University Press, 1952.

Gisbert, Teresa. "El arte misional, el barroco, y las tierras altas." In *Memoria: III Festival Internacional de Música Renacentista y Barroca Americana Misiones de Chiquitos 2000; Del 26 de julio al 13 de agosto de 2000*, 79–90. Santa Cruz de la Sierra, Bolivia: Asociación Pro Arte y Cultura, 2000.

González Navarro, Moisés. "Instituciones indígenas del México independiente." In *La política indigenista en México: Métodos y resultados*, ed. Gonzalo Aguirre Beltrán and Ricardo Pozas Arciniegas, 217–313. 1954. Mexico City: Instituto Nacional Indigenista, 1973.

González Ruiz, Luis, ed. *Etnología y misión en la Pimería Alta, 1715–1740: Informes y relaciones misioneras de Luis Xavier Velarde, Giuseppe Maria Genovese, Daniel Januske, José Agustín de Campos y Cristóbal de Cañas*. Mexico City: Universidad Nacional Autónoma de México, Instituto de Investigaciones Históricas, 1977.

Gould, Jeffrey L. *To Die in This Way: Nicaraguan Indians and the Myth of Mestizaje, 1880–1965*. Durham, NC: Duke University Press, 1998.

Gow, Peter. "Land, People, and Paper in Western Amazonia." In *The Anthropology of Landscape: Perspectives on Place and Space*, ed. Eric Hirsch and Michael O'Hanlon, 43–62. Oxford: Clarendon, 1995.

Gramsci, Antonio. *Prison Notebooks*. Ed. and trans. Joseph A. Buttigieg. New York: Columbia University Press, 1991.

Grandin, Greg. *The Blood of Guatemala: A History of Race and Nation*. Durham, NC: Duke University Press, 2000.

Greever, Janet Goff. *José Ballivián y el Oriente boliviano*. Trans. José Luis Roca. La Paz: Empresa Editorial Siglo, 1987.

Griffen, William B. *The Apaches at War and Peace: The Janos Presidios, 1750–1858*. Albuquerque: University of New Mexico Press, 1988.

——. *Culture Change and Shifting Populations in Central Northern Mexico*. Tucson: University of Arizona Press, 1969.

——. *Indian Assimilation in the Franciscan Area of Nueva Vizcaya*. Tucson: University of Arizona Press, 1979.

Griffith, James. *Beliefs and Holy Places: A Spiritual Geography of the Pimería Alta*. Tucson: University of Arizona Press, 1992.

Griffiths, Nicholas. *The Cross and the Serpent: Religious Repression and Resurgence in Colonial Peru*. Norman: University of Oklahoma Press, 1996.

Griffiths, Nicholas, and Fernando Cervantes, eds. *Spiritual Encounters: Interactions between Christianity and Native Religion in Colonial America*. Lincoln: University of Nebraska Press, 1999.

Grove, Richard. *Green Imperialism: Colonial Expansion, Tropical Island Edens, and the Origins of Environmentalism, 1600–1860*. Cambridge: Cambridge University Press, 1995.

Gruzinski, Serge. *The Conquest of Mexico: The Incorporation of Indian Societies into the Western World, 16th–18th Centuries*. Trans. Eileen Corrigan. 1989. Cambridge: Polity, 1993.

——. *La guerra de las imagines: De Cristóbal Colón a "Blade Runner" (1492–2019)*. Trans. Juan José Utrilla. Mexico City: Fondo de Cultura Económica, 1994.

Guardino, Peter. *Peasants, Politics, and the Formation of Mexico's National State: Guerrero, 1800–1857*. Stanford, CA: Stanford University Press, 1996.

Guerra, François-Xavier. *Modernidad e independencias: Ensayos sobre las revoluciones hispánicas*. Madrid: Editorial Mapfre, 1992.

Guerrero, Andrés. "El proceso de identificación: Sentido común ciudadano, ventriloquia y transescritura." In *Pueblos, comunidades y municipios frente a los proyectos modernizadores en América Latina, siglo XIX*, ed. Antonio Escobar Ohmstede, Romana Falcón, and Raymond Buve, 29–64. Amsterdam: Centrum voor Studie en Documentatie van Latijns-Amerika, 2002.

Guha, Ramachandra. *The Unquiet Woods: Ecological Change and Peasant Resistance in the Himalaya*. Berkeley: University of California Press, 1989.

——, ed. *Social Ecology*. Dehli: Oxford University Press, 1994.

Gumerman, George J., ed. *Exploring the Hohokam: Prehistoric Desert Peoples of the American Southwest*. Dragoon, AZ: Amerind Foundation, 1991.

Gutiérrez, Ramón A. *When Jesus Came, the Corn Mothers Went Away: Marriage, Sexuality, and Power in New Mexico, 1500–1846*. Stanford, CA: Stanford University Press, 1991.

Gutiérrez, Ramón A., and Richard J. Orsi, eds. *Contested Eden: California before the Gold Rush*. Berkeley: University of California Press, 1998.

Gutiérrez da Costa, Ramón, and Rodrigo Gutiérrez Viñuales. "Territorio, urbanismo y arquitectura en Moxos y Chiquitos." In *Las misiones jesuíticas de Chiquitos*, ed. Pedro Querejazu and Plácido Molina Barbery, 303–81. La Paz: Fundación BHN, 1995.

Guy, Donna, and Thomas E. Sheridan, eds. *Contested Ground: Comparative Frontiers on the Northern and Southern Edges of the Spanish Empire*. Tucson: University of Arizona Press, 1998.

Haas, Lisbeth. *Conquests and Historical Identities in California, 1769–1936*. Berkeley: University of California Press, 1995.

Haefeli, Evan. "American Historical Review Forum Essay: Responses; A Note on the Use of North American Borderlands." *American Historical Review* 104.4 (1999): 1222–25.

Hall, Thomas D. "The Río de la Plata and the Greater Southwest: A View from World System Theory." In *Contested Ground: Comparative Frontiers on the Northern and Southern Edges of the Spanish Empire*, ed. Donna Guy and Thomas E. Sheridan, 150–66. Tucson: University of Arizona Press, 1998.

——. *Social Change in the Southwest, 1350–1880*. Lawrence: University Press of Kansas, 1989.

Hämäläinen, Pekka. "The Comanche Empire: A Study of Indigenous Power." Ph.D. diss., University of Helsinki, 2001.

——. "The Western Comanche Trade Center: Rethinking the Plains Indian Trade System." *Western Historical Quarterly* 29.4 (1998): 485–513.

Hardt, Michael, and Antonio Negri. *Empire*. Cambridge, MA: Harvard University Press, 2000.

Harris, Olivia, Brooke Larson, and Enrique Tandeter, eds. *La participación indígena en los mercados surandinos: Estrategias y reproducción social, siglos XVI a XX*. La Paz: Centro de Estudios de la Realidad Económico y Social, 1987.

Harvey, Herbert R., ed. *Land and Politics in the Valley of Mexico: A Two Thousand–Year Perspective*. Albuquerque: University of New Mexico Press, 1991.

Haskett, Robert. *Indigenous Rulers: An Ethnohistory of Town Government in Colonial Cuernavaca*. Albuquerque: University of New Mexico Press, 1991.

Haury, Emil W. *The Hohokam, Desert Farmers and Craftsmen: Excavations at Snaketown, 1964–1965*. Tucson: University of Arizona Press, 1976.

Hausberger, Bernd. "La violencia en la conquista espiritual: Las misiones jesuitas de Sonora." *Jahrbuch für Geschichte von Staat, Wirtschaft und Gesellschaft Lateinamerikas* 30 (1991): 27–54.

Hayden, Julian D. *The Sierra Pinacate*. Tucson: University of Arizona Press, 1998.

Hedrick, Basil C., J. Charles Kelley, and Carroll L. Riley, eds. *The North Mexican Frontier: Readings in Archaeology, Ethnohistory, and Ethnography*. Carbondale: Southern Illinois University Press, 1971.

Hefner, Robert W., ed. *Conversion to Christianity: Historical and Anthropological Perspectives on a Great Transformation*. Berkeley: University of California Press, 1989.

Hendricks, Margo, and Patricia Parker, eds. *Women, "Race," and Writing in the Early Modern Period*. London: Routledge, 1994.

Hernández Chávez, Alicia. *Anenecuilco: Memoria y vida de un pueblo*. 2d ed. Mexico City: Colegio de México, 1993.

Hers, Marie-Areti, et al., eds. *Nómadas y sedentarios en el norte de México: Homenaje a Beatriz Braniff*. Mexico City: Universidad Nacional Autónoma de México, 2000.

Hickerson, Nancy Parrott. *The Jumanos: Hunters and Traders of the South Plains*. Austin: University of Texas Press, 1994.

Hill, Jonathan D. *Rethinking History and Myth: Indigenous South American Perspectives on the Past*. Urbana: University of Illinois Press, 1988.

——, ed. *History, Power, and Identity: Ethnogenesis in the Americas, 1492–1992*. Iowa City: University of Iowa Press, 1996.

Hilton, Rodney, and Paul Sweezy, eds. *The Transition from Feudalism to Capitalism*. London: Verso, 1978.

Hirsch, Eric. "Landscape: Between Place and Space." In *The Anthropology of Landscape: Perspectives on Place and Space*, ed. Hirsch and Michael O'Hanlon, 1–30. Oxford: Clarendon Press, 1995.

Hirsch, Eric, and Michael O'Hanlon, eds. *The Anthropology of Landscape: Perspectives on Place and Space*. Oxford: Clarendon Press, 1995.

Hobsbawm, E. J. *The Age of Capital, 1848–1875*. London: Weidenfeld and Nicolson, 1975.

Hoffmann, Werner. *Las misiones jesuíticas entre los Chiquitanos*. Buenos Aires: Fundación para la Educación, la Ciencia y la Cultura, 1979.

Hu-Dehart, Evelyn. *Missionaries, Miners, and Indians: Spanish Contact with the Yaqui Nation of Northwestern New Spain*. Tucson: University of Arizona Press, 1981.

——. *Yaqui Resistance and Survival: The Struggle for Land and Autonomy, 1821–1910*. Madison: University of Wisconsin Press, 1984.

Hulme, Peter. "Including America." *Ariel* 26.1 (1995): 117–23.

Illari, Bernardo, ed. *Música barroca del Chiquitos Jesuítico: Trabajos leídos en el encuentro de musicólogos; I Festival Internacional de Música Renacentista y Barroca*

Americana "Misiones de Chiquitos"; Santa Cruz de la Sierra, abril de 1996. Santa Cruz de la Sierra, Bolivia: Agencia Española de Cooperación Internacional, 1998.

Irurozqui, Marta. "El bautismo de la violencia: Indígenas patriotas en la revolución de 1870 en Bolivia." In *Identidad, ciudadanía y participación popular desde la colonia al siglo XX*, ed. Josefa Salmón and Guillermo Delgado, 115–52. La Paz: Plural, 2003.

——. "La vecindad y sus promesas: De vecino a ciudadano, Bolivia, 1810–1930." In *Anuario ABNB 2000*, 203–27. Sucre: Archivo y Biblioteca Nacionales de Bolivia, 2000.

Ives, Ronald L. "The Pinacate Region, Sonora, Mexico." *Occasional Papers of the California Academy of Sciences* 47 (1964): 1–43.

Jackson, Robert H. *Indian Population Decline: The Missions of Northwestern New Spain*. Albuquerque: University of New Mexico Press, 1994.

Jacobsen, Nils. *Mirages of Transition: The Peruvian Altiplano, 1780–1930*. Berkeley: University of California Press, 1993.

——. "Montoneras, la comuna de Chalaco y la revolución de Piérola: La sierra piurana entre el clientelismo y la sociedad civil, 1868–1895." With Alejandro Díez Hurtado. In *Los ejes de la disputa: Movimiento sociales y actores colectivos en América Latina, siglo XIX*, ed. Antonio Escobar Ohmstede and Ramona Falcón, 57–132. Madrid: Iberoamericana, 2002.

Jardim, Antony, Timothy J. Killeen, and Alfredo Fuentes. *Guía de los árboles y arbustos del bosque seco chiquitano, Bolivia*. Ed. Damián Rumiz. Santa Cruz de la Sierra, Bolivia: Museo de Historia Natural, Fundación Amigos de la Naturaliza Noel Kempff (FAN-Bolivia), 2003.

John, Elizabeth A. H. *Storms Brewed in Other Men's Worlds: The Confrontation of Indians, Spanish, and French in the Southwest, 1540–1795*. College Station: Texas A&M University Press, 1975.

Johnson, Jay K. "From Chiefdom to Tribe in Northeast Mississippi: The Soto Expedition as a Window on a Culture in Transition." In *The Hernando de Soto Expedition: History, Historiography, and "Discovery" in the Southeast*, ed. Patricia Galloway, 295–312. Lincoln: University of Nebraska Press, 1997.

Johnson, Lyman L., and Enrique Tandeter, eds. *Essays on the Price History of Eighteenth-Century Latin America*. Albuquerque: University of New Mexico Press, 1990.

Jones, Kristine L. "Comparative Raiding Economies: North and South." In *Contested Ground: Comparative Frontiers on the Northern and Southern Edges of the Spanish Empire*, ed. Donna Guy and Thomas E. Sheridan, 97–114. Tucson: University of Arizona Press, 1998.

Joseph, Gilbert M., and Daniel Nugent, eds. *Everyday Forms of State Formation: Revolution and the Negotiation of Rule in Modern Mexico*. Durham, NC: Duke University Press, 1994.

Joseph, Gilbert M., and Susan Deans-Smith, eds. "Mexico's New Cultural History: *¿Una Lucha Libre?*" Special issue, *Hispanic American Historical Review* 79.2 (1999).

Kelley, J. Charles. "The C. de Berghes Map of 1833." In *The North Mexican Frontier: Readings in Archaeology, Ethnohistory, and Ethnography*, ed. Basil C. Hedrick, J. Charles Kelley, and Carroll L. Riley, xiv–xvi. Carbondale: Southern Illinois University Press, 1971.

Kessell, John L. *Friars, Soldiers, and Reformers: Hispanic Arizona and the Sonora Mission Frontier, 1767–1856*. Tucson: University of Arizona Press, 1976.

Killeen, Timothy J., Emilia García E., and Stephan G. Beck. *Guía de árboles de Bolivia*. La Paz: Herbario Nacional de Bolivia, 1993.

Kino, Eusebio Francisco. *Las misiones de Sonora y Arizona: Comprendiendo la crónica titulada "Favores celestiales" y la "Relación diaria de la entrada al noroeste."* Ed. Francisco Fernández del Castillo, with notes by Dr. Emilio Bose. Mexico City: Editorial Porrúa, S. A., 1989.

Klein, Herbert. *African Slavery in Latin America and the Caribbean*. New York: Oxford University Press, 1986.

Klor de Alva, J. Jorge. "The Postcolonization of the (Latin) American Experience: A Reconsideration of 'Colonialism,' 'Postcolonialism,' and 'Mestizaje.' " In *After Colonialism: Imperial Histories and Postcolonial Displacements*, ed. Gyan Prakash, 241–75. Princeton, NJ: Princeton University Press, 1995.

Kolaz, Thomas M. "Tohono O'odham *Fariseos* at the Village of Kawari'k." *Journal of the Southwest* 39.1 (1997): 59–77.

Kozak, David L., and David I. López, *Devil Sickness and Devil Songs: Tohono O'odham Poetics*. Washington, DC: Smithsonian Institution Press, 1999.

Kraniauskas, John. "Hybridity in a Transnational Frame: Latin-Americanist and Postcolonial Perspectives on Cultural Studies." In *Hybridity and Its Discontents: Politics, Science, and Culture*, ed. Avtar Brah and Annie E. Coombes, 235–56. London: Routledge, 2000.

Kühne, Eckart. "Semana Santa y Fiesta Patronal en San José de Chiquitos: Topografía de ritos, con notas explicatorias del esquema urbanístico." In *Las misiones jesuíticas de Chiquitos*, ed. Pedro Querejazu and Plácido Molina Barbery, 557–64. La Paz: Fundación BHN, 1995.

Kula, Witold. *Problemas y métodos de la historia económica*. 1963. Barcelona: Ediciones Península, 1973.

Kupperman, Karen Ordall, ed. *America in European Consciousness, 1493–1750*. Chapel Hill: University of North Carolina Press, 1994.

Kuznesof, Elizabeth Anne. "Ethnic and Gender Influences on 'Spanish' Creole Society in Colonial Spanish America." *Colonial Latin American Review* 4.1 (1995): 153–76.

Lafaye, Jacques. *Quetzalcóatl y Guadalupe: La formación de la consciencia nacional en México*. Trans. Ida Vitale y Fulgencio López Vidarte. 1974. Mexico City: Fondo de Cultura Económica, 1977.

Lahmeyer Lobo, Eulalia Maria. "Caminho de Chiquitos às Missões Guaranis de 1690–1718." *Revista de História* 10.39 (1959): 67–79.

——. "A importância estratégica e económica da Provincia de Santa Cruz de la Sierra durante a guerra da Tríplice Aliança—1865/1870." *Boletim de História* 6.6 (1961): 11–22.

Langer, Erick. "El liberalismo y la abolición de la comunidad indígena en el siglo XIX." *Historia y Cultura* 14 (1988): 59–95.

——. "Missions and the Frontier Economy: The Case of the Franciscan Missions among the Chiriguanos, 1845–1930." In *The New Latin American Mission History*, ed. Langer and Robert H. Jackson, 49–76. Lincoln: University of Nebraska Press, 1995.

Langer, Erick, and Robert H. Jackson, eds. *The New Latin American Mission History*. Lincoln: University of Nebraska Press, 1995.

Larson, Brooke. *Cochabamba, 1550–1900: Colonialism and Agrarian Transformation in Bolivia*. 2d ed. Durham, NC: Duke University Press, 1998.

Larson, Brooke, and Olivia Harris, eds. *Ethnicity, Markets, and Migration in the Andes: At the Crossroads of History and Anthropology*. With Enrique Tandeter. Durham, NC: Duke University Press, 1995.

Laurie, Nina, Sarah Radcliffe, and Robert Andolina. "The Excluded 'Indigenous'? The Implications of Multi-ethnic Policies for Water Reform in Bolivia." In *Multiculturalism in Latin America: Indigenous Rights, Diversity, and Democracy*, ed. Rachel Sieder, 252–76. Houndmills, Basingstoke, UK: Palgrave Macmillan, 2002.

Lavrín, Asunción, ed. *Sexuality and Marriage in Colonial Latin America*. Lincoln: University of Nebraska Press, 1989.

Le Roy Ladurie, Emmanuel. *Montaillou: The Promised Land of Error*. Trans. Barbara Bray. 1975. New York: Vintage, 1978.

Lears, T. J. Jackson. "The Concept of Cultural Hegemony: Problems and Possibilities." *American Historical Review* 90.3 (1985): 567–93.

Lehmann, David, ed. *Ecology and Exchange in the Andes*. Cambridge: Cambridge University Press, 1982.

León Portilla, Miguel. *Visión de los vencidos: Relaciones indígenas de la conquista*. 1959. Mexico City: Universidad Nacional Autónoma de México, 1992.

Lewis, G. Malcolm. "Rhetoric of the Western Interior: Modes of Environmental Description in American Promotional Literature of the Nineteenth Century." In *The Iconography of Landscape: Essays on the Symbolic Representation, Design, and Use of Past Environments*, ed. Stephen Daniels and Denis E. Cosgrove, 179–93. Cambridge: Cambridge University Press, 1988.

Limerick, Patricia Nelson. *The Legacy of Conquest: The Unbroken Past of the American West*. New York: Norton, 1987.

Limón, José E. *Dancing with the Devil: Society and Cultural Poetics in Mexican-American South Texas*. Madison: University of Wisconsin Press, 1994.

Lockart, James, and Stuart B. Schwartz, eds. *Early Latin America: A History of Colonial Spanish America and Brazil*. New York: Cambridge University Press, 1983.

Lockhart, James. *The Nahuas after the Conquest: A Social and Cultural History of the Indians of Central Mexico, Sixteenth through Eighteenth Centuries*. Stanford, CA: Stanford University Press, 1992.

——, ed. and trans. *We People Here: Nahuatl Accounts of the Conquest of Mexico*. Berkeley: University of California Press, 1993.

Lorenzana Durán, Gustavo. "Tierra, agua y mercado en el Distrito de Alamos, Sonora, 1769–1915." PhD diss., Universidad Veracruzana, 2001.

López Austin, Alfredo. *Hombre-dios: Religión y política en el mundo nahuatl*. Mexico City: Universidad Nacional Autónoma de México, Instituto de Investigacioned Históricas, 1989.

Lovejoy, Paul E., and Nicholas Rogers, eds. *Unfree Labour in the Development of the Atlantic World*. London: Frank Cass, 1994.

Lovell, W. George, and Christopher H. Lutz. *Demography and Empire: A Guide to the Population History of Spanish Central America, 1500–1821*. Boulder, CO: Westview, 1995.

Lugar, Catherine. "Rice Industry." In *Encyclopedia of Latin American History and Culture*, ed. Barbara A. Tenenbaum, 4:559–60. New York: Scribner's, 1996.

Lynch, Thomas F. "Earliest South American Lifeways." In *The Cambridge History of the Native Peoples of the Americas*, vol. 3, *South America*, ed. Frank Salomon and Stuart B. Schwartz, Part I: 188–263. Cambridge: Cambridge University Press, 1999.

MacCameron, Robert. "Environmental Change in Colonial New Mexico." In *Out of the Woods: Essays in Environmental History*, ed. Char Miller and Hal Rothman, 79–98. Pittsburgh, PA: University of Pittsburgh Press, 1997.

MacCormack, Sabine. "Ethnography in South America: The First Two Hundred Years." In *The Cambridge History of the Native Peoples of the Americas*, vol. 3, *South America*, ed. Frank Salomon and Stuart B. Schwartz, 96–187. Cambridge: Cambridge University Press, 1999.

——. " 'The Heart Has Its Reasons': Predicaments of Missionary Christianity in Early Colonial Peru." *Hispanic American Historical Review* 65.3 (1985): 443–66.

——. *Religion in the Andes: Vision and Imagination in Early Colonial Peru*. Princeton, NJ: Princeton University Press, 1991.

Magaña, Mario Alberto. *Población y misiones de Baja California: Estudio histórico demográfico de la misión de Santo Domingo de la Frontera, 1775–1850*. Tijuana, Mexico: El Colegio de la Frontera Norte, 1998.

Mallon, Florencia E. *The Defense of Community in Peru's Central Highlands: Peasant Struggle and Capitalist Transition, 1860–1940*. Princeton, NJ: Princeton University Press, 1983.

——. *Peasant and Nation: The Making of Postcolonial Mexico and Peru*. Berkeley: University of California Press, 1995.

——. "The Promise and Dilemma of Subaltern Studies: Perspectives from Latin American History." *American Historical Review* 99.5 (1995): 1491–515.

Malvido, Elsa. "¿El arca de Noé o la caja de Pandora? Suma y recopilación de pandemias, epidemias y endemias en Nueva España, 1519–1810." In *Temas médicos de la Nueva España*, ed. Enrique Cárdenas de la Peña, 45–87. Mexico City: Instituto Mexicana del Seguro Social, Instituto Cultural Domecq, 1992.

Mann, Kristin Dutcher. "The Power of Song in the Missions of Northern New Spain (Mexico, California, Texas)." PhD diss., Northern Arizona University, 2002.

María Mercado, Melchor. *Album de paisajes: Tipos humanos y costumbres de Bolivia, 1841–1869*. Ed. Gunnar Mendoza L. La Paz: Archivo y Biblioteca Nacionales de Bolivia, 1991.

Mariluz Urquijo, José M. "Las escuelas de dibujo y pintura de Mojos y Chiquitos." *Anales del Instituto de Arte Americano e Investigaciones Estéticas* 9 (1956): 37–51.

Martin, Cheryl English. *Governance and Society in Colonial Mexico: Chihuahua in the Eighteenth Century*. Stanford, CA: Stanford University Press, 1996.

Martinez, Oscar. *Border Peoples: Life and Society in the U.S.-Mexico Borderlands*. Tucson: University of Arizona Press, 1994.

Martínez-Alier, Joan. "Ecology and the Poor: A Neglected Dimension of Latin American History." *Journal of Latin American Studies* 23.2 (1991): 621–39.

Marx, Karl. *The Eighteenth Brumaire of Louis Bonaparte*. Ed. C. P. Dutt. New York: International Publishers, 1963.

Masuda, Shozo, Izumi Shimada, and Craig Morris, eds. *Andean Ecology and Civilization: An Interdisciplinary Perspective on Andean Ecological Complementarity*. Tokyo: Tokyo University Press, 1985.

Mathewson, Kent. "Tropical Riverine Regions: Locating the 'People without History.'" *Journal of the Steward Anthropological Society* 20.1–2 (1992): 167–80.

Matson, Daniel S., and Bernard L. Fontana, eds. *Friar Bringas Reports to the King: Methods of Indoctrination on the Frontier of New Spain, 1796–97*. Tucson: University of Arizona Press, 1977.

Maybury-Lewis, David, ed. *Dialectical Societies: The Gê and Bororo of Central Brazil*. Cambridge, MA: Harvard University Press, 1979.

Mayer, Enrique. *The Articulated Peasant: Household Economies in the Andes*. Boulder, CO: Westview, 2002.

McCaa, Robert. "Marriageways in Mexico and Spain, 1500–1900." *Continuity and Change* 9.1 (1994): 11–43.

McCarty, Kieran R. *A Spanish Frontier in the Enlightened Age: Franciscan Beginnings in Sonora and Arizona, 1767–1770*. Washington, DC: Academy of American Franciscan History, 1981.

McNeill, William H. *Plagues and Peoples*. 1976. New York: Anchor Books, 1998.

Melville, Elinor G. K. *A Plague of Sheep: The Environmental Consequences of the Conquest of Mexico*. Cambridge: Cambridge University Press, 1994.

Menacho, Antonio. *Por tierras de Chiquitos: Los jesuitas en Santa Cruz y en las misiones de Chiquitos en los siglos 16 a 18*. San Javier, Bolivia: Vicariato Apostólico de Ñuflo de Chávez, 1991.

Mercado, Melchor María. Album de paisajes. La Paz, Bolivia: Archivo y Biblioteca Nacionales, 1991.

Merrill, William L. "Conversion and Colonialism in Northern Mexico: The Tarahumara Response to the Jesuit Mission Program, 1601–1767." In *Conversion to Christianity: Historical and Anthropological Perspectives on a Great Transformation*, ed. Robert W. Hefner, 129–63. Berkeley: University of California Press, 1989.

——. *Rarámuri Souls: Knowledge and Social Process in Northern Mexico*. Washington, DC: Smithsonian Institution Press, 1988.

Métraux, Alfred. "The Guaraní." In *Handbook of South American Indians*, vol. 3, *The Tropical Forest Tribes*, ed. Julian H. Steward, 69–94. Washington, DC: Smithsonian Institution Press, 1949.

——. "The Social Organization and Religion of the Mojo and Manasí." *Primitive Man: Quarterly Bulletin of the Catholic Anthropological Conference* 16.1–2 (1943): 1–30.

——. "Tribes of Eastern Bolivia and the Madeira Headwaters," In *Handbook of South American Indians*, vol. 3, *The Tropical Forest Tribes*, ed. Julian H. Steward, 381–424. Washington, DC: Smithsonian Institution Press, 1949.

Mignolo, Walter. *The Darker Side of the Renaissance: Literacy, Territoriality, and Colonization*. Ann Arbor: University of Michigan Press, 1995.

Miller, Char, and Hal Rothman, eds. *Out of the Woods: Essays in Environmental History*. Pittsburgh, PA: University of Pittsburgh Press, 1997.

Mills, Kenneth. *An Evil Lost to View? An Investigation of Post-Evangelisation Andean Religion in Mid-colonial Peru*. Liverpool: University of Liverpool, Institute of Latin American Studies, 1994.

Mirafuentes Galván, José Luis. "Agustin Ascuhul, el profeta de Moctezuma: Milenarismo y aculturación en Sonora (Guaymas, 1737)." *Estudios de Historia Novohispana* 12 (1992): 123–41.

Molina del Villar, América. *La Nueva España y el matlazáhuatl, 1736–1739*. Mexico City: Centro de Investigaciones y Estudios Superiores en Antropología Social, 2001.

Molina Molina, Flavio. *Historia de Hermosillo antiguo: En memoria del aniversario doscientos de haber recibido el título de Villa del Pitic*. Hermosillo, Mexico: F. Molina Molina, 1983.

Montané Martí, Julio César. "De *nijoras* y 'españoles a medias.'" In *XV Simposio de Historia de Sonora*, 105–23. Hermosillo, Mexico: Universidad de Sonora, 1990.

——. "Desde los orígenes hasta 3000 años a.p." In *Historia general de Sonora*, vol. 1, *Período prehistórico y prehispánico*, ed. Sergio Calderón Valdés, 177–221. Hermosillo, Mexico: Gobierno del Estado de Sonora, 1985.

Monteiro, John. "Invaded Societies: Sixteenth-Century Coastal Brazil." In *The Cambridge History of the Native Peoples of the Americas*, vol. 3, *South America*, ed. Frank Salomon and Stuart B. Schwartz, 973–1024. Cambridge: Cambridge University Press, 1999.

Morin, Claude. *Michoacán en la Nueva España del siglo XVIII: Crecimiento y desigualdad en una economía colonial*. Trans. Roberto Gómez Ciriza. Mexico City: Fondo de Cultura Económica, 1979.

Mukerjee, Radkahamal. "Ecological Contributions to Sociology." *Sociological Review* 22.4 (1930): 281–91.

——. *Social Ecology*. New York: Longman, Green, 1942.

Murra, John. *Formaciones económicas y políticas del mundo andino*. Trans. Daniel Wagner. Lima: Instituto de Estudios Peruanos, 1975.

Nabhan, Gary P. "*Ak-ciñ* and the Environment of Papago Indian Fields." *Applied Geography* 6.1 (1986): 61–76.

——. "Amaranth Cultivation in the U.S. Southwest and Northwest Mexico." *Amaranth Proceedings* (1979): 129–33.

——. *Cultures of Habitat: On Nature, Culture, and Story*. Washington, DC: Counterpoint, 1997.

——. *The Desert Smells like Rain: A Naturalist in Papago Indian Country*. San Francisco: North Point, 1982.

——. *Gathering the Desert*. Tucson: University of Arizona Press, 1985.

——, ed. *Counting Sheep: Twenty Ways of Seeing Desert Bighorn*. Tucson: University of Arizona Press, 1993.

Nabhan, Gary P., and Thomas Sheridan. "Living with a River: Traditional Farmers of the Río San Miguel." *Journal of Arizona History* 19.1 (1978): 1–16.

Nawrot, Piotr. *Indígenas y cultura musical de las reducciones jesuíticas*. Vols. 1 and 3. Cochabamba, Bolivia: Editorial Verbo Divino, 2000.

——. "Producciones de opera en las reducciones jesuíticas." In *Festival Internacional "Misiones de Chiquitos": III Reunión Científica; 10 y 11 de Agosto de 2000*, 65–78. Santa Cruz de la Sierra, Bolivia: Asociación Pro Arte y Cultura, 2000.

Newell, Gillian E., and Emiliano Gallaga, eds. *Surveying the Archaeology of Northwest Mexico*. Salt Lake City: University of Utah Press, 2004.

Officer, James E. *Hispanic Arizona, 1536–1856*. Tucson: University of Arizona Press, 1987.

Officer, James E., Mardith K. Schuetz-Miller, and Bernard L. Fontana, eds. *The Pimería Alta: Missions and More*. Tucson, AZ: Southwestern Mission Research Center, 1996.

O'Hanlon, Rosalind. "Recovering the Subject: *Subaltern Studies* and Histories of Resistance in Colonial South Asia." *Modern Asian Studies* 22.1 (1988): 189–224.

O'Leary, Daniel Florencio. *Memorias*. Madrid: Biblioteca Ayacucho, 1919.
Orlove, Benjamin. "Ecological Anthropology." *Annual Review of Anthropology* 9 (1980): 235–73.
——. *Lines in the Water: Nature and Culture at Lake Titicaca*. Berkeley: University of California Press, 2002.
Ortega Noriega, Sergio. *Un ensayo de historia regional: El noroeste de México, 1530–1880*. Mexico City: Universidad Nacional Autónoma de México, Instituto de Investigaciones Históricas, 1993.
——. "El sistema de misiones jesuíticas: 1591–1699." In *Tres siglos de historia sonorense (1530–1830)*, ed. Ortega Noriega and Ignacio del Río, 41–94. Mexico City: Universidad Nacional Autónoma de México, 1993.
Ortega Noriega, Sergio, and Ignacio del Río, eds. *Tres siglos de historia sonorense (1530–1830)*. Mexico City: Universidad Nacional Autónoma de México, 1993.
Ortiz, Alfonso. *The Tewa World: Space, Time, Being, and Becoming in a Pueblo Society*. Chicago: University of Chicago Press, 1969.
Ouweneel, Arij. *Shadows over Anáhuac: An Ecological Interpretation of Crisis and Development in Central Mexico, 1730–1800*. Trans. Peter Mason. Albuquerque: University of New Mexico Press, 1996.
Ovando Sanz, Jorge Alejandro. *La invasión brasileña a Bolivia (una de las causas del Congreso de Panamá)*. La Paz: Ediciones ISLA, 1977.
Pacquier, Alain. *Las chemins du baroque dans le Nouveau Monde: De la Terre de Feu à l'embouchure du Saint-Laurent*. Paris: Librairie Arthème Fayard, 1996.
Parejas Moreno, Alcides. "Chiquitos: Historia de una utopia." In *Las misiones jesuíticas de Chiquitos*, ed. Pedro Querejazu and Plácido Molina Barbery, 253–302. La Paz: Fundación BHN, 1995.
——. "Marco geográfico." In *Santa Cruz: Tiempo y espacio*, ed. Parejas Moreno and Bernd Fischermann, 13–24. Santa Cruz de la Sierra, Bolivia: Cooperativa Rural de Electrificación, 2000.
——. "Organización misionera." In *Las misiones jesuíticas de Chiquitos*, ed. Pedro Querejazu and Plácido Molina Barbery, 275–82. La Paz: Fundación BHN, 1995.
Parejas Moreno, Alcides, and Bernd Fischermann, eds. *Santa Cruz: Tiempo y espacio*. Santa Cruz de la Sierra, Bolivia: Cooperativa Rural de Electrificación, 2000.
Parejas Moreno, Alcides, and Virgilio Suárez Salas. *Chiquitos: Historia de una utopía*. Santa Cruz de la Sierra, Bolivia: CORDECRUZ, Universidad Privada de Santa Cruz de la Sierra, 1992.
Peet, Richard, and Michael Watts. "Liberation Ecology: Development, Sustainability, and Environment in an Age of Market Triumphalism." In *Liberation Ecologies: Environment, Development, Social Movements*, ed. Peet and Watts, 1–45. London: Routledge, 1996.

——, eds. *Liberation Ecologies: Environment, Development, Social Movements*. London: Routledge, 1996.

Pennington, Campbell W. *The Pima Bajo of Central Sonora*. Salt Lake City: University of Utah Press, 1979.

——. *The Tepehuan of Chihuahua: Their Material Culture*. Salt Lake City: University of Utah Press, 1979.

——. *Vocabulario en la Lengua Névome: The Pima Bajo of Central Sonora, Mexico*. Vol. 2. Salt Lake City: University of Utah Press, 1979.

——, ed. *Arte y vocabulario de la lengua dohema, heve o eudeve: Anónimo (siglo XVII)*. Mexico City: Universidad Nacional Autónoma de México, 1981.

Penry, S. Elizabeth. "The Rey Común: Indigenous Political Discourse in Eighteenth-Century Alto Perú." In *The Collective and the Public in Latin America: Collective Identities and Political Order*, ed. Luis Roniger and Tamar Herzog, 219–37. Brighton, UK: Sussex Academic Press, 2000.

Pérez Bedolla, Raúl Gerardo. "Geografía de Sonora." In *Historia general de Sonora*, vol. 1, *Período prehistórico y prehispánico*, ed. Sergio Calderón Valdés, 111–24. Hermosillo, Mexico: Gobierno del Estado de Sonora, 1985.

Pickens, Buford, ed. *The Missions of Northern Sonora: A 1935 Field Documentation*. Tucson: University of Arizona Press, 1993 [1935].

Platt, Tristan. "Ethnic Calendars and Market Interventions among the *Ayllus* of Lipes during the Nineteenth Century." In *Ethnicity, Markets, and Migration in the Andes: At the Crossroads of History and Anthropology*, ed. Brooke Larson and Olivia Harris, with Enrique Tandeter, 259–96. Durham, NC: Duke University Press, 1995.

——. "Fronteras imaginarias en el sur andino (siglos XV–XVII)." In *Anuario ABNB 1995*, 329–44. Sucre: Archivo y Biblioteca Nacionales de Bolivia, 1995.

——. *La persistencia de los ayllus en el norte de Potosí de la invasión europea a la República de Bolivia*. La Paz: Fundación Diálogo con el Apoyo de la Embajada del Reino de Dinamarca en Bolivia, 1999.

Polzer, Charles W. *Rules and Precepts of the Jesuit Missions of Northwestern New Spain*. Tucson: University of Arizona Press, 1976.

Prakash, Gyan. *After Colonialism: Imperial Histories and Postcolonial Displacements*. Princeton, NJ: Princeton University Press, 1995.

Presta, Ana María, ed. *Espacio, etnías, frontera: Atenuaciones políticas en el sur del Tawantinsuyu, siglos XV–XVIII*. Sucre: Editorial Antropólogos del Surandino, 1995.

Querejazu, Pedro, and Plácido Molina Barbery, eds. *Las misiones jesuíticas de Chiquitos*. La Paz: Fundación BHN, 1995.

Quiroga, Rosa María, and Bernd Fischermann. "Pueblo viejo–pueblo nuevo: Tradición y modernización en San Javierito." In *Las misiones de ayer para los días de mañana*, ed. Bolcato Custodio and Juan Carlos Ruiz, 119–25. Santa Cruz de la Sierra, Bolivia: Colegio de Arquitectos de Santa Cruz, 1993.

Rabasa, José. *Writing Violence on the Northern Frontier: The Historiography of Sixteenth-Century New Mexico and Florida and the Legacy of Conquest*. Durham, NC: Duke University Press, 2000.

Radding, Cynthia. "Crosses, Caves, and *Matachinis:* Divergent Appropriations of Catholic Discourse in Northwestern New Spain." *Americas* 55.2 (1998): 177–201.

——. "Domesticating Rural Space: Communities, Culture, and Nature in Sonora." Paper presented at the "Environment of Greater Mexico: History, Culture, Economy, and Politics" symposium, University of California–San Diego, March 1999.

——. *Entre el desierto y la sierra: Las naciones o'odham y Tegüima de Sonora, 1530–1840*. Mexico City: Instituto Nacional Indigenista, Centro de Investigaciones y Estudios Superiores en Antropología Social, 1995.

——. "From the Counting House to the Field and Loom: Ecologies, Cultures, and Economies in the Missions of Sonora (Mexico) and Chiqiuitanía (Bolivia)." *Hispanic American Historical Review* 81.1 (2002): 45–88.

——. "Historical Perspectives on Gender, Security, and Technology: Gathering, Weaving, and Subsistence in Colonial Mission Communities of Bolivia." *International Journal of Politics, Culture, and Society* 15.1 (2001): 107–23.

——. "Peasant Resistance on the Yaqui Delta: An Historical Inquiry into the Meaning of Ethnicity." *Journal of the Southwest* 31.3 (1989): 330–61.

——. "Voces chiquitanas: Entre la encomienda y la misión en el oriente de Bolivia (siglo XVIII)." In *Anuario ABNB 1997*, 123–38. Sucre: Archivo y Biblioteca Nacionales de Bolivia, 1997.

——. *Wandering Peoples: Colonialism, Ethnic Spaces, and Ecological Frontiers in Northwestern Mexico, 1700–1850*. Durham, NC: Duke University Press, 1997.

Rafael, Vicente. *Contracting Colonialism: Translation and Christian Conversion in Tagalog Society under Early Spanish Rule*. Ithaca, NY: Cornell University Press, 1988.

Ramos, Alcida Rita. *Sanumá Memories: Yanomami Ethnography in Times of Crisis*. Madison: University of Wisconsin Press, 1995.

Rappaport, Joanne. *The Politics of Memory: Native Historical Interpretation in the Colombian Andes*. 1990. Durham, NC: Duke University Press, 1998.

Real, Leslie A., and James H. Brown, eds. *Foundations of Ecology: Classic Papers with Commentaries*. Chicago: University of Chicago Press, 1991.

Reed, Nelson A. *The Caste War of Yucatán*. Rev. ed. Stanford, CA: Stanford University Press, 2001.

Reff, Daniel T. *Disease, Depopulation, and Culture Change in Northwestern New Spain, 1518–1764*. Salt Lake City: University of Utah Press, 1991.

——. "Text and Context: Cures, Miracles, and Fear in the *Relación* of Alvar Núñez Cabeza de Vaca." *Journal of the Southwest* 38.2 (1996): 115–38.

Regalsky, Pablo. *Etnicidad y clase: El estado boliviano y las estrategias andinas de manejo de su espacio*. La Paz: Plural, 2003.

Renard Casevitz, France-Marie, Thierry Saignes, and Anne C. Taylor, eds. *El este de los Andes: Relaciones entre las sociedades amazónicas y andinas entre los siglos XV y XVII*. 2 vols. Trans. Juan Carrera Colin. 1986. Lima: Instituto Francés de Estudios Andinos, 1988.

René-Moreno, Gabriel. *Catálogo del Archivo de Mojos y Chiquitos*. 1888. La Paz: Librería Editorial "Juventud," 1973.

Restall, Matthew. *The Maya World: Yucatec Culture and Society, 1550–1850*. Stanford, CA: Stanford University Press, 1997.

Riester, Jürgen. "Los Chiquitanos." In *En busca de la loma santa*, ed. Riester and Bernd Fischermann, 121–82. La Paz: Editorial Los Amigos del Libro, 1976.

Riester, Jürgen, and Bernd Fischermann, eds. *En busca de la loma santa*. La Paz: Editorial Los Amigos del Libro, 1976.

Riley, Carroll L. *The Frontier People: The Greater Southwest in the Protohistoric Period*. Rev. and exp. ed. Albuquerque: University of New Mexico Press, 1987.

Río, Ignacio del. *La aplicación regional de las reformas borbónicas en Nueva España: Sonora y Sinaloa, 1768–1787*. Mexico City: Universidad Nacional Autónoma de México, 1995.

——. *Conquista y aculturación en la California jesuítica, 1697–1768*. Mexico City: Universidad Nacional Autónoma de México, 1984.

——. "Las efímeras 'ciudades' del desierto sonorense." In *La ciudad y el campo en la historia de México: Memoria de la VIII Reunión de Historiadores Mexicanos y Norteamericanos, Oaxaca, Oax., 1985*/Papers Presented at the VIII Conference of Mexican and United States Historians. 2 vols. Ed. Ricardo Sánchez, Eric Van Young, and Gisela von Wobeser, 2:673–86. Mexico City: Universidad Nacional Autónoma de México, 1992.

Rodríguez, François, and Nelly Silva. *Etnoarqueología de Quitovac, Sonora*. Mexico City: Centro de Estudios Mexicanos y Centroamericanos and Instituto Nacional de Antropología e Historia, 1985–86.

Rojas Rabiela, Teresa, ed. *Agricultura indígena: Pasado y presente*. México City: Centro de Investigaciones y Estudios Superiores en Antropología Social, 1994.

Romero, Mary, Pierrette Hondagneu-Sotelo, and Vilma Ortiz, eds. *Challenging Fronteras: Structuring Latina and Latino Lives in the U.S.; An Anthology of Readings*. New York: Routledge, 1997.

Romero, Saúl Jerónimo. *De las misiones a los ranchos y haciendas: La privatización de la tenencia de la tierra en Sonora, 1740–1860*. Hermosillo, Mexico: Gobierno del Estado de Sonora, 1991.

Roniger, Luis, and Tamar Herzog, eds. *The Collective and the Public in Latin America: Cultural Identities and Political Order*. Brighton, UK: Sussex Academic Press, 2000.

Roosevelt, Anna C. "The Maritime, Highland, Forest Dynamic and the Origins of Complex Culture." In *The Cambridge History of the Native Peoples of the Americas*, vol. 3, *South America*, ed. Frank Salomon and Stuart B. Schwartz, 264–349. Cambridge: Cambridge University Press, 1999.

Roseberry, William. *Anthropologies and Histories: Essays in Culture, History, and Political Economy*. New Brunswick, NJ: Rutgers University Press, 1989.

Roth, Hans, and Eckart Kühne. "'Esta nueva y hermosa iglesia.' La construcción y restauración de las iglesias de M. Schmid." In *Las misiones jesuíticas de Bolivia: Martin Schmid, 1694–1772; Misionero, músico y arquitecto entre los Chiquitanos*, ed. Eckart Kühne, 89–102. Santa Cruz de la Sierra, Bolivia: Asociación Suiza por la Cultura "Pro Helvetia," 1996.

Ruibal Corella, Juan Antonio. "*¡Y Caborca se cubrió de gloria!*": *La expedición filibustera de Henry Alexander Crabb a Sonora*. Mexico City: Editorial Porrúa, 1976.

Saeger, James Schofield. *The Chaco Mission Frontier: The Guaycuruan Experience*. Tucson: University of Arizona Press, 2000.

Sahlins, Marshall. "Cosmologies of Capitalism: The Trans-Pacific Sector of 'The World System.'" In *Culture in Practice: Selected Essays*, 415–69. New York: Zone, 2000.

——. *Culture in Practice: Selected Essays*. New York: Zone, 2000.

——. *Islands of History*. Chicago: University of Chicago Press, 1985.

Said, Edward W. *Culture and Imperialism*. New York: Knopf, 1993.

——. *Orientalism*. New York: Pantheon, 1978.

Saignes, Thierry. *Los Andes orientales: Historia de un olvido*. Lima: Instituto Francés de Estudios Andinos, 1985.

——. "De la filiation à la résidence: Las ethnies dans les vallées de Larecaja." *Annales Économies, Sociétés, Civilisations* 33 (1978): 1160–81.

Saldívar, José David. *The Dialectics of Our America: Genealogy, Cultural Critique, and Literary History*. Durham, NC: Duke University Press, 1991.

Salmón, Josefa, and Guillermo Delgado. *Identidad, ciudadanía y participación popular desde la colonia al siglo XX*. La Paz: Plural, Asociación de Estudios Bolivianos, 2003.

Salomon, Frank. "Testimonies: The Making and Reading of Native South American Historical Sources." In *The Cambridge History of the Native Peoples of the Americas*, vol. 3, *South America*, ed. Salomon and Stuart B. Schwartz, 1:19–95. Cambridge: Cambridge University Press, 1999.

Salomon, Frank, and Stuart B. Schwartz, eds. *The Cambridge History of the Native Peoples of the Americas*. Vol. 3, *South America*. Cambridge: Cambridge University Press, 1999.

Sánchez, Ricardo, Eric Van Young, and Gisela von Wobeser, eds. *La ciudad y el campo en la historia de México: Memoria de la VIII Reunión de Historiadores Mexicanos y Norteamericanos, Oaxaca, Oax., 1985*/Papers Presented at the VIII Conference of Mexican and United States Historians. 2 vols. Mexico City: Universidad Nacional Autónoma de México, 1992.

Santamaría, Daniel. "Fronteras indígenas del oriente boliviano: La dominación colonial en Moxos y Chiquitos, 1675–1810." *Boletín Americanista* 36 (1986): 197–228.

——. " 'Que reconozcan el triste estado de sus almas': Los métodos misioneros de los jesuitas en Chiquitos." In *Las misiones jesuíticas de Bolivia: Martin Schmid, 1694–1772; Misionero, músico y arquitecto entre los Chiquitanos*, ed. Eckart Kühne, 25–29. Santa Cruz de la Sierra, Bolivia: Asociación Suiza por la Cultura "Pro Helvetia," 1996.

Sauer, Carl O. *Aboriginal Population of Northwestern Mexico*. Berkeley: University of California Press, 1935.

——. *The Distribution of Aboriginal Tribes and Languages in Northwestern Mexico*. Berkeley: University of California Press, 1934.

——. "The Morphology of Landscape." In *Land and Life: A Selection of the Writings of Carl Sauer*, 315–50. Ed. John Leighly. Berkeley: University of California Press, 1963.

Schroeder, Susan, Stephanie Wood, and Robert Haskett, eds. *Indian Women of Early Mexico*. Norman: University of Oklahoma Press, 1997.

Schwartz, Stuart B., ed. *Implicit Understandings: Observing, Reporting, and Reflecting on the Encounters between Europeans and Other Peoples in the Early Modern Era*. Cambridge: Cambridge University Press, 1994.

Schwarz, Burkhar. *Yabaicürr, yabaitucürr, chiyabaiturrüp: Estrategias neocoloniales de "desarrollo" versus territorialidad chiquitana*. La Paz: Ediciones Fondo Editorial FIA-SEMILLA-CEBIAE, 1995.

Scott, James C. *Domination and the Arts of Resistance: Hidden Transcripts*. New Haven, CT: Yale University Press, 1990.

——. *Seeing like a State: How Certain Schemes to Improve the Human Condition Have Failed*. New Haven, CT: Yale University Press, 1998.

——. "The State and People Who Move Around: How the Valleys Make the Hills in Southeast Asia." Keynote address to "Peasants in Comparative and Interdisciplinary Perspective: Landscapes of Identity, Nature, and Power" conference, University of Illinois–Urbana-Champaign, 9–10 April 1998.

——. *Weapons of the Weak: Everyday Forms of Peasant Resistance*. New Haven, CT: Yale University Press, 1985.

Seed, Patricia. *Ceremonies of Possession in Europe's Conquest of the New World, 1492–1640*. Cambridge: Cambridge University Press, 1995.

——. "Colonial and Postcolonial Discourse." *Latin American Research Review* 26.3 (1991): 181–200.

Seoane Urioste, Carlos. "Música en las misiones jesuíticas de Moxos y Chiquitos." *Historia y Cultura* 12 (1987): 111–20.

Sewell, William. "The Concept(s) of Culture." In *Beyond the Cultural Turn: New Directions in the Study of Society and Culture*, ed. Victoria E. Bonnell and Lynn Hunt, 35–61. Berkeley: University of California Press, 1999.

Shaw, Carolyn Martin. *Colonial Inscriptions: Race, Sex, and Class in Kenya*. Minneapolis: University of Minnesota Press, 1995.

Sheridan, Cecilia. *Anónimos y desterrados: La contienda por el "sitio que llaman de Quauyla," siglos XVI–XVIII*. Mexico City: Centro de Investigaciones y Estudios Superiores en Antropología Social, 2000.

Sheridan, Thomas E. *Los Tucsonenses: The Mexican Community in Tucson, 1854–1941*. Tucson: University of Arizona Press, 1986.

——. *Where the Dove Calls: The Political Ecology of a Peasant Corporate Community in Northwestern Mexico*. Tucson: University of Arizona Press, 1988.

——, ed. *Empire of Sand: The Seri Indians and the Struggle for Spanish Sonora, 1645–1803*. Tucson: University of Arizona Press, 1999.

Shimada, Izeumi. "Evolution of Andean Diversity: Regional Formations, ca. 500 B.C.E.–C.E. 600." In *The Cambridge History of the Native Peoples of the Americas*, vol 3, *South America*, ed. Frank Salomon and Stuart B. Schwartz, 1:350–517. Cambridge: Cambridge University Press, 1999.

Shohat, Ella. "Notes on the 'Postcolonial.'" *Social Text* 31–32 (1992): 99–113.

Sieder, Rachel, ed. *Multiculturalism in Latin America: Indigenous Rights, Diversity, and Democracy.* Houndmills, Basingstoke, UK: Palgrave Macmillan, 2002.

Siemens, Alfred H. *A Favored Place: San Juan River Wetlands, Central Veracruz, A.D. 500 to the Present*. Austin: University of Texas Press, 1998.

Sigal, Pete. "Gendered Power, the Hybrid Self, and Homosexual Desire in Late Colonial Yucatan." In *Infamous Desire: Male Homosexuality in Colonial Latin America*, ed. Sigal, 102–33. Chicago: University of Chicago Press, 2003.

——, ed. *Infamous Desire: Male Homosexuality in Colonial Latin America*. Chicago: University of Chicago Press, 2003.

Sluyter, Andrew. *Colonialism and Landscape: Postcolonial Theory and Applications*. Lanham, MD: Rowman and Littlefield, 2002.

Socolow, Susan Migden. "Spanish Captives in Indian Societies: Cultural Contact along the Argentine Frontier, 1600–1835." *Hispanic American Historical Review* 72.1 (1992): 73–99.

Soja, Edward W. *Postmodern Geographies: The Reassertion of Space in Critical Social Theory*. London: Verso, 1989.

Solano, Francisco de, and Salvador Bernabeu, eds. *Estudios (nuevos y viejos) sobre la frontera*. Madrid: Centro Superior de Investigaciones Científicas, Centro de Estudios Históricos, Departamento de Historia de América, 1991.

Spalding, Karen. *Huarochirí: An Andean Society under Inca and Spanish Rule*. Stanford, CA: Stanford University Press, 1984.

Spicer, Edward H. *Cycles of Conquest: The Impact of Spain, Mexico, and the United States on the Indians of the Southwest, 1533–1960*. Tucson: University of Arizona Press, 1962.

——. *The Yaquis: A Cultural History*. Tucson: University of Arizona Press, 1980.

Spicer, Rosamond, "People on the Desert." In *The Desert People: A Study of the Papago Indians*, ed. Alice Joseph, Spicer, and Jane Chesky, 3–59. Chicago: University of Chicago Press, 1949.

Stanfield, Michael E. *Red Rubber, Bleeding Trees: Violence, Slavery, and Empire in Northwest Amazonia, 1850–1933*. Albuquerque: University of New Mexico Press, 1998.

Stauffer, Robert Clinton. "Ecology in the Long Manuscript Version of Darwin's *Origin of Species* and Linnaeus' *Oeconomy of Nature*." *Proceedings of the American Philosophical Society* 104.2 (1960): 235–41.

Stavig, Ward. *The World of Túpac Amaru: Conflict, Community, and Identity in Colonial Peru*. Lincoln: University of Nebraska Press, 1999.

Stein, Stanley J., and Barbara H. Stein. *The Colonial Heritage of Latin America: Essays on Economic Dependence in Perspective*. New York: Oxford University Press, 1970.

Stern, Alexandra Minna. "Buildings, Boundaries, and Blood: Medicalization and Nation-Building on the U.S.-Mexico Border, 1910–1930." *Hispanic American Historical Review* 79.1 (1999): 41–82.

Stern, Steve J. "Feudalism, Capitalism, and the World-System in the Perspective of Latin America and the Caribbean." *American Historical Review* 93.4 (1988): 829–73.

——. *Peru's Indian Peoples and the Challenge of Spanish Conquest: Huamanga to 1640*. Madison: University of Wisconsin Press, 1982.

Steward, Julian H. *Theory of Culture Change: A Methodology of Multilinear Evolution*. Urbana: University of Illinois Press, 1955.

Stoler, Ann Laura. *Carnal Knowledge and Imperial Power: Race and the Intimate in Colonial Rule*. Berkeley: University of California Press, 2002.

——. *Race and the Education of Desire: Foucault's History of Sexuality and the Colonial Order of Things*. Durham, NC: Duke University Press, 1995.

——."Rethinking Colonial Categories: European Communities and the Boundaries of Rule." *Comparative Studies in Society and History* 31.1 (1989): 134–61.

Strack, Peter. " 'De tradiciones españolas y costumbres raras': Ritos religiosos en Chiquitos vistos por europeos." In *Las misiones jesuíticas de Bolivia: Martin Schmid, 1694–1772; Misionero, músico y arquitecto entre los Chiquitanos*, ed. Eckart Kühne, 39–46. Santa Cruz de la Sierra, Bolivia: Asociación Suiza por la Cultura "Pro Helvetia," 1996.

——. "Forasteros matizando recuerdos." *Posthumo Hans Roth: Presencia en la cultura* 1.5 (2000): 18–19.

——. *Frente a Dios y los pozokas: Las tradiciones culturales y sociales de las reducciones jesuíticas desde la conquista hasta el presente; Fiesta patronal y Semana Santa en Chiquitos*. Bielefeld, Germany: Verlag für Regionalgeschichte, 1992.

Szuter, Christine, Frank S. Crosswhite, and Carol D. Crosswhite. "The Sonoran Desert." In *Reference Handbook on the Deserts of North America*, ed. Gordon L. Bender, 163–319. Westport, CT: Greenwood, 1982.

Taussig, Michael T. *Shamanism, Colonialism, and the Wild Man: A Study in Terror and Healing*. Chicago: University of Chicago Press, 1987.

Tenenbaum, Barbara A., ed. *Encyclopedia of Latin American History and Culture*. 5 vols. New York: Scribner's, 1996.

Tenenti, Alberto. *La formación del mundo moderno, siglos XIV–XVII*. Trans. Pedro Roqué Ferrer. 1980. Barcelona: Editorial Crítica, 1985.

Thomas, David H., ed. *Columbian Consequences*. Vol. 1, *Archaeological and Historical Perspectives on the Spanish Borderlands West*. Washington, DC: Smithsonian Institution Press, 1989.

Thomson, Guy P. C. "Bulwarks of Patriotic Liberalism: The National Guard, Philharmonic Corps, and Patriotic Juntas in Mexico, 1847–88." *Journal of Latin American Studies* 22.1 (1990): 31–68.

——. "Los indios y el servicio militar en el México decimonónico: ¿Leva o ciudadanía?" In *Indio, nación y comunidad en el México del siglo XIX*, ed. Antonio Escobar Ohmstede and Patricia Lagos Preisser, 207–52. Mexico City: Centro de Investigaciones y Estudios Superiores en Antropología Social and Centro Francés de Estudios Mexicanos y Centroamericanos, 1993.

——. "Movilización conservadora, insurrección liberal y rebeliones indígenas, 1854–1876." In *America Latina: Dallo stato coloniale allo stato nazione*, ed. Antonio Annino, 2:592–614. Milan: Franco Angeli, 1987.

Thurner, Mark. *From Two Republics to One Divided: Contradictions of Postcolonial Nationmaking in Andean Peru*. Durham, NC: Duke University Press, 1997.

Todorov, Tzvetan. *The Conquest of America: The Question of the Other*. Trans. Richard Howard. 1982. New York: Harper and Row, 1984.

——. *La conquista de América: La cuestión del otro*. Trans. Flora Botton Burlá. 1982. Mexico City: Siglo Veinteuno Editores, 1987.

Toledo, Marisol. "Inventario de una unidad doméstica." In *Las plantas útiles de Lomerío*, ed. Teresa R. Centurión and Ivo J. Kraljevic, 429–34. Santa Cruz de la Sierra, Bolivia: Proyecto de Manejo Forestal Sostenible, Herbario, CICOL, 1996.

Tomichá Charupá, Roberto. *La primera evangelización en las reducciones de Chiquitos, Bolivia, 1691–1767: Protagonistas y metodología misional*. Cochabamba, Bolivia: Editorial Verbo Divino, 2002.

Tracy, James D., ed. *The Rise of Merchant Empires: Long-Distance Trade in the Early Modern World, 1350–1750*. Cambridge: Cambridge University Press, 1990.

Trigger, Bruce G. *Natives and Newcomers: Canada's "Heroic Age" Reconsidered*. 1985. Kingston, ON: McGill-Queen's University Press, 1986.

Truett, Samuel. "Neighbors by Nature: Rethinking Region, Nation, and Environmental History in the U.S.-Mexico Borderlands." *Environmental History* 2.2 (1997): 160–79.

Turner, Frederick Jackson. *The Frontier in American History*. 1920. Tucson: University of Arizona Press, 1986.

——. "The Significance of the Frontier in American History." In *Annual Report of the American Historical Association for the Year 1893*, 199–227. Washington, DC: Government Printing Office, 1894.

Twinam, Ann. *Public Lives, Private Secrets: Gender, Honor, Sexuality, and Illegitimacy in Colonial Spanish America*. Stanford, CA: Stanford University Press, 1988.

Underhill, Ruth. *Papago Indian Religion*. 1946. New York: AMS Press, 1969.

——. *Papago Woman*. 1936. Prospect Heights, IL: Waveland, 1979.

Valiñas Coalla, Leopoldo. "Lo que la lingüística yutoazteca podría aportar en la reconstrucción histórica del norte de México." In *Nómadas y sedentarios en el norte de México: Homenaje a Beatriz Braniff*, ed. Marie-Areti Hers et al., 175–205. Mexico City: Universidad Nacional Autónoma de México, 2000.

Vanderwood, Paul. *The Power of God against the Guns of Government: Religious Upheaval in Mexico at the Turn of the Nineteenth Century*. Stanford, CA: Stanford University Press, 1998.

Van Kirk, Sylvia. *"Many Tender Ties": Women in Fur-Trade Society in Western Canada, 1670–1870*. Winnipeg, MB: Watson and Dwyer, 1980.

Van Young, Eric. "Cities, Hinterlands, and Marches: Incommensurable New World Colonial Experiences Compared." Paper presented at the "Greater American Histories?" conference, Huntington Library, San Marino, California, 2001.

——. *Hacienda and Market in Eighteenth-Century Mexico: The Rural Economy of the Guadalajara Region, 1675–1820*. Berkeley: University of California Press, 1981.

——. *The Other Rebellion: Popular Violence, Ideology, and the Struggle for Mexican Independence, 1810–1825*. Stanford, CA: Stanford University Press, 2001.

——. "La otra rebelión: Un perfil social de la insurgencia popular en México, 1810–1815," In *Los ejes de la disputa: Movimientos sociales y actores collectivos en América Latina, siglo XIX*, ed. Antonio Escobar Ohmstede and Ramona Falcón, 25–55. Madrid: Iberoamericana, 2002.

——, ed. *Mexico's Regions: Comparative History and Development*. San Diego: Center for U.S.-Mexican Studies, University of California–San Diego, 1992.

Vaughan, Mary Kay. *Cultural Politics in Revolution: Teachers, Peasants, and Schools in Mexico, 1930–1940*. Tucson: University of Arizona Press, 1997.

Villalpando, Elisa C. "Los que viven en las montañas: Correlación arqueológico-etnográfica en Isla San Estéban, Sonora, México." In *Noroeste de México*, 8:9–95. Hermosillo: INAH, Centro Sonora, 1989.

Voss, Stuart F. *Latin America in the Middle Period, 1750–1929*. Wilmington, DE: Scholarly Resources, 2002.

——. *On the Periphery of Nineteenth-Century Mexico: Sonora and Sinaloa, 1810–1877*. Tucson: University of Arizona Press, 1982.

Wachtel, Nathan. *The Vision of the Vanquished: The Spanish Conquest of Peru through*

Indian Eyes, 1530–1570. 1971. Trans. Ben Reynolds and Siân Reynolds. New York: Barnes and Noble Imports, 1977.

Waisman, Leonardo. " 'Soy misionero porque canto, taño, y danzo': Martin Schmid, músico." In *Las misiones jesuíticas de Bolivia: Martin Schmid, 1694–1772; Misionero, músico y arquitecto entre los Chiquitanos*, ed. Eckart Kühne, 55–64. Santa Cruz de la Sierra, Bolivia: Asociación Suiza por la Cultura "Pro Helvetia," 1996.

Walker, Charles F. *Smoldering Ashes: Cuzco and the Creation of Republican Peru, 1780–1840*. Durham, NC: Duke University Press, 1999.

Wallerstein, Immanuel. *The Modern World-System*. 2 vols. New York: Academic Press, 1974–80.

Warman, Arturo. *Y venimos a contradecir: Los campesinos de Morelos y el estado nacional.* Mexico City: Centro de Estudios Superiores del Instituto Nacional de Antropología e Historia, 1976.

Weber, David J. *The Mexican Frontier, 1821–1846: The American Southwest under Mexico*. Albuquerque: University of New Mexico Press, 1982.

——. *The Spanish Frontier in North America*. New Haven, CT: Yale University Press, 1992.

Weber, David J., and Jane M. Rausch. Introduction to *Where Cultures Meet: Frontiers in Latin American History*, ed. Weber and Rausch, xiii–xli. Wilmington, DE: Scholarly Resources, 1994.

Weiner, Annette B., and Jane Schneider, eds. *Cloth and Human Experience*. Washington, DC: Smithsonian Institution Press, 1989.

West, Robert. *Sonora: Its Geographical Personality*. Austin: University of Texas Press, 1993.

Whigham, Thomas. "Paraguay's *Pueblos de Indios:* Echoes of a Missionary Past." In *The New Latin American Mission History*, ed. Erick Langer and Robert H. Jackson, 157–88. Lincoln: University of Nebraska Press, 1995.

White, Richard. *The Middle Ground: Indians, Empires, and Republics in the Great Lakes Region, 1650–1815*. Cambridge: Cambridge University Press, 1991.

——. *The Roots of Dependency: Subsistence, Environment, and Social Change among the Choctaws, Pawnees, and Navajos*. Lincoln: University of Nebraska Press, 1983.

White, Richard, Patricia Nelson Limerick, and James Grossman. *The Frontier in American Culture: An Exhibition at the Newberry Library, August 26, 1994–January 7, 1995*. Berkeley: University of California Press, 1994.

Whitehead, Neil L. "Ethnic Transformation and Historical Discontinuity in Native Amazonia and Guayana, 1500–1900." *L'Homme* 33.2–4 (1993): 185–305.

Whitten, Norman E. "Commentary: Historical and Mythic Evocations of Chthonic Power in South America." In *Rethinking History and Myth: Indigenous South American Perspectives on the Past*, ed. Jonathan D. Hill, 282–306. Urbana: University of Illinois Press, 1988.

——. *Sicuanga Runa: The Other Side of Development in Amazonian Ecuador*. Urbana: University of Illinois Press, 1985.

Whitten, Norman E., Dorothea Scott Whitten, and Alfonso Chango. "Return of the Yumbo: The Indigenous Caminata from Amazonia to Andean Quito." *American Ethnologist* 24.2 (1997): 355–91.

Williams, Jerry M., and Robert E. Lewis, eds. *Early Images of the Americas: Transfer and Invention*. Tucson: University of Arizona Press, 1993.

Williams, Raymond. *The Country and the City*. London: Chatto and Windus, 1973.

Williams, Robert. A., Jr. *Linking Arms Together: American Indian Treaty Visions of Law and Peace, 1600–1800*. New York: Routledge, 1999.

Wolf, Eric R. *Europe and the People without History*. Berkeley: University of California Press, 1982.

Worster, Donald. *Nature's Economy: A History of Ecological Ideas*. 2d ed. Cambridge: Cambridge University Press, 1994.

Wunder, John R., and Pekka Hämäläinen. "Of Lethal Places and Lethal Essays." *American Historical Review* 104.4 (1999): 1229–34.

Zepeda, Ofelia. "Where the Wilderness Begins." *Interdisciplinary Studies in Literature and the Environment* 4.1 (1997): 85–107.

Zimmerer, Karl S. "Discourses on Soil Loss in Bolivia: Sustainability and the Search for Socioenvironmental 'Middle Ground.'" In *Liberation Ecologies: Environment, Development, Social Movements*, ed. Richard Peet and Michael Watts, 110–24. London: Routledge, 1996.

Zimmerer, Karl S., and Thomas J. Bassett. "Approaching Political Ecology: Society, Nature, and Scale in Human-Environment Studies." In *Political Ecology: An Integrative Approach to Geography and Environment-Development Studies*, ed. Zimmerer and Bassett, 1–28. New York: Guilford, 2003.

——, eds. *Political Ecology: An Integrative Approach to Geography and Environment-Development Studies*. New York: Guilford, 2003.

Index

Acequias, 56, 316; definition of, 375
Adelantados, 125–26; definition of, 375
Agriculture: in Chiquitos, 17, 31–39; in Sonora, 7, 19, 26, 255
Alamos (Sonora), 85, 262, 264–65, 274
Alcabala: definition of, 375; in Sonora, 85
Aldeas, 63; definition of, 375
Almela, Miguel, 70, 71
Altar: presidio of, 276–77; villa of Guadalupe and, 276–77
Altar Desert: Pinacate, 31. *See also* Sonoran Desert
Altar Magdalena river, 31, 312
Altar Valley. *See* Sonora
Amazonia, 43–44; tribal peoples of, 282
Ancheta, Antonio, 100, 177
Andes, 43–44, 230, 240, 322; Andean highlands, 279, 281; *chacras* in, 249
Anza, Juan Bautista de, 178, 184, 202, 205, 237
Apache, 155, 194, 296; bands of, 17, 164; Captain Valdés, 186–88; capture of, 127; frontier wars and, 144, 182, 185–88, 207; peace encampments of, 260–61; raiding by, 259–60, 262, 264, 276, 312, 317, 320
Apoderados: in Chiquitos, 79–80
Arawak, 44, 140, 143
Architecture: baroque, 235; in Chiquitos, 65; domestic, 29, 32, 41–42; public, 29; sacred, 198, 211, 215, 217–19, 222–26; in Sonora, 25, 33, 223. *See also* landscapes
Areñenos: "sand people," 31, 318; Hiach-ed, S-ohbmakam, 32, 254
Arivechi, 105, 145
Arizpe, 61, 263; intendancy of 85, 92, 256, 317; village of, 99, 105
Artachu, Joaquín, 76, 79
Asunción de Paraguay, 33, 45–46, 53, 96, 127
Athapaskan. *See* Apache.
Audiencia: definition of, 375; of Guadalajara, 170; of La Plata, 191, 241
Ayllu, 137; definition of, 375; tribute and, 248, 281. *See also* Tribute
Ayoreo, (Ayoreóde), 124, 143, 325; cosmology of, 199–201, 206, 215; as Zamuco, 200

Bach, Moritz, 6, 302
Bacoachi (Sonora), 105, 176, 260–61, 316
Baja California, 61, 72, 144, 235, 262
Ballivián, José (President of Bolivia), 299; Battle of Ingavi and, 299
Bandeiras, 130; or *bandeirantes*, 130, 132; definition of, 375
Barbastro, Francisco, 219, 227–29. *See also* Franciscans
Baroyeca (Sonora), 264–65
Bartlett, John Russell, 316–21
Basan, Gregorio, 111–12
Batuc, 102, 147
Batucari, 63; *monte* for, 63
Bautista de Escalante, Alférez, 68–69
Baviácora, 100, 105, 263, 267
Bavispe Valley (Sonora), 104, 147; garrison, 176, 260
Beni, department of, 299–300
Binational Boundary Commission, 294, 297. *See also* Frontier
Bisanig (Sonora), 25
Bolivia, 227, 240, 295, 298, 302–3, 306–9, 325; Bolivian state, 248, 267, 282; eastern lowlands of, 33, 35, 43, 10, 98, 124, 215, 283, 288, 292, 296, 324; La Plata region of, 76–77; state, 110
Borderlands, 240, 292, 322; in Chiquitos, 282; cultural, 162; ecological, 162; international, 297; regional, 295; social and ethnic, 271; U.S./Mexican, 311–12, 324–25

Bororós, 304, 307
Bosque: gardens and, 38, 114, 216; uncultivated forestland, 37
Boturiquis River, 96, 299
Bourbon administration, 58, 79, 81, 84, 104, 261; economic policies of, 244, 249
Brazil, 252, 283–84, 294, 297, 304–5, 307, 324; Afro-Brazilians, 189; Atlantic rainforest of, 17; port of Coimbra, 301; port of Corumbá, 308
Buenavista: estate of, 99–101, 150, 155, 177, 182, 184
Buenos Aires, 45, 242, 248, 284
Buingoechea, Manuel, 92, 101–2

Caballero, Lucas, 141, 202–4; Quiriquicas and, 203–4
Cabeceras: in Sonora, 64
Cabeza de Vaca, Alvar Núñez, 9, 18–23, 32, 45, 48, 127, 302; as *adelantado* of Río de la Plata (1541–1545), 127; contact with Chanís tribe, 49, 51; expedition through Sonora of, 19–32; failure in greater Río de la Plata of, 50, 96; Pedro de Hernandez chronicles of, 45–51, 65, 66; *Relación* by, 19–20; relations with Indians of, 46, 51; southern expedition of, 33–48
Cabildo, 162, 172, 197, 219–20, 231–32, 234, 269, 273, 279, 292; *actas* of, 169; in Chiquitos, 171, 174–75, 189, 283–84, 286–88; *corregidor* of, 226; definition of, 376; hierarchy of offices, 173; native officers of, 165, 168; political identity and, 164; in Sonora, 169, 176, 180, 312
Caborca (Sonora), 145, 254–55, 266, 276
Cacicazgo, 165; in Chiquitos, 176; in Sonora, 181
Cacique, 69, 84, 140, 162, 173, 223, 231, 234, 270; actions and words of, 166; in Chiquitos, 170, 202; definition of, 337 n.5, 376; political role of, 171; in Sonora, 59, 63–64, 184, 186
Caguillona, 187–88
Cahitan peoples, 59, 204, 212, 215, 271, 312, 356; Pima revolt of, 181; religious festivals of, 226; revolt of 1739–41, 237; shamanism and, 207–9, 217. *See also* Yaqui–Mayo
Carvajal, Antonio: Governor of Chiquitos, 232
Casas grandes (Paquimé), 22, 27, 323, 333 n.10, 376
Castas, 157, 254, 271; definition of, 121, 376; as stereotypes, 122. See also *Mestizaje*
Castelnau, Francis de, 6, 107, 158
Catholicism, 14, 165, 167, 241; Catholic missionization and, 199–200, 205, 207, 214, 220; Christian rituals of, 207–10, 217, 226; *fiesta–cargo* system and, 269–70, 281; folk, 222, 269; imagery and vocabulary, 197, 202, 213, 229, 236; institutional presence of, 288; multiple interpretations of, 229–30; rituals of, 207–10, 217, 226, 237–38, 279–80; symbols of, 187, 194, 236–37; theatrical elements and, 235–37
Cattle ranching: in Chiquitos, 110–12; in Sonora, 99–100
Caudillo: definition of, 376; in Sonora, 312, 317
Cercado (Chiquitos), 108, 115
Cerro Prieto, 182–85, 188, 237
Chaco, 171, 293, 307; agricultural plot, 38, 43, 65, 84, 98, 302; definition of, 377
Chaco Boreal (northern), 3, 45, 200, 299, 307; Gran Chaco of, 33, 52, 119, 127, 323, 325; people of, 170–71, 216
Chapacura, 140, 143
Charcas: *audiencia* of, 8, 62, 76, 78, 80, 93, 130, 132–33, 135, 190, 283, 296, 323
Cháves, Ñulfo de, 53, 131–32
Chicha (fermented drink), 41, 203, 234
Chiefdom: native polity, 50, 52; *principales* of, 46, 50, 52
Chihuahua, 176, 271
Chiquitanía, 223, 281, 286, 292, 309; cosmology of, 217–18; ethno-botanical knowledge of, 217; dialects in, 124; drought in, 105–7; geography of, 8, 34–38; Jesuit presence in, 58; language of, 118; *monte* of, 37, 68; origins of term, 138–40; Serranía de Santiago in, 35; territory of, 98. *See also* Jesuits
Chiquitanos, 197, 216, 230–31, 239, 241–49, 302, 308–9; agriculture and, 37; architecture of, 41–43; communities of, 41; cosmology of, 201, 211, 239; diet of, 40; division of labor among, 41; ethnic identities of, 57; human geography of, 38; Iberian contact and, 45; Jesuit missions and, 75; language of, 189, 196, 232; *minca*, 41; political culture of, 162; pre-Hispanic culture of, 44; social life of, 43, 44

Chiquitos, 33, 108, 165–67, 189, 195, 232–35, 268, 305, 325–26; Bañados de Izózog in, 35, 45; *bosque alto* of, 37; *campo cerrado* of, 37; Chaco borderlands of, 194; indigenous peoples of, 52–54, 211, 215–17, 227, 298; Jesuit *reducciones* in, 222; province of, 284–85, 290, 302, 307; salt flats of, 297; Sonora vs., 35, 38, 223, 235, 260; territorial and political configuration of, 279, 288; Valle Grande of, 106
Chiriguano, 44, 110, 130, 132, 139, 292, 296
Citizenship, 284, 292–93; constitutional definitions of, 268–69, 279–81; ideologies of, 271, 277, 282; Indians as citizens, 287
Coahuíla, 126, 144
Cochabamba, 8, 131, 300
Cócorit, 184–85
Cocóspera (Sonora), 254–55, 259
Colegio (Chiquitos), 167, 173, 193, 219, 222–23, 237
Colegio de la Santa Cruz de Querétaro, 254, 259
Colonial regimes, 14; power relation of, 117
Colorado River, 31, 212
Comanche, 17, 180, 194
Commandancy General of the Internal Provinces, 61, 176, 270
Communal land holdings: in Chiquitos, 107, 109; division of, 105; restitution of, 272; in Sonora, 100–104. *See also* Land tenure
Compadrazgo, 146; definition of, 377
Composición: definition of, 101, 377; use of, 102
Común, 102, 105, 161, 193, 254, 312; definition of, 72, 169, 377; indigenous villagers of, 100, 177; landed patrimony as, 105; native elite and, 179
Concepción (Chiquitos), 106, 247, 287, 298; fire and epidemic in, 251; *parcialidades of*, 138, 141, 143, 189; population of, 252; rebellion in, 193, 236–37
Cordillera. See Sonora
Cordillera: province of, 45, 108–9, 115
Coronado, Francisco Vásquez de: expedition of 1540–42, 23, 53
Cosmology, 211–18, 227, 229–33, 238–39
Creole society, 14, 223, 299, 302, 312
Cruceño: definition of, 377; households, 131; labor demands, 136; society, 106, 115; raiding expeditions by, 130–31
Cruceros de día, 82, 172; definition of, 377
Culiacán, San Miguel de, 21, 126
Cumpas, Señora de la Asunción de (Sonora), 99, 188
Cumuripa, 27, 100, 177, 182–84
Cunca'ac, 32–33, 92, 103, 144, 164, 176; Seri as, 181, 215, 237, 260, 271
Curavés, 96, 298

Denuncia: in Chiquitos, 107–8; definition of, 101, 377; in Sonora, 102
Despoblado, 100; definition of, 23, 377
Devil, 207–8, 213; devil way (*jiawul himdag*), 205–7, 210, 214, 233, 295, 318, 379
Disease: in Chiquitos, 67–69, 96, 131, 138, 143; epidemics as, 203–4, 238–39, 251–52, 254; fear of, 168; in Sonora, 66–67, 143, 181
Drought: in Chiquitos, 251–52; in Sonora, 255–56, 260

Ecology: cultural ecology, 5; ecological diversity, 43, 275; environment, 15, 277–78, 292–93; human ecology, 54; human geographies and, 22, 244–45; social ecology, 5, 89, 114, 293; theorists of, 4
Ecónomos, 298; definition of, 106, 377
Emory, William H., 317–21
Encomienda, 62, 124, 160, 202, 280, 290, 336; in Chiquitos, 67–68, 98, 127–36; definition of, 47, 125, 378; *encomenderos* and, 53, 58, 161; in northern Mexico, 126
Entradas: in Chiquitos, 58, 65, 141, 171; definition of, 378; in Sonora, 59, 69
Environmental history, 15–17. *See also* Ecology
Escalante y Arvizu, Manuel, 259
Escudero, José Agustín de, 261–64, 267
Ethnic classifications: colonial institutions and, 125; in Sonora, 149, 254
Ethnic identities, 117, 163; amalgamations of, 123–24, 143, 159, 181; in Chiquitos, 118, 124, 130, 292; differences and, 121; gender roles and, 124; languages and, 119; in Sonora, 118, 143, 146, 149, 154–56, 175
Ethnicity, 159–60, 162, 196, 288, 292; in Chiquitos, 124; definition of, 117; partial societies and, 122–23; territories and, 141

Ethnic mosaics, 118, 136, 160; in eastern Bolivia, 120, 174
Ethnic polity, 164, 176, 193, 324; indigenous polity as, 269, 279, 292
Eudeve: Cucurpe, 99; Tuape–Eudeve, 99; villages of, 61, 99, 105, 264. *See also* Opata

Fernández, Juan Patricio, 1, 3, 58, 170–71, 203, 207; descriptions of Chiquitos and, 40, 141
Filibusteros, 314–15
Franciscans, 73–74, 177, 255, 309; expulsion of Spaniards and, 259; missionaries, 220
Fronteras (Sonora), 186, 188, 260–61, 316
Frontier: as boundary, 24, 206–7, 292, 218; between Chiquitos and Brazil, 283–84; between culture and nature, 198, 200, 208, 321, 323; defense and, 176; ecological and cultural, 48, 314; Guaycurúan–Chiquitano, 162; shamans and, 198–201, 208; social and spatial, 279; in Sonora, 312

Gadsden Purchase, 297,
Ganado de secciones, 306, 309; definition of, 378
Gécori: estate of, 150, 152, 153
Gender, 288, 292; captivity and, 51–52, 160–61; definition of, 10, 117, 196, 302; division of labor and, 86–87, 117, 241–43; family formation and, 117; gendered violence, 216; hierarchies and, 165, 195; identity and, 118, 121; inequalities of, 160; language and, 123, 230–31, 234; masculinity and, 210–11; sexuality and, 117
Gente de razón, 146–47; definition of, 378
Gila River valley, 204, 323
Guaraní, 17, 44–48, 50–51, 123, 227, 292, 302; concept of *Tumpa*, 200; frontier with Guaycurú, 110, 337 n.61; language of, 119; 140, 143; Spaniards allied with, 51–53, 126–27
Guarayo, 124, 292, 296, 302, 309
Guaycurú, 44, 46, 96, 127, 194, 302; Guaycurúan–Chiquitano frontier and, 162. *See also* Guaraní
Guayma Indians, 103, 262–63, 265
Gulf of California, 31–32, 145, 212, 318

Hanansaya, 137; definition of, 378
Hardy, Robert, 6, 262, 264–65, 267
Hermosillo (Sonora), 261–63, 274
Hernandez, Pedro de, author of *Comentarios*, 45–48. *See also* Cabeza de Vaca, Alvar Nuñez de
Herrera, José Manuel, 297–98
Hohokam, 204, 323
Hybridity, xviii, 118, 121, 322; definition of, 117; in Sonora, 147

Iberia: Iberian imperialism, 165; Ibero-American world, xviii; legal traditions of, 14
Indios encomenderos, 133–35
Indios laboríos, 148–52, 156, 161; definition of, 379. See also *Mestizos*
Intendancy of Arizpe. *See* Arizpe
Isózog, Bañados de. *See* Chiquitos: Bañados de Isózog in
Itonamas, 132–33

Jamaica: estate of, 150–51, 155–56
Jaurú river, 297, 304, 306
Jesuits, 13, 249–50; administrative system of, 56; *annuas*, 120; Chiquitanos and, 65; expulsion from Chiquitos of, 78, 188; expulsion from Sonora of, 74, 247; indigenous agrarian systems affected by, 61–62; mission enterprise of, 59, 62, 77, 264; missions in Chiquitos, 200, 217, 222–23; origins in San Lorenzo, 58; shamans and, 201–4; in Sonora, 59, 61, 64, 207–9, 220
Jichis, 38, 216, 295; spiritual guardians, 43, 201
Jova, 30, 144
Junta Provincial de Temporalidades, 76, 78; definition of, 379
Jusacamea, Juan Ignacio (Banderas), 273–74
Justiniano Mercado, Manuel José, 108

Knogler, Julian, 58, 65, 77, 171, 231, 236

Labor: in Sonora, 87.
Landscapes, xvi, 63–64, 84, 118, 219, 321; agrarian, 17; agricultural architecture as, 29, 43; colonial, 167, 181; constructed, 6–7; cultivated, 316, 319; cultural, 116, 164, 230, 295–97, 314; post-colonial, 240–41, 292; of power, 167; of refuge, 188; in Sonora, 316,

321; spiritual, 196–97, 211–15, 227, 217; western notions of, 5–6
Land tenure, 89; changes in, 114; in Chiquitos, 289; politics of, 115–16, 270, 274; in Sonora, 293
Languages: codification and translation of, 227–32, 238; gendered, 230–31; sermons and, 231–34; Tagalog translation, 230
La Paz: department of, 300
López, David I., 205

Magdalena River valley (Sonora), 145, 182, 254–55, 275, 312
Malocas, 130–32
Mamelucos, 59, 81, 87, 133; definition of, 379. *See also* Brazil; Mato Grosso.
Manapeca, 124, 189, 236–37
Manazica, 124, 141, 170–71, 201–3; Quiriquicas as, 203
Mancomún: definition of, 110
Manioc, 37, 39, 85, 111
Marcos: definition of, 71
Markets, 15, 22, 199
Martínez, Antonio, 178–79
Martinez de Irala, Domingo, 45, 48, 132
Martínez de la Torre, Diego Antonio (first governor of Chiquitos), 79–80
Mary, Holy Virgin, 196–97, 200, 202, 204; as symbol of Catholic Church, 208, 214, 228–29, 232. *See also* Catholicism
Masavi family, 134–35
Matachines, 227, 238; definition of, 380
Mátape, 102, 183
Matlazáhuatl scourge of 1737 (New Spain), 143; definition of, 380. *See also* Disease
Mato Grosso, 48, 81, 189, 302, 304; border with Chiquitos, 45, 79, 82, 283–84; Portuguese province of, 8; slave hunters and, 53, 130. See also *Mamelucos*
Mayo (Sonora), 237. *See also* Cahitan
Mbayás, 162, 299
Mendez, Pedro, 62–63
Mendoza, Pedro de, 45, 49
Mesoamerica, 17, 21–22, 52, 136–37, 166, 269; definition of, 380
Mestizaje: in Chiquitos, 158, 290; definition of, 117, 380; in Sonora, 271
Mestizo, 200, 220; in Chiquitos, 163; definition of, 380; in Sonora, 147–50, 153–57, 254, 260
Mexico, 10, 240, 256, 274, 292, 295, 311; Mexican Republic, 270–71; U.S. invasion of, 207, 294
Migratory patterns, 57; in Chiquitos, 65
Milpa, 27–29, 70, 100, 104; definition of, 380
Mining *reales*, 59, 70; abandonment of, 182–83; definition of, 13; link to colonial enterprise and, 99; operations of, 73; in Sonora, 144, 157
Missions, 280, 284; economy of, 70, 71, 75, 93, 245–48, 252–53, 260, 296; enterprise in, 59, 76, 254; indigenous alternatives to, 87; *memorias* of, 71; outlays of, 73; in Pimería Alta, 74; secularization and, 253–54; weaving in, 242–45. *See also* Jesuits; *Reducciones*
Mission towns. See *Reducciones*
M'Oñeyca (Chiquitos), 118, 139, 345–46 n.5
Montero, Juan Miguel, vicar of southern cordillera, 109–10
Moxos, 107, 171, 279, 290, 302, 305; *llanos* (plains) of, 35–36 (map 2), 43–44; missions of, 62, 81, 132, 235, 249; renamed department of Beni, 299–300

Nácori, 102, 145, 183
Nacameri mission, 145, 177–79, 188
Nahuatl, 23, 119, 168–69, 229–30, 333 n.12
Narváez, Panfílio, 19
Nava, Pedro de, 104, 158
Nebome. *See* Pima
Nentvig, Juan, 91, 169, 170, 334 n.20
New Mexico, 176, 180, 185; pueblos of, 21
New Spain, 240
Nijoras (nixoras: captives), 145, 155
Nizza, Fray Marcos de: expedition of 1539, 23, 53
Northern New Spain. *See* Sonora
Nueva Galicia, 21; definition of, 380
Nueva Vizcaya, 61, 180, 185, 188, 323; definition of, 380; diocese of, 150; *encomienda* in, 125–26; mining frontier of, 59, 73, 99; nomadic tribes in, 144
Nuevo León (Mexico), 126, 144
Núñez Cabeza de Vaca, Alvar. *See* Cabeza de Vaca, Alvar Núñez

Occidente: state of, 105, 256–57, 260, 263, 272–73, 312
Ocuca, 90–93, 99, 101, 275
Olíden, José León, 301–2, 304, 308
Olíden, Manuel Luis de, 297, 300–301
Oliva, Juan María, 184–85
Ónavas, 146
O'odham, 239, 254–55, 266, 279, 293, 318, 323, 356; cosmology of, 204, 212–13, 215, 226, 229; definition of, 380; *himdag* and, 209; missions and, 219–20; oral traditions of, 205–7, 233; Papagos as, 32–33, 207, 222, 275, 317; Papawi Ko'odham as, 32; seasonal calendar of, 34; Tohono O'odham, 145, 176, 207, 211, 222, 228, 254, 271, 292; uprisings of 1840–43, 272, 274–78; *wi' igita* and, 217, 233. *See also* Areñenos; Pima
Opata, 188, 195, 276, 292, 312, 317–18, 323; auxiliary troops as, 176, 178–79, 193; census of Oposura and, 154; Eudeve and, 25, 30, 57, 144, 169, 227–29; language of, 118, 196; military garrisons of, 260–61; native polity of, 279; peasant cultivators as, 263–264; Pima and, 87; rebellion of, 271–74, 277
Opatería, 228, 312
Opodepe, 178–79; Eudeve and, 99; mission of, 145, 177
Opodepe (Sonora): church of, 220
Oposura valley (Sonora), 104, 147, 187, 263–64, 267, 271; renamed Moctezuma, 262
Oquitoa (Sonora), 254–55
Orbigny, Alcides d', 97, 250–51, 253, 282, 299; as author, 6; French naturalist, 96, 158, 174; tour of Chiquitanía by, 106, 120
Orientalist debate, 13
Ostimuri, 67, 145, 193, 195, 237, 256, 260, 312; Cahita revolt in, 181
Otukés, 96, 298
Otuquis, 35, 45, 140, 143, 300–301, 304, 308, 324

Paiconeca, 124
Pantanal, 162, 300, 302; definition of, 381
Papago Indian Reservation, 207, 222. *See Also* O'odham
Paraguay: province of, 45, 127, 130, 235, 252, 294, 307–8, 324; river, 96, 119, 299–301, 304, 307–8, 324; watershed, 162
Parapetí River, 119
Paraná River, 119
Parcialidades, 57, 59, 163, 172–75, 223; definition of, 381; *Cah* and, 137, 376; *Calpulli* and, 137, 376; in Chiquitos, 86, 111, 113, 137–43, 141–42, 189, 191–93, 195, 293, 298, 309; definition of, 136–37, 381; in Sonora, 59, 144
Paurús. *See* Water
Peña, Marcelino de la, 73, 111, 173–74, 282, 297–98; report of 1831, 251–52
Peonage, 312, 321
Pérez de Ribas, Andrés: as author, 66–67; as missionary, 63–64, 65, 207
Pérez Llera, Jóse María, 255, 259
Pfefferkorn, Ignaz, 91, 169
Pima, 57, 176, 188, 193, 237, 271, 276, 312; Akimel as, 32; auxiliary troops, 270; metal tools and, 53; military garrison of, 260; missions of, 61, 100, 264; Nebome as, 30; nomadic groups of, 69, 185; Piato as, 181–82; rebellion of, 181; Tepehuán and, 25, 27, 99–100, 119, 144–46, 227. *See also* O'odham.
Pimería Alta, 59, 72, 97, 101–2, 105, 145; arid plains of, 311; Francisco Barbastro in, 228; Francisco Eusebio Kino and, 205; missions of, 169, 254–61, 275, 312; mines in, 265–66; rebellion in, 182
Pineda, Juan Claudio, 146, 185
Piñoca, 124; languages of, 143
Pitic (Sonora), 254. *See also* Hermosillo
Pitiquito mission, 145
Pivipa, estate of, 150, 154
Political culture, 162, 193; in Chiquitos, 288; hegemony and, 58, 180, 270, 366 n.55; in Sonora, 312; spatial dimensions of, 293
Political economy: definition of, 55
Pónida, 105
Pópulo mission, 145
Porongo, 108, 189
Potam, 184
Potosí, 77, 323
Power: concept of, 197, 236–38, 239; spiritual, 211, 216, 218, 220, 226, 230–31, 234; symbols of, 236–37; water and, 216
Prestación vial, 175; definition of, 381
Prestamero, Juan de, 178–79

Priest: priesthood, Judeo-Christian meaning of, 199
Principales: definition of, 381. *See also* Chiefdom
Private property, 321; evolution of, 93; expansion of, 100–104, 110
Puerto de los Reyes, 48–50
Punitive expeditions, 180, 182

Quechua, 44
Quibichicocíes, 132
Quitobac (Sonora), 211, 265, 275–76, 296

Ramos, Sebastián, 283–84, 297, 304, 306–7, 324
Rancherias, 41, 53, 55, 61, 200–203, 219, 296; Apaches and, 180–86, 188, 193; as center of social and cultural life, 43; in Chiquitos, 141, 170, 302, 307; consolidation of, 63, 175; definition of, 381; missions and, 144; settlements as, 156; in Sonora, 143, 156, 177, 234, 255, 261, 271, 276–79, 318
Rarámuri (Tarahumara), 27, 204, 295; cosmology of, 212–15, 230, 239; religious festivals of, 226, 234; sermons and, 233
Realengas, 92, 101–2; definition of, 381
Reciprocity, 17, 124, 164
Redondo, Santiago, 102
Redondo, Xavier, 92, 101
Reducciones, 61, 86, 165, 167–68, 193, 219; in Chiquitos, 58, 65, 69, 76, 78, 81, 84, 132, 140–41, 162–63, 171, 202, 222–23; definition of, 381; religious life of, 137; in Sonora, 63, 66, 75, 181, 208, 277. *See also* Missions
Repartimiento, 68, 125–26, 168; definition of, 382
República de Indios, 164; definition of, 382
Riberto, Domingo, 112–13
Río Chico, 71, 146
Río Conchas, 19
Río de la Plata, 240, 282; *encomiendas* in, 125; Iberian contact in, 45–46; province of 33, 45, 52, 304; warfare in, 50
Río Grande (Chiquitos), 106
Río Grande (Mexico), 19
Río Guapay, 35
Roxas, Manuel, 190–92
Roza: definition of, 382; *roce y quema*, 38. *See also* Chaco; Swidden cultivation
Saenz de Rico, José, 92
Saguaripa, 105, 145
Saint way (*sa:nto himdag*), 205–7, 214, 295; definition of, 382. *See also* O'odham
Salazar, Alejo, 90, 91
Salazar, Prudencio, 90, 91
Salazar family: land holdings of, 91, 101
Salinero, 69
San Antonio de la Huerta (Sonora), 71, 146, 183, 264
Sánchez Labrador, Joseph, 162, 194, 324
San Felipe, villa de (Sinaloa), 209
San Francisco: in Alta California, 180
San Ignacio de Cabúrica (Sonora), 74
San Ignacio de Chiquitos, 78, 143, 226, 245, 287, 299; population of, 252–53; tumult in, 232, 237
San Ignacio de Onavas (Sonora), 157, 158, 254–55
San Ignacio de Tubac, 92
San Ignacio de Velasco (Chiquitos), 174–75
San Ignacio de Zamucos (Chiquitos), 143, 189–93, 201
San José (Chiquitos), 226
San José de Chiquitos, 111, 241–45, 247, 298; electoral assembly in, 286; population of, 253
San José de Pimas (Sonora), 183–84
San José de Tumacácori (Sonora), 100
San Juan, pueblo of (Chiquitos), 226, 247, 304; population of, 252–53
San Lorenzo de la Barranca (Chiquitos). *See* Santa Cruz de la Sierra
San Lorenzo ranch (Sonora), 91, 182
San Miguel de Bavispe (Sonora), 104
San Miguel de Culiacán (Sinaloa), 126
San Miguel de Horcasitas (Sonora), 71, 103, 145, 178, 180, 183
San Miguel de los Ures, 100
San Miguel de Oposura (Sonora), 99 147, 150, 152, 156–57; demographic profile of, 150, 154–55
San Miguel pueblo (Chiquitos), 226, 243–44, 247, 287, 299; population of, 252–53
San Miguel River (Sonora), 31, 73, 103, 145, 182; valley of, 177, 320
San Pedro de Aconchi (Sonora), 71, 72, 74, 105, 263

San Pedro de la Conquista de Pitic (Sonora), 102–4, 182; mixed population of 102; *suertes* assigned at, 103
San Rafael de Buenavista (Sonora), 260
San Rafael pueblo (Chiquitos), 226, 232, 243–44, 287, 289
Santa Ana de Nuri (Ostimuri), 100, 146
Santa Ana de Tepache (Sonora), 78, 90–91, 147–50, 156–57, 263, 267, 271; demographic profile of, 147–49; mines of, 147
Santa Ana pueblo (Chiquitos), 226, 236, 243, 283; caciques in, 171; electoral assembly in, 286; Francis de Castelnau in, 107; founding of, 250; Marcelino de la Peña in, 174; population of, 249–53
Santa Cruz: department of, 300, 304, 306
Santa Cruz de la Sierra (Chiquitos), 59, 62, 127, 131, 167, 247, 279, 284, 289, 299; bishop of, 58, 80, 189, 249; Chiquitos missions and, 62; description of, 106; elections in, 287–88; *encomiendas* and, 127; founding of, 53–54; *mestizaje* in, 290–92; San Lorenzo de la Barranca as, 53, 130, 135, 202, 337 n.64; villa of, 59. *See also* Orbigny, Alcides d'
Santa Cruz river (Sonora), 92, 145
Santa Gertrudís del Altar, 92, 182
Santa María Bacerác, 104
Santa María de Mobas, 100
Santa María Magdalena, 132
Santa Rosa, 184–85
Santiago de Cocóspera, 74
Santiago pueblo (Chiquitos), 226, 247, 298, 300, 301, 308; population of, 252–53
Santo Corazón (Chiquitos), 93–98, 111, 162, 297–301, 304–5, 308; population of, 252–53
San Xavier (Chiquitos), 201, 226, 249–51, 298; Alcides d'Orbigny in, 106; ethnic groups in, 143; indigenous officers of, 75, 79–80; Manazicas and, 170–71
San Xavier del Bac (Sonora), 145, 254–55, 259
Sáric (Sonora), 255, 259
Savannas, 26, 35, 322
Seri, 182, 184, 188, 194, 317. *See also* Cunca'ac
Sermons, 231–36. *See also* Languages
Shamanism: healing rituals of, 20, 33, 66; Mapono, Manazica, 201–4, 217, 223; mediators, 64, 66; in Sonora, 204–6; sorcerers and healers as, 199, 208, 210–11, 217, 233; spiritual powers and, 197–98, 207, 238
Sibilat, Joseph, 241–45
Sibubapas (or Suvbapas), 146, 181
Sierra Bacatete (Sonora), 212. *See also* Yaqui–Mayo.
Sierra Madre Occidental, 19, 176, 180, 204, 212, 323; geography of, 23; Pima–Tepehuán and, 119; provinces west of, 119
Sin: concept of, 228–29
Sinaloa, 21, 59, 61, 63, 67, 70, 181, 193, 235, 237, 256, 260, 321
Slash and burn. *See* Swidden Cultivation
Social ecology, 5, 89, 114, 293
Sociedad Progresista de Bolivia, 297, 300, 308–9
Society of Jesus. *See* Jesuits
Sonora, 217, 239, 241, 256, 268, 311, 325–26; *alcaldes* of, 61; Altar desert in, 31, 99, 211, 274–76; Altar valley of, 22; changing natural environment of, 99; coalescing of ethnic groups in, 57; language amalgamation in, 123–24; Mesoamerican influence in, 22; missions of, 197, 235, 237; *monte*, 99–100; post-colonial administrator of, 260, 269–71; pre-Hispanic settlements in, 29; province of, 19–52; *tenientes* of, 61; three major regions of, 23; tribes of, 32; Zona Serrana of, 27–30
Sonoran desert, 31–34, 181, 204, 210, 215, 230, 311, 320, 323; Areñenos of, 31
Sonoran highlands (*cordillera*), 25–27, 84, 181; state formation of, 15
Sonora River, 73, 103, 182, 228
Soyopa, 183
Spanish state: Castilian crown of, 19; Constitution of Cádiz, 241, 268, 285; Cortes of 1812, 241, 280; imperial project of, 164
Suaqui, 146, 183
Suárez, José Ramón, 108–9
Sumas, 126
Swidden cultivation, xvii, 93; definition of, 382; techniques, 25–27, 114, 133

Tao, 141, 143
Tapacuras, 132
Tapisques, 73
Tapuymiri, 139; definition of, 382

Tarahumara, 27. *See also* Rarámuri
Tarija, 131
Tawantinsuyo, 8
Technology: in Chiquitos, 244–45
Tecoripa, 183
Tepache. *See* Santa Ana de Tepache
Tepoca, 69
Terrenate, 185, 187
Territory, 294; in Chiquitos, 98; Native American expansion of, 17; in social ecology, 89; in Sonora, 98–99
Tesgüino (fermented drink, Sonora), 234
Teuricachi, 101
Tienda de Cuervo, Joseph, 183, 188
Tierras realengas: definition of, 382; in Sonora, 90
Toiserobavi: estate of, 150–52, 154, 156
Transculturation, 228, 322; transculturative processes and, 206
Tremedal (Chiquitos), 297, 304, 306–7, 324
Tribute, 166, 168, 300
Trincheras, 29–30; definition of, 383
Trinidad (Sonora), 146
Tubac (Sonora), 180, 182, 312; presidio of, 255, 260
Tubutama (Sonora), 145, 228, 255, 259, 276
Tucson (Sonora), 260, 312
Tucumán, 62, 96, 127
Tumacácori (Sonora), 254–55, 259

United States, 297
Upper Peru. *See* Bolivia
Ures (Sonora), 99, 188, 261, 263, 274
Uruguay River, 119

Valle Grande (Chiquitos), 29
Vecinos, 299, 302; in Chiquitos, 279, 284, 288–89; definition of, 382; land claims by, 91–92, 97, 102, 105, 259–60; *mestizos*, 156; missions and, 99; population growth and, 181; in Santa Cruz, 68, 93, 98, 108–9; in Sonora, 75, 144, 146, 183, 185–88, 237, 255, 263, 276–77, 293
Velarde, Luis, 1, 3
Velasco, José Francisco, 261–62, 264–67
Velasco, Maria Guadalupe, 92, 101
Velasco, Miguel Antonio, 90, 92, 101
Verdugo, Juan Barthelemí, 81–83, 96
Viedma, Francisco de (intendant governor), 244
Vildósola, Gabriel Antonio, 186–88, 261
Villa Rica de Potosí, 296
Visita: in Sonora, 64, 71

Wahia, 91
Warfare, 15, 29, 33, 38, 162–63, 166, 168, 201, 295, 299; captivity and, 20, 22, 50, 69, 183, 195; history of, 50; interethnic, 124, 126; labor supply and, 130; raiding and, 53, 127; *rescate* and, 62, 130–31, 134, 181, 183; in Sonora, 144–45, 311
War of the Triple Alliance, 307–8
Water: access to, 41; *jichis* as guardians of, 43, spiritual power of, 216. See also *Paurús*; *Jichis*
White, Richard, 15, 16
Wilderness, 219; concept of, 211; Rarámuri and, 212, 295; Tohono O'odham and, 211–13
Women: captivity and, 50–51; in Chiquitos, 76–77; marital alliances and, 50; reproductive labor and, 50–51; warfare and, 20. *See also* gender
World systems theory, 13

Xamares, 132
Xarayes, 49, 132
Xecatacari, 177

Yaqui–Mayo, 25, 188, 279, 292, 295, 317–19; armed uprisings by, 260, 273–74; captives as, 184–85; laborers as, 265–67, 277; Oposura census and, 154; *rancherías* of, 67; Sierra Bacatete and, 212. *See also* Cahitan peoples
Yaqui River, 182–83, 262, 275
Yaqui valley, 105, 177
Ybañez, Francisco, 108–9
Yuca, 37, 39, 85, 111
Yuma, 145
Yuracarés, 132, 296

Zamucos, 96, 124, 143, 298; language of, 140
Zanjón: arroyo of, 145
Zea, Juan Bautista de, 200–201
Zona serrana. *See* Sonora
Zuñiga, Ignacio, 4, 312, 320

CYNTHIA RADDING is Professor of History and Director of the Latin American and Iberian Institute at the University of New Mexico.

LIBRARY OF CONGRESS CATALOGING-IN-PUBLICATION DATA
Radding Murrieta, Cynthia.
Landscapes of power and identity : comparative histories in the Sonoran desert and the forests of Amazonia from colony to republic / Cynthia Radding.
p. cm.
Includes bibliographical references and index.
ISBN 0-8223-3652-9 (cloth : alk. paper)
ISBN 0-8223-3689-8 (pbk. : alk. paper)
1. Landscape changes—Mexico—Sonora (State) 2. Landscape changes—Mexico—Chiquitos (Province) 3. Land use—Mexico—Sonora (State) 4. Land use—Mexico—Chiquitos (Province) 5. Landscape ecology—Mexico—Sonora (State) 6. Landscape ecology—Mexico—Chiquitos (Province) 7. Sonora (Mexico : State)—Social conditions. 8. Sonora (Mexico : State)—Environmental conditions. 9. Chiquitos (Bolivia : Province)—History. 10. Chiquitos (Bolivia : Province)—Environmental conditions. I. Title.
GF91.M6R33 2005
304.2′0972′17—dc22 2005021135